T0094080

# Encyclopedia of
# Artificial Intelligence

# Encyclopedia of Artificial Intelligence

## The Past, Present, and Future of AI

Philip L. Frana and Michael J. Klein, Editors

ABC-CLIO®

An Imprint of ABC-CLIO, LLC
Santa Barbara, California • Denver, Colorado

**Library of Congress Cataloging-in-Publication Data**

Names: Frana, Philip L., editor. | Klein, Michael Joseph, editor.
Title: Encyclopedia of artificial intelligence : the past, present, and
    future of AI / Philip L. Frana and Michael J. Klein, editors.
Description: Santa Barbara, California : ABC-CLIO, LLC, 2021. | Includes
    bibliographical references and index.
Identifiers: LCCN 2020035798 (print) | LCCN 2020035799 (ebook) | ISBN
    9781440853265 (hardcover) | ISBN 9781440853272 (ebook)
Subjects: LCSH: Artificial intelligence—Encyclopedias.
Classification: LCC Q334.2 .K54 2021  (print) | LCC Q334.2  (ebook) | DDC
    006.303—dc23
LC record available at https://lccn.loc.gov/2020035798
LC ebook record available at https://lccn.loc.gov/2020035799

ISBN: 978-1-4408-5326-5 (print)
      978-1-4408-5327-2 (ebook)

25  24  23  22  21     1  2  3  4  5

This book is also available as an eBook.

ABC-CLIO
An Imprint of ABC-CLIO, LLC

ABC-CLIO, LLC
147 Castilian Drive
Santa Barbara, California 93117
www.abc-clio.com

This book is printed on acid-free paper ∞

Manufactured in the United States of America

# Contents

List of Entries    vii

Introduction    xi

Chronology    xvii

**Entries    1**

Bibliography    353

About the Editors    359

List of Contributors    361

Index    365

# List of Entries

AARON

Accidents and Risk Assessment

Advanced Soldier Sensor Information Systems and Technology (ASSIST)

AI Winter

Air Traffic Control, AI and

Alchemy and Artificial Intelligence

Algorithmic Bias and Error

Animal Consciousness

Asimov, Isaac

Automated Machine Learning

Automated Multiphasic Health Testing

Automatic Film Editing

Autonomous and Semiautonomous Systems

Autonomous Weapons Systems, Ethics of

Autonomy and Complacency

Battlefield AI and Robotics

Bayesian Inference

Beneficial AI, Asilomar Meeting on

Berger-Wolf, Tanya

Berserkers

Biometric Privacy and Security

Biometric Technology

*Blade Runner*

Blue Brain Project

Bostrom, Nick

Brooks, Rodney

Brynjolfsson, Erik

Calo, Ryan

Campaign to Stop Killer Robots

Caregiver Robots

Chatbots and Loebner Prize

Cheng, Lili

Climate Crisis, AI and

Clinical Decision Support Systems

Cognitive Architectures

Cognitive Computing

Cognitive Psychology, AI and

Computational Creativity

Computational Neuroscience

Computer-Assisted Diagnosis

Cybernetics and AI

Dartmouth AI Conference

de Garis, Hugo

Deep Blue

Deep Learning

DENDRAL

Dennett, Daniel

Diamandis, Peter

Digital Immortality

Distributed and Swarm Intelligence

Driverless Cars and Trucks

Driverless Vehicles and Liability

ELIZA

Embodiment, AI and

Emergent Gameplay and Non-Player Characters

Emily Howell

*Ex Machina*

Expert Systems

Explainable AI

Farmer, J. Doyne

Foerst, Anne

Ford, Martin

Frame Problem, The

Gender and AI

General and Narrow AI

General Problem Solver

Generative Design

Generative Music and Algorithmic Composition

Giant Brains

Goertzel, Ben

Group Symbol Associator

Hassabis, Demis

Human Brain Project

Intelligent Sensing Agriculture

Intelligent Transportation

Intelligent Tutoring Systems

Interaction for Cognitive Agents

INTERNIST-I and QMR

Ishiguro, Hiroshi

Knight, Heather

Knowledge Engineering

Kurzweil, Ray

Lethal Autonomous Weapons Systems

Mac Hack

Machine Learning Regressions

Machine Translation

Macy Conferences

McCarthy, John

Medicine, Artificial Intelligence in

Minsky, Marvin

Mobile Recommendation Assistants

MOLGEN

Monte Carlo

Moral Turing Test

Moravec, Hans

Musk, Elon

MYCIN

Natural Language Generation

Natural Language Processing and Speech Understanding

Newell, Allen

Nissenbaum, Helen

Nonhuman Rights and Personhood

Omohundro, Steve

PARRY

Pathetic Fallacy

*Person of Interest*

Post-Scarcity, AI and

Precision Medicine Initiative

Predictive Policing

Product Liability and AI

Quantum AI

Reddy, Raj

Robot Ethics

RoboThespian

Rucker, Rudy

Simon, Herbert A.

Sloman, Aaron

Smart Cities and Homes

Smart Hotel Rooms

"Software Eating the World"

Spiritual Robots

Superintelligence

Symbol Grounding Problem, The

Symbol Manipulation

Symbolic Logic

SyNAPSE

Tambe, Milind

Technological Singularity

*The Terminator*

Tilden, Mark

Trolley Problem

Turing, Alan

Turing Test

Turkle, Sherry

*2001: A Space Odyssey*

Unmanned Ground and Aerial Vehicles

Warwick, Kevin

Workplace Automation

Yudkowsky, Eliezer

# Introduction

Artificial intelligence is a fast-growing branch of computer science, a discipline famous for its interdisciplinarity and free-spirited interest in next-generation challenges of automation. Within universities, it encompasses the study of intelligent agents, autonomous programs that sense and actively respond to data collected in an environment. The virtual assistants Alexa and Siri are examples of intelligent agents. Agents may be embedded in smart robots, such as driverless cars and unmanned aerial vehicles.

More broadly, artificial intelligence describes real or imagined efforts to simulate cognition and creativity. The term distinguishes machine and code from the natural intelligence of animals and people. But artificial intelligence (AI) researchers often view the brain as a natural computer and the mind as a human-made computer program. AI pioneer Herbert Simon distinguished them this way: natural occurrences possess an "air of 'necessity' about them" as they observe physical laws; artificial phenomena have an "air of 'contingency' in their malleability" (Simon 1981, ix–xi). Even so, humans have only a handful of natural characteristics that limit their ability to learn, improve, and invent. Indeed, they are very good at adding complexity to their environment. The way forward, Simon reasoned, is not to retreat to the natural state, but to create more of the artificial world that people cannot do without.

A lot of AI has disappeared into the applications and devices we use every day. It's become so routine we don't even recognize it anymore. Bank software and Internet search engines are examples. Journalist Pamela McCorduck has identified this as the AI effect: "[I]t's part of the history of the field of artificial intelligence that every time somebody figured out how to make a computer do something—play good checkers, solve simple but relatively informal problems—there was a chorus of critics to say, but 'that's not thinking'" (McCorduck 2004, 204).

Instead, it might be helpful to frame artificial intelligence analogous to the way computer science generally has been defined by the "new new" and "next big" things—as transformative smart systems and technologies that haven't been invented yet. Mathematician John McCarthy coined the term "artificial intelligence" in 1956 to attract participants to the Dartmouth Summer Research Project, the Constitutional Convention of AI, and also boldly mark the divide between new and old approaches to understanding brain and machine. This liberating view is embodied today in Silicon Valley scientist and business leader Andrew Ng's unabashed pronouncement that AI is "the new electricity."

Since the 1950s, researchers have explored two major avenues to artificial intelligence: connectionism and symbolic representation. Interest in one or the other of these approaches has waxed and waned over the decades. Connectionist approaches nurtured by cybernetics, the study of neural pathways, and associative learning dominated in the 1950s and 1960s. This was followed by a surge in interest in symbolic lines of attack from the 1970s to the mid-1990s. In this period of "Good Old-Fashioned AI" (GOFAI), scientists in growing artificial intelligence programs and laboratories began programming chatbots such as ELIZA and PARRY and expert systems for capturing the knowledge and emulating the skill of the organic chemist (DENDRAL), physician (INTERNIST-I), and artist (AARON).

Since 2000, more improvements in performance have been wrung out of connectionist neural network approaches. Progress is being made in machine translation, computer vision, generative design, board and video game playing, and more. Among more audacious efforts are the Blue Brain Project and Human Brain Project, which attempt computational reconstructions of whole brains. Connectionist and symbolic AI are often described as rivals, but together they account for a large share of the spectacular progress in the field. Efforts are currently underway to integrate the approaches in a structured way under the banner of neural-symbolic learning and reasoning. When brought together, neural-symbolic computation, systems, and applications may (or may not) begin to approach General AI of the sort commonly depicted in science fiction.

## IN THE POPULAR IMAGINATION

Over the past hundred years, the concept of intelligent artificial beings has been a mainstay of literature and film. Beginning with Czech writer Karel Čapek's 1921 play *R.U.R.* (Rossum's Universal Robots), the human imagination has been sparked by the idea of creating artificial life that has independent thought. Similar to Čapek's depiction, German filmmaker Fritz Lang's *Metropolis* (1927) depicts these artificial beings as slaves to humans with little autonomy.

A watershed moment for these depictions came in 1968 with the release of Arthur C. Clarke's book and Stanley Kubrick's film, both entitled *2001: A Space Odyssey*. The book/film's antagonist, HAL 9000, serves humans in their exploration of the source of an alien signal coming from the outer planets of our solar system. But HAL does more than serve his human masters, instead taking the initiative to kill the human astronauts on board the ship *Discovery*. While logical to HAL, his actions clearly depicted a concept that would figure more prominently in subsequent works—AI rebellion against humans.

For example, *Blade Runner* (1982) and *The Terminator* (1984) both depict the effects of machines having free will. In *Blade Runner*, replicants serve as sex slaves and soldiers banned from Earth because of their predisposition to gain empathy, a defining human trait in this work. The cyborg soldier of *The Terminator*, under orders from the AI Skynet, is simply a killing machine on a mission to terminate a human. And in *The Matrix* (1999), humans live in an artificial cyber world, functioning as batteries for their robot overlords.

More recent films have been more nuanced in their treatment of AI. For example, *Ex Machina* (2014) depicts a scientist's attempts at creating a self-aware female android. His previous failures, kept as living art pieces or servants, attest to his misogynistic impulses. When his most recent creation escapes his house/laboratory, leaving him for dead in the process, the audience is clearly expected to recognize her humanity, something the scientist clearly lacks.

Smart robots in the real world, while impressive, have fought an uphill battle against their imaginary counterparts. The Roomba robotic vacuum cleaner is impressive, but cannot compete against Rosey the Robot housekeeper of *The Jetsons*. Caregiver robots are designed to be physically and socially assistive, but do not come close to replacing human nurses and nannies. Battlefield AI, however, is another matter. Autonomous weapons systems are not yet Terminators, but they are lethal enough to inspire campaigns to stop them, laws to govern them, and ethical frameworks to guide them.

## TECHNICAL, PHILOSOPHICAL, AND ETHICAL CRITICISMS

One of the main criticisms of AI is that while computers can be trained to do a limited number of things very well, it is much harder to teach a computer to be adaptive in its learning. For example, machines can be programmed with a series of algorithms, allowing them to classify patterns using a neural network—in other words, learning by example. This type of programming is called deep learning and is found in numerous applications including automated driving, medical diagnosis, and image and text recognition. Deep learning is often used in Natural Language Processing (NLP), a means of programming computers to analyze and recognize large chunks of natural language. For example, email filters are able to recognize key terms as specified by a user and follow specific rules to organize or classify messages. Other examples include predictive text in word processing programs, speech understanding, and data analysis.

However, while computers can store vast quantities of information, they can only utilize what has actually been programmed into them. In other words, they cannot adapt to situations that programmers didn't foresee. Some critics see this limitation as a significant barrier to real intelligence as it precludes machines from understanding causation, an integral component of human intelligence (Bergstein 2020).

In addition to technological criticisms, philosophical criticisms of AI abound. For example, the Turing Test, a foundational analysis to determine if responses to questions are created by a human or machine, is implicitly biased toward language use as a measure of intelligence. Along similar lines, philosophers ask if being like a human means that something is intelligent. Critics contend that the Turing Test doesn't measure intelligence; rather, it measures how well a machine can simulate being a human. Additional questions include whether intelligent behavior requires a mind, how to define sentience, what is the connection between consciousness and intelligence, and whether creativity can be programmed.

AI is increasingly used to monitor our daily lives. As AI surveillance and secu-rity systems become more commonplace, critics have raised questions regarding its potential to be biased against certain people, especially those who come from traditionally underrepresented groups. For example, a recent study showed that systems used by custom and border agents misidentifies the faces of Black people five to ten times more than white people (Simonite 2019). This could lead to racial disparities in policing as facial recognition systems become more embedded in law enforcement agencies around the world.

The reason for these disparities is because the programming of the algorithms to recognize faces is inaccurate. Often, the datasets of faces used to train these systems are not representative of the culture producing them. Another factor may be the actual photo quality, as photographic technology has been optimized for lighter skin since its inception (Lewis 2019). This is just one example of how the biases inherent in a culture may often become embedded in the AI systems that human culture designs and produces.

A larger issue that ethicists and philosophers raise is whether it is morally acceptable to create sentient life for the sole purpose of having it serve the needs of humans. If we were successful in creating true intelligence, one that was con-scious and creative, then how would our laws regarding the treatment of people apply? Would these intelligent machines be afforded the protections of the Consti-tution (or a similar document), or would they be treated as second-class citizens, with limited or no rights? Some scholars contend that these legal issues need to be considered and, if possible, resolved before AI becomes a reality.

## AI FOR GOOD

Artificial intelligence is helping people create and embrace new forms of art, perform unique forms of theater and dance, and make distinctive kinds of music. Quantum AI may help us understand the origins of life and the ultimate shape of the universe. Contrastingly, it is precipitating a gut-wrenching Fourth Industrial Revolution. AI threatens to automate people out of jobs, upend systems of wealth creation, and blur relied-upon boundaries between the biological and digital worlds. Nations and firms are rushing in to hack human civilization itself using AI.

The great paradox is that artificial intelligence is maximizing our preferences and simultaneously making us vulnerable to global catastrophes and existential risks. Artificial intelligence is fueling the reinvention of ourselves in a new com-putational universe. That process engenders a full range of emotions and outcomes from euphoria, anxiety, and—potentially—misery. AI's potential for dislocation and disorder is energizing movements to use the technology for common advan-tage. These movements go by various names: AI for Social Good, Beneficial AI, Trustworthy AI, Friendly AI. Together, they embody the wishes of scientists and policymakers that artificial intelligence balance its benefits against the risks and costs involved and wherever possible avoid harms.

Humanity is interested in making intelligent machines our caregivers, compan-ions, guides, and gods. And yet we have done a far better job turning humans into intelligent cogs in society's smart machine than in transforming artifacts into carbon copies of human beings. This is not for lack of interest. Several prominent artificial intelligence researchers argue for the inevitability of a Technological Singularity, beyond which alterations to artificial and human intelligence are unforeseeable and uncontrollable. Accelerating change and recursive self-improvement, these boosters say, could produce a superintelligent machine with its own unfathomable hypergoals.

The purpose of the present volume is to help the reader more carefully evaluate claims of "successes" and "failures" in artificial intelligence; assess the real impact of smart technologies in society; and understand the historical, literary, cultural, and philosophical significance of machine intelligence. Our machines are highly polished mirrors that reflect and magnify human feeling and ambition. AI opens us up to another way in which the world might be imagined and also sensitizes us to the richness of the human search for meaning.

*Philip L. Frana and Michael J. Klein*

**Further Reading**

Bergstein, Brian. 2020. "What AI Still Can't Do." *MIT Technology Review*, February 19, 2020. https://www.technologyreview.com/2020/02/19/868178/what-ai-still-cant-do/.

Cardon, Dominique, Jean-Philippe Cointet, and Antoine Mazieres. 2018. "Neurons Spike Back: The Invention of Inductive Machines and the Artificial Intelligence Controversy." *Reseaux* 36, no. 211: 173–220.

Lewis, Sarah. 2019. "The Racial Bias Built into Photography." *The New York Times*, April 25, 2019. https://www.nytimes.com/2019/04/25/lens/sarah-lewis-racial-bias-photography.html.

McCorduck, Pamela. 2004. *Machines Who Think: A Personal Inquiry into the History and Prospects of Artificial Intelligence.* Natick, MA: CRC Press.

Simon, Herbert. 1981. *The Sciences of the Artificial.* Second edition. Cambridge, MA: MIT Press.

Simonite, Tom. 2019. "The Best Algorithms Struggle to Recognize Black Faces Equally." *Wired*, July 22, 2019. https://www.wired.com/story/best-algorithms-struggle-recognize-black-faces-equally/.

# Chronology

**1942**

Science fiction author Isaac Asimov's Three Laws of Robotics appear in the short story "Runaround."

**1943**

Mathematician Emil Post writes about "production systems," a concept borrowed for the General Problem Solver of 1957.

**1943**

Publication of Warren McCulloch and Walter Pitts' paper on a computational theory of neural networks, entitled "A Logical Calculus of the Ideas of Immanent in Nervous Activity."

**1944**

John von Neumann, Norbert Wiener, Warren McCulloch, Walter Pitts, and Howard Aiken form the Teleological Society to study, among other things, communication and control in the nervous system.

**1945**

George Polya highlights the value of heuristic reasoning as a means of solving problems in his book *How to Solve It*.

**1946**

The first of ten Macy Conferences on Cybernetics begins in New York City. The theme of the first meeting is "Feedback Mechanisms and Circular Causal Systems in Biological and Social Systems."

**1948**

Publication of *Cybernetics, or Control and Communication in the Animal and the Machine* by mathematician Norbert Wiener.

**1949**

Psychologist Donald Hebb proposes an explanation for neural adaptation in human education in *The Organization of Behavior*: "neurons that fire together wire together."

**1949**

Publication of *Giant Brains, or Machines That Think* by mathematician Edmund Berkeley.

**1950**

The Turing Test, attributing intelligence to any machine capable of exhibiting intelligent behavior equivalent to that of a human, is described in Alan Turing's "Computing Machinery and Intelligence."

**1950**

Claude Shannon publishes a pioneering technical paper on "Programming a Computer for Playing Chess," sharing search algorithms and techniques.

**1951**

Mathematics student Marvin Minsky and physics student Dean Edmonds design an electric rat capable of learning how to negotiate a maze utilizing Hebbian theory.

**1951**

Mathematician John von Neumann publishes "General and Logical Theory of Automata," reducing the human brain and central nervous system to a computing machine.

**1951**

Christopher Strachey writes a checkers program and Dietrich Prinz creates a chess routine for the University of Manchester's Ferranti Mark 1 computer.

**1952**

*Design for a Brain: The Origin of Adaptive Behavior*, on the logical mechanisms of human cerebral function, is published by cyberneticist W. Ross Ashby.

**1952**

Physiologist James Hardy and physician Martin Lipkin begin devising a McBee punched card system for mechanical diagnosis of patients at Cornell University Medical College.

**1954**

Groff Conklin publishes the theme-based anthology *Science-Fiction Thinking Machines: Robots, Androids, Computers.*

**1954**

The Georgetown-IBM experiment demonstrates the potential of machine translation of text.

**1955**

Artificial intelligence research begins at Carnegie Tech (now Carnegie Mellon University) under economist Herbert Simon and graduate student Allen Newell.

**1955**

Mathematician John Kemeny writes "Man Viewed as a Machine" for *Scientific American.*

**1955**

Mathematician John McCarthy coins the term "artificial intelligence" in a Rockefeller Foundation proposal for a Dartmouth University conference.

**1956**

Logic Theorist, an artificial intelligence computer program for proving theorems in Alfred North Whitehead and Bertrand Russell's *Principia Mathematica*, is created by Allen Newell, Herbert Simon, and Cliff Shaw.

**1956**

The Dartmouth Summer Research Project, the "Constitutional Convention of AI," brings together experts in cybernetics, automata, information theory, operations research, and game theory.

**1956**

Electrical engineer Arthur Samuel demonstrates his checkers-playing AI program on television.

**1957**

The General Problem Solver AI program is written by Allen Newell and Herbert Simon.

**1957**

The Rockefeller Medical Electronics Center demonstrates an RCA Bizmac computer program to aid the physician in the differential diagnosis of blood diseases.

**1958**

Publication of John von Neumann's unfinished *The Computer and the Brain.*

**1958**

Firmin Nash gives a first public demonstration of the Group Symbol Associator at the "Mechanisation of Thought Processes" conference at UK's Teddington National Physical Laboratory.

**1958**

Frank Rosenblatt introduces the single layer perceptron, including a neural network and supervised learning algorithm for linear classification of data.

**1958**

John McCarthy at the Massachusetts Institute of Technology (MIT) specifies the high-level programming language LISP for AI research.

**1959**

Physicist Robert Ledley and radiologist Lee Lusted publish "The Reasoning Foundations of Medical Diagnosis," which introduces Bayesian inference and symbolic logic to problems of medicine.

**1959**

John McCarthy and Marvin Minsky start what becomes the Artificial Intelligence Laboratory at MIT.

**1960**

The Stanford Cart, a remote control vehicle equipped with a television camera, is constructed by engineering student James L. Adams.

**1962**
Science fiction and fantasy author Fred Saberhagen introduces intelligent killer machines called Berserkers in the short story "Without a Thought."

**1963**
The Stanford Artificial Intelligence Laboratory (SAIL) is founded by John McCarthy.

**1963**
The U.S. Department of Defense's Advanced Research Projects Agency begins funding artificial intelligence projects at MIT under Project MAC.

**1964**
ELIZA, the first program for natural language communication with a machine ("chatbot"), is programmed by Joseph Weizenbaum at MIT.

**1965**
British statistician I. J. Good publishes his "Speculations Concerning the First Ultraintelligent Machine" about a coming intelligence explosion.

**1965**
Philosopher Hubert L. Dreyfus and mathematician Stuart E. Dreyfus release a paper critical of artificial intelligence entitled "Alchemy and AI."

**1965**
The Stanford Heuristic Programming Project, with the twin goals of modeling scientific reasoning and creating expert systems, is initiated by Joshua Lederberg and Edward Feigenbaum.

**1965**
Donald Michie organizes the Department of Machine Intelligence and Perception at Edinburgh University.

**1965**
Georg Nees establishes in Stuttgart, West Germany, the first generative art exhibit, called Computer Graphic.

**1965**
Computer scientist Edward Feigenbaum begins a ten-year effort to automate the molecular analysis of organic compounds with the expert system DENDRAL.

**1966**
The Automatic Language Processing Advisory Committee (ALPAC) releases its skeptical report on the current state of machine translation.

**1967**
Richard Greenblatt completes work on Mac Hack, a program that plays competitive tournament chess, on a DEC PDP-6 at MIT.

**1967**
Ichiro Kato at Waseda University initiates work on the WABOT project, which unveils a full-scale anthropomorphic intelligent robot five years later.

**1968**

Director Stanley Kubrick turns Arthur C. Clarke's science fiction book *2001: A Space Odyssey*, about the HAL 9000 artificially intelligent computer, into one of the most influential and critically acclaimed movies of all time.

**1968**

Terry Winograd at MIT begins work on the natural language understanding program SHRDLU.

**1969**

The First International Joint Conference on Artificial Intelligence (IJCAI) is held in Washington, DC.

**1972**

Artist Harold Cohen creates AARON, an AI program to create paintings.

**1972**

Ken Colby reports on his experiments simulating paranoia with the software program PARRY.

**1972**

Hubert Dreyfus publishes his critique of the philosophical foundations of artificial intelligence in *What Computers Can't Do*.

**1972**

The MYCIN expert system, designed to diagnose bacterial infections and recommend treatment options, is begun by doctoral student Ted Shortliffe at Stanford University.

**1972**

The Lighthill Report on Artificial Intelligence is released by the UK Science Research Council, highlighting failures of AI technology and difficulties of combinatorial explosion.

**1972**

Arthur Miller publishes *The Assault on Privacy: Computers, Data Banks, and Dossiers*, an early work on the social impact of computers.

**1972**

University of Pittsburgh physician Jack Myers, medical student Randolph Miller, and computer scientist Harry Pople begin collaborating on INTERNIST-I, an internal medicine expert system.

**1974**

Social scientist Paul Werbos finishes his dissertation on a now widely used algorithm for backpropagation used in training artificial neural networks for supervised learning tasks.

**1974**

Marvin Minsky releases MIT AI Lab memo 306 on "A Framework for Representing Knowledge." The memo details the concept of a frame, a "remembered framework" that fits reality by "changing detail as necessary."

**1975**
John Holland uses the term "genetic algorithm" to describe evolutionary strategies in natural and artificial systems.

**1976**
Computer scientist Joseph Weizenbaum publishes his ambivalent views of work on artificial intelligence in *Computer Power and Human Reason.*

**1978**
EXPERT, a generalized knowledge representation scheme for creating expert systems, becomes operational at Rutgers University.

**1978**
The MOLGEN project at Stanford is begun by Joshua Lederberg, Douglas Brutlag, Edward Feigenbaum, and Bruce Buchanan to solve DNA structures derived from segmentation data in molecular genetics experiments.

**1979**
The Robotics Institute is established by computer scientist Raj Reddy at Carnegie Mellon University.

**1979**
The first human is killed while working with a robot.

**1979**
The Stanford Cart, evolving over almost two decades into an autonomous rover, is rebuilt and equipped with a stereoscopic vision system by Hans Moravec.

**1980**
The First National Conference of the American Association of Artificial Intelligence (AAAI) is held at Stanford University.

**1980**
Philosopher John Searle makes his Chinese Room argument that a computer's simulation of behavior does not in itself demonstrate understanding, intentionality, or consciousness.

**1982**
Release of the science fiction film *Blade Runner,* which is broadly based on Philip K. Dick's story *Do Androids Dream of Electric Sheep?* (1968).

**1982**
Physicist John Hopfield popularizes the associative neural network, first described by William Little in 1974.

**1984**
Tom Alexander publishes "Why Computers Can't Outthink the Experts" in *Fortune Magazine.*

**1984**
Computer scientist Doug Lenat starts the Cyc project to build a massive commonsense knowledge base and artificial intelligence architecture at the Microelectronics and Computer Consortium (MCC) in Austin, TX.

**1984**
The first *Terminator* film, with android assassins from the future and an AI called Skynet, is released by Orion Pictures.

**1986**
Honda opens a research center for the development of humanoid robots to coexist and collaborate with human beings.

**1986**
MIT roboticist Rodney Brooks introduces the subsumption architecture for behavior-based robotics.

**1986**
Marvin Minsky publishes *The Society of Mind,* which describes the brain as a set of cooperating agents.

**1989**
Rodney Brooks and Anita Flynn of the MIT Artificial Intelligence Lab publish "Fast, Cheap, and Out of Control: A Robot Invasion of the Solar System," about the potential for launching tiny robots on missions of interplanetary discovery.

**1993**
Rodney Brooks, Lynn Andrea Stein, Cynthia Breazeal, and others launch the Cog interactive robot project at MIT.

**1995**
Musician Brian Eno coins the term generative music to describe systems that create ever-changing music by altering parameters over time.

**1995**
The General Atomics MQ-1 Predator unmanned aerial vehicle enters U.S. military and reconnaissance service.

**1997**
IBM's Deep Blue supercomputer defeats reigning chess champion Garry Kasparov under regular tournament conditions.

**1997**
The first RoboCup, an international competition with over forty teams fielding robot soccer players, is held in Nagoya, Japan.

**1997**
Dragon Systems releases NaturallySpeaking, their first commercial speech recognition software product.

**1999**
Sony releases AIBO, a robotic dog, to the consumer market.

**2000**
Honda introduces its prototype ASIMO, the Advanced Step in Innovative Mobility humanoid robot.

**2001**
Visage Corporation debuts the FaceFINDER automated face-recognition system at Super Bowl XXXV.

**2002**
The iRobot Corporation, founded by Rodney Brooks, Colin Angle, and Helen Greiner, begins marketing the Roomba autonomous home vacuum cleaner.

**2004**
DARPA sponsors its first autonomous car Grand Challenge in the Mojave Desert around Primm, NV. None of the cars finish the 150-mile course.

**2005**
The Swiss Blue Brain Project to simulate the mammalian brain is established under neuroscientist Henry Markram.

**2006**
Netflix announces a $1 million dollar prize to the first programming team that develops the best recommender system based on a dataset of previous user ratings.

**2007**
DARPA launches its Urban Challenge, an autonomous vehicle competition meant to test merging, passing, parking, and negotiating traffic and intersections.

**2009**
Google begins its Self-Driving Car Project (now called Waymo) in the San Francisco Bay Area under Sebastian Thrun.

**2009**
Stanford University computer scientist Fei-Fei Li presents her work on ImageNet, a collection of millions of hand-annotated images for training AIs to visually recognize the presence or absence of objects.

**2010**
A "flash crash" of the U.S. stock market is triggered by human manipulation of automated trading software.

**2011**
UK artificial intelligence start-up DeepMind is founded by Demis Hassabis, Shane Legg, and Mustafa Suleyman to teach AIs to play and excel at classic video games.

**2011**
IBM's natural language computing system Watson defeats past *Jeopardy!* champions Ken Jennings and Brad Rutter.

**2011**
Apple releases the mobile recommendation assistant Siri on the iPhone 4S.

**2011**
An informal Google Brain deep learning research collaboration is started by computer scientist Andrew Ng and Google researchers Jeff Dean and Greg Corrado.

**2013**

The Human Brain Project of the European Union is launched to understand how the human brain works and also emulate its computational capabilities.

**2013**

Human Rights Watch begins a campaign to Stop Killer Robots.

**2013**

*Her*, a science fiction drama directed by Spike Jonze, is released. The film features a romance between a man and his AI mobile recommendation assistant Samantha.

**2014**

Ian Goodfellow and collaborators at the University of Montreal introduce Generative Adversarial Networks (GANs) for use in deep neural networks, which prove useful in creating realistic images of fake people.

**2014**

The chatbot Eugene Goostman, portraying a thirteen-year-old boy, is controversially said to have passed a Turing-like test.

**2014**

Physicist Stephen Hawking predicts the development of AI could result in the extinction of humanity.

**2015**

Facebook releases DeepFace deep learning facial recognition technology on its social media platform.

**2016**

DeepMind's AlphaGo program defeats 9th dan Go player Lee Sedol in a five-game match.

**2016**

Microsoft's artificial intelligence chatbot Tay is released on Twitter, where users train it to make offensive and inappropriate tweets.

**2017**

The Future of Life Institute organizes the Asilomar Meeting on Beneficial AI.

**2017**

The Way of the Future church is founded by AI self-driving start-up engineer Anthony Levandowski, who is motivated to create a superintelligent robot deity.

**2018**

Google announces Duplex, an AI application for scheduling appointments over the phone using natural language.

**2018**

The European Union publishes its General Data Protection Regulation (GDPR) and "Ethics Guidelines for Trustworthy AI."

**2019**

Google AI and Northwestern Medicine in Chicago, IL, collaborate on a lung cancer screening AI that outperforms specialist radiologists.

**2019**

OpenAI, cofounded by Elon Musk, develops an artificial intelligence text generation system that creates realistic stories and journalism. It is initially deemed "too dangerous" to use because of its potential to generate fake news.

**2020**

Google AI in collaboration with the University of Waterloo, the "moonshot factory" X, and Volkswagen announce TensorFlow Quantum, an open-source library for quantum machine learning.

# A

## AARON

AARON is computer software created by Harold Cohen to create paintings. Cohen himself dates the creation of the first version to "around 1972." Since AARON is not open source, it can be said that its development ended in 2016 when Cohen died. AARON was still producing new images in 2014, and its functionality was evident even in 2016. AARON is not an acronym. The name was given because it is at the beginning of the alphabet, and Cohen imagined he would subsequently create other programs later on, which he never did.

During its four decades of evolution, AARON had several versions with different capabilities. The earlier versions were able to create black-and-white line drawings, while the later versions were also able to paint in color. Some versions of AARON were configured to create abstract paintings while others painted scenes with objects and people in it.

The primary purpose of AARON was to create not only digital images but also tangible, large-sized images or paintings. In Cohen's exhibition at The San Francisco Museum of Modern Art, the lines drawn by AARON, a program written in C at the time, were traced directly on the wall. In later artistic installments of AARON, the program was coupled with a machine that had a robotic arm and was able to apply paint on canvas. For instance, the version of AARON on exhibit in The Computer Museum in Boston in 1995, which was implemented in LISP by this time and ran on a Silicon Graphics computer, created a file with a set of instructions. This file was then transferred to a PC that ran a C++ program. This computer had a robotic arm attached to it. The C++ code interpreted the instructions and controlled the movement of the arm, the mixing of the dyes, and the application of them on the canvas. The machines built by Cohen to draw and paint were also important innovations. Even later versions used industrial inkjet printers. Cohen held this configuration of AARON the most advanced because of the colors these new printers could produce; he believed that when it came to colors, the inkjet was the biggest invention since the industrial revolution.

While Cohen mostly focused on tangible images, around 2000, Ray Kurzweil created a version of AARON that was a screensaver program. By 2016, Cohen himself had created a version of AARON that generated black-and-white images that the user could color using a large touch screen. He called this "Fingerpainting." Cohen always believed that AARON is neither a "fully autonomous artist" nor is it truly creative. He believed though that AARON exhibits one condition of autonomy: a form of emergence, which in Cohen's terms means that the paintings generated are genuinely surprising and novel. Cohen never ventured very far into the philosophical implications of AARON. Based on the amount of time he dedicates to the coloring problem in almost all of the interviews made with him, it is safe to assume that he regarded AARON's performance as a colorist his biggest achievement.

*Mihály Héder*

*See also:* Computational Creativity; Generative Design.

**Further Reading**

Cohen, Harold. 1995. "The Further Exploits of AARON, Painter." *Stanford Humanities Review* 4, no. 2 (July): 141–58.

Cohen, Harold. 2004. "A Sorcerer's Apprentice: Art in an Unknown Future." Invited talk at Tate Modern, London. http://www.aaronshome.com/aaron/publications/tate -final.doc.

Cohen, Paul. 2016. "Harold Cohen and AARON." *AI Magazine* 37, no. 4 (Winter): 63–66.

McCorduck, Pamela. 1990. *Aaron's Code: Meta-Art, Artificial Intelligence, and the Work of Harold Cohen.* New York: W. H. Freeman.

## Accidents and Risk Assessment

The most important property of many computer-based systems is dependability. Mechanical and software failures can lead to physical damage, information loss, economic dislocation, or human casualties. Robotics, automation, and artificial intelligence are now in control of many critical systems. They monitor nuclear power plants, financial markets, social security disbursements, traffic signals, and military radar installations.

High technologies may be made deliberately dangerous to humans—as in the case of Trojan horses, viruses, and spyware—or they can be dangerous because of mistakes of the human programmer or operator. They may become unsafe in response to unexpected environmental factors or in the future because of intentional or unintentional decisions taken by the machines themselves.

In 1979, the first human was killed while working with a robot. Robert Williams, an engineer at Ford Motor Company, was struck in the head by a one-ton parts-retrieval robot manufactured by Litton Industries. Two years later, Japanese engineer Kenji Urada was killed after failing to completely turn off a malfunctioning robot on the factory floor at Kawasaki Heavy Industries. The robot's arm pushed Urada into a grinding machine. Not all accidents result in fatalities. In 2016, for instance, a 300-lb. Knightscope K5 security robot on patrol at a shopping center in Northern California knocked down a toddler and ran over his foot. The child escaped with only a few scrapes and some swelling.

The history of the Cold War is replete with examples of nuclear close-calls triggered by malfunctioning computer equipment. In 1979, a computer error at the North American Aerospace Defense Command (NORAD) briefly convinced the Strategic Air Command that the Soviet Union had launched more than 2,000 nuclear missiles at the United States. An investigation led to the discovery that a training scenario had accidentally been uploaded to an active defense computer.

In 1983, a Soviet military early warning system detected a nuclear attack by a single U.S. intercontinental ballistic missile. The operator of the missile defense system, Stanislav Petrov, rightly dismissed the warning as a false alarm. This and other false alarms were later determined to be caused by sunlight striking high-altitude clouds. Despite averting global thermonuclear war, Petrov later suffered punishment for embarrassing his superiors by exposing errors.

Stock market trading software triggered the so-called "2010 Flash Crash." On May 6, 2010, the S&P 500, Dow Jones, and NASDAQ stock indexes lost—and then mostly regained—a trillion dollars in value in just over a half hour. After five years of investigations by the U.S. Department of Justice, the U.K. trader Navinder Singh Sarao was arrested for allegedly modifying an automated system to generate and then cancel large numbers of sell orders, which allowed his company to reap stocks at temporarily deflated prices. The flash crash was followed by two software-induced market flash crashes in 2015 and by flash crashes in the gold futures market and digital cryptocurrency in 2017.

In 2016 a Microsoft Corporation artificial intelligence social media chatterbot named Tay (after the acronym for "Thinking about you") went horribly awry. Microsoft researchers designed Tay to mimic a nineteen-year-old American girl and to learn from conversations with other Twitter users. Instead, online trolls taught Tay to use offensive and inflammatory language, which it repeated in tweets. Microsoft removed Tay's account after only sixteen hours.

In the future, more AI-related accidents may occur in motor vehicle operation. The first fatal crash involving a self-driving automobile occurred in 2016, when the driver of a Tesla Model S in autopilot mode crashed through a semi-trailer crossing the highway. Witnesses report that the driver may have been watching a Harry Potter movie on a portable DVD player when the accident occurred. Tesla's software does not yet permit fully autonomous driving and requires an alert human operator. Despite such risks, one management consulting firm suggests that up to 90 percent of roadway accidents could be prevented by automated cars.

Artificial intelligence safety is becoming an important subfield in cybersecurity research. Militaries around the world are engaged in development of prototype lethal autonomous weapons systems. Automatic weapons, such as drones that now depend on a human operator to make decisions about lethal force against targets, may be replaced by life and death decisions initiated by automated systems. On the battlefield, robotic decision-makers may one day outclass humans in pulling patterns from the fog of war and responding rapidly and rationally to new or perplexing situations.

Contemporary society is increasingly dependent on high technology, and those technologies are sometimes brittle and susceptible to failure. In 1987, an inquisitive squirrel caused the crash of the NASDAQ's main computer, bringing one of the largest stock exchanges to its knees. In another case, discovery of the ozone hole over Antarctica was delayed for years because the extremely low levels found in data-processed satellite observations were thought to be errors. It may be that the complexity of autonomous systems, and society's dependency upon them under rapidly changing conditions, will make fully testable artificial intelligence impossible. Artificial intelligence operates by using software capable of adjusting to and interacting with its environment and user. The results of changes to variables, individual actions, or events are sometimes unpredictable and may even be catastrophic.

One of the dark secrets of advanced artificial intelligence is that it relies on mathematical methods and deep learning algorithms so complex that even its makers cannot understand how it makes reliable decisions. Autonomous vehicles, for instance, usually rely on instructions written solely by the computer as it observes humans driving under real conditions. But how can a driverless car come to expect the unexpected? And additionally, will efforts to tweak AI-generated code to reduce perceived errors, omissions, and impenetrability reduce the risk of accidental negative outcomes or simply amplify errors and generate new ones? It remains unclear how to mitigate the risks of artificial intelligence, but it is likely that society will use proven and presumably trustworthy machine-learning systems to automatically provide rationales for their behavior, and even examine newly invented cognitive computing systems on our behalf.

*Philip L. Frana*

*See also:* Algorithmic Bias and Error; Autonomy and Complacency; Beneficial AI, Asilomar Meeting on; Campaign to Stop Killer Robots; Driverless Vehicles and Liability; Explainable AI; Product Liability and AI; Trolley Problem.

### Further Reading

De Visser, Ewart Jan. 2012. "The World Is Not Enough: Trust in Cognitive Agents." Ph.D. diss., George Mason University.

Forester, Tom, and Perry Morrison. 1990. "Computer Unreliability and Social Vulnerability." *Futures* 22, no. 5 (June): 462–74.

Lee, John D., and Katrina A. See. 2004. "Trust in Automation: Designing for Appropriate Reliance." *Human Factors* 46, no. 1 (Spring): 50–80.

Yudkowsky, Eliezer. 2008. "Artificial Intelligence as a Positive and Negative Factor in Global Risk." In *Global Catastrophic Risks*, edited by Nick Bostrom and Milan M. Ćirković, 308–45. New York: Oxford University Press.

# Advanced Soldier Sensor Information Systems and Technology (ASSIST)

Soldiers are often asked to perform missions that can last many hours and be under great stress. After a mission is complete, the soldiers are asked to provide a report describing the most important things that happened. This report is used to gather intelligence about the environment and local/foreign personnel to allow for more informed planning for future missions. Soldiers usually provide this report based solely on their memory, still pictures, and GPS data from handheld units. Due to the extreme stress that they encounter, there are undoubtedly many instances in which important information is missed and not available for the planning of future missions.

The ASSIST (Advanced Soldier Sensor Information Systems and Technology) program addressed this challenge by instrumenting soldiers with sensors that they could wear directly on their uniforms. During the mission, sensors continuously captured what was going on around the soldiers. When soldiers returned from their mission, the sensor data was run through a series of AI-based software systems that indexed the data and created an electronic chronicle of the events that happened while the ASSIST system was recording. With this information, soldiers could give more accurate reports without relying solely on their memory. The AI-based algorithms had enabled numerous functionalities, including:

- "Image/Video Data Analysis Capabilities"
  - Object Detection/Image Classification—the ability to recognize and identify objects (e.g., vehicles, people, and license plates) through analysis of video, imagery, and/or related data sources
  - Arabic Text Translation—the ability to detect, recognize, and translate written Arabic text (e.g., in imagery data)
  - Change Detection—the ability to identify changes over time in related data sources (e.g., identify differences in imagery of the same location at different times)
- "Audio Data Analysis Capabilities"
  - Sound Recognition/Speech Recognition—the ability to identify sound events (e.g., explosions, gunshots, and vehicles) and recognize speech (e.g., keyword spotting and foreign language identification) in audio data
  - Shooter Localization/Shooter Classification—the ability to identify gunshots in the environment (e.g., through analysis of audio data), including the type of weapon producing the shots and the location of the shooter
- "Soldier Activity Data Analysis Capabilities"
  - Soldier State Identification/Soldier Localization—the ability to identify a soldier's path of movement around an environment and characterize the actions taken by the soldier (e.g., running, walking, and climbing stairs)

For AI systems such as these (often termed *autonomous* or *intelligent* systems) to be successful, they must be comprehensively and quantitatively evaluated to ensure that they will function appropriately and as expected in a wartime environment. The National Institute of Standards and Technology (NIST) was tasked with evaluating these AI systems based on three metrics:

1. The accuracy of object/event/activity identification and labeling
2. The system's ability to improve its classification performance through learning
3. The utility of the system in enhancing operational effectiveness

NIST developed a two-part test methodology to produce its performance measures. Metrics 1 and 2 were evaluated through component- and system-level technical performance evaluations and metric 3 was evaluated through system-level utility assessments. The technical performance evaluations were designed to measure the progressive development of ASSIST system technical capabilities, and the utility assessments were designed to predict the impact these technologies will have on warfighter performance in a variety of missions and job functions. In specifying the detailed procedures for each type of evaluation, NIST attempted to define evaluation strategies that would provide a reasonable level of difficulty for system and soldier performance.

At the component level, the ASSIST systems were broken into components that implemented particular capabilities. For example, to test the Arabic translation capability, the system was broken down into an Arabic text identification component, an Arabic text extraction component (to localize individual text characters), and a text translation component. Each aspect was tested separately to assess their individual impact on the system.

At the system level, each ASSIST system was evaluated as a black box, where the overall performance of the system was evaluated, independent of the performance of the individual components. A single score was provided for the overall system, which represented the capability of the system to accomplish the overall task.

Also, at the system level, a test was performed to assess the utility of the system in enhancing operational effectiveness for after-mission reporting. All the systems that were evaluated under this program were in relatively early stages of development; therefore, a formative evaluation approach was appropriate. NIST was particularly interested in assessing the utility of the system for the war fighter. As such, we were concerned with impacts on both their processes and products. To reflect this perspective, user-centered metrics were employed. NIST attempted to identify metrics that would help assess questions such as: What information do infantry soldiers want and/or need after completing a mission in the field? How well are information needs met, both from the soldiers' perspective and the S2's (Staff 2—Intelligence Officer) perspective? What were the ASSIST contributions to mission reporting with respect to user-stated information needs?

The Aberdeen Test Center Military Operations in Urban Terrain (MOUT) site at the Aberdeen Test Center in Aberdeen, Maryland, was chosen to carry out this evaluation. The site was chosen for the following reasons:

- Ground truth—Aberdeen was able to provide ground truth at selected locations to two-centimeter accuracy. This allowed the evaluation team to gain a very good representation of what actually happened in the environment, thus allowing us to provide a good benchmark against which to compare the system output.
- Realism—The MOUT site contained approximately twenty buildings set up to resemble an Iraqi village.
- Infrastructure for testing—The site was instrumented with numerous cameras (both indoors and outdoors) to allow us to gain a better understanding of the environment during testing.
- Soldier availability—The site was able to provide a small squad of active duty soldiers to participate in the evaluation.

To further facilitate an operationally relevant test environment, the MOUT site was augmented with objects, persons, and background sounds, whose placement and behavior were scripted. The purpose was to provide an environment that would exercise the different ASSIST systems' capabilities as they detected, identified, and/or captured various types of information. NIST included the following elements in the utility assessments: foreign language speech detection and classification, Arabic text detection and recognition, detection of shots fired and vehicular sounds, classification of soldier states and tracking their locations (both inside and outside of buildings), identifying objects of interest including vehicles, buildings, people, etc. The soldiers' actions were not scripted as they moved through each exercise because the tests required the soldiers to act according to their training and experience.

*Craig I. Schlenoff*

*Portions of this entry adapted from Schlenoff, Craig, Michelle Potts Steves, Brian A. Weiss, Mike Shneier, and Ann Virts. 2007. "Applying SCORE to Field-Based Performance Evaluations of Soldier Worn Sensor Techniques."* Journal of Field Robotics, *24: 8–9, 671– 698. Copyright © 2007 Wiley Periodicals, Inc., A Wiley Company. Used by permission.*

*See also:* Battlefield AI and Robotics; Cybernetics and AI.

**Further Reading**

Schlenoff, Craig, Brian Weiss, Micky Steves, Ann Virts, Michael Shneier, and Michael Linegang. 2006. "Overview of the First Advanced Technology Evaluations for ASSIST." In *Proceedings of the Performance Metrics for Intelligence Systems Workshop*, 125–32. Gaithersburg, MA: National Institute of Standards and Technology.

Steves, Michelle P. 2006. "Utility Assessments of Soldier-Worn Sensor Systems for ASSIST." In *Proceedings of the Performance Metrics for Intelligence Systems Workshop*, 165–71. Gaithersburg, MA: National Institute of Standards and Technology.

Washington, Randolph, Christopher Manteuffel, and Christopher White. 2006. "Using an Ontology to Support Evaluation of Soldier-Worn Sensor Systems for ASSIST." In *Proceedings of the Performance Metrics for Intelligence Systems Workshop*, 172– 78. Gaithersburg, MA: National Institute of Standards and Technology.

Weiss, Brian A., Craig I. Schlenoff, Michael O. Shneier, and Ann Virts. 2006. "Technology Evaluations and Performance Metrics for Soldier-Worn Sensors for ASSIST." In *Proceedings of the Performance Metrics for Intelligence Systems Workshop*, 157–64. Gaithersburg, MA: National Institute of Standards and Technology.

# AI Winter

The phrase AI Winter was coined at the 1984 annual meeting of the American Association of Artificial intelligence (now the Association for the Advancement of Artificial Intelligence or AAAI). Two leading researchers, Marvin Minsky and Roger Schank, used the expression to refer to the then-impending bust period in research and commercial development in artificial intelligence. Canadian AI researcher Daniel Crevier has documented how angst over a coming AI Winter triggered a domino effect that began with cynicism in the AI research community, trickled into mass media, and finally led to adverse reactions by funding bodies. The result was a freeze in serious AI research and development. Initial pessimism is now mainly attributed to the overly ambitious promises made at the time—AI's actual results being far humbler than expectations.

Other factors such as insufficient computing power available during early days of AI research also contributed to the opinion that an AI Winter was at hand. This was particularly true of neural network research, which required vast computing resources. Similarly, economic factors, particularly during overlapping periods of economic crisis, resulted in restricted focus on more tangible investments.

Several periods throughout the history of AI can be described as AI Winters, with two of the major periods spanning from 1974 to 1980 and from 1987 to 1993. Although the dates of AI Winters are contentious and source-dependent, periods of overlapping trends mark these periods as prone to research abandonment and defunding.

Similar to the hype and eventual bust of emerging technologies such as nanotechnology, the development of AI systems and technologies has advanced nonetheless. The current boom period is marked by not only an unprecedented amount of funding toward fundamental research but also unparalleled progress in the development of machine learning. Motivations behind the investment boom differ, as they depend on the various stakeholders who engage in artificial intelligence research and development. Industry, for example, has wagered large sums on the promise that breakthroughs in AI will yield dividends by revolutionizing whole market sectors. Governmental bodies such as the military, on the other hand, invest in AI research to make both defensive and offensive technologies more efficient and remove soldiers from immediate harm.

Because AI Winters are fundamentally caused by a perceptual loss of faith in what AI can yield, the current hype surrounding AI and its promises has led to concern that another AI Winter will be triggered. Conversely, arguments have been made that the current technological advances in applied AI research have solidified the growth in future innovation in this area. This argument stands in stark contrast with the so-called "pipeline problem," which argues that the lack of fundamental research in AI will lead to finite amounts of applied results. The pipeline problem is often cited as one of the contributing factors to previous AI Winters. If the counterargument is correct, however, a feedback loop between applied innovations and fundamental research will provide the pipeline with enough pressure for continued progress.

*Steven Umbrello*

*See also:* Minsky, Marvin.

**Further Reading**

Crevier, Daniel. 1993. *AI: The Tumultuous Search for Artificial Intelligence.* New York: Basic Books.

Kurzweil, Ray. 2005. *The Singularity Is Near: When Humans Transcend Biology.* New York: Viking.

Muehlhauser, Luke. 2016. "What Should We Learn from Past AI Forecasts?" https://www .openphilanthropy.org/focus/global-catastrophic-risks/potential-risks-advanced -artificial-intelligence/what-should-we-learn-past-ai-forecasts.

# Air Traffic Control, AI and

Air Traffic Control (ATC) refers to the ground-based air navigation service of directing the aircraft on the ground and in controlled airspace. Air traffic controllers occasionally provide advisory services through the noncontrolled airspace as well. Controllers organize the safe flow of air traffic by coordinating the movement of commercial and private jets and ensuring a safe separation of the traffic in the air and on the ground. They typically give directional instructions to pilots with real-time traffic and weather alerts.

According to the Federal Aviation Administration (FAA), the main objectives of the ATC are to organize and expedite the flow of air traffic, as well as to prevent aircraft collisions and provide real-time information and other navigational support for pilots. The ATC is a risk-averse and safety critical service. In addition to the visual observation of air traffic controllers, the ATC also benefits from a diverse set of technologies such as computer systems, radars, and transmitters.

The volume and density of air traffic have been on the rise globally. As global air traffic becomes denser, it pushes the operational limits of contemporary ATC systems. Systems of air navigation and air traffic management need to become more sophisticated to keep pace with the increasing demand for accommodating future growth of air traffic. For safer, more efficient, and effective management of increasing air traffic, artificial intelligence (AI) offers the possibility of a variety of applications. According to the Global Air Navigation Plan (GANP) of the International Civil Aviation Organization (ICAO), AI-based air traffic management systems can provide solutions to the operational challenges of the increasing volume and diversity of air traffic. The training of human air traffic controllers already benefits from simulation programs with AI that can monitor and guide the actions of trainee controllers.

Operationally, the capacity of machine learning-based AI systems to digest vast volumes of data can be employed to overcome the complexities and challenges of traffic management operations. Such systems can be used during planning phases to analyze traffic data for flight planning and route choices. AI is also able to provide reliable traffic predictions by detecting wide-ranging flight patterns. For en route operations, AI-based ATC systems can be utilized for trajectory prediction and decision-making, especially in uncertain situations with limited data. AI can improve taxiing strategies and runway configurations. Also, AI-applied speech recognition systems can improve the communication between the controllers and the pilots.

With such diverse applications, AI systems can enhance the overall performance of human air traffic controllers by providing them with in-depth information and fast decision-making processes. It is also important to note that rather than replacing human air traffic controllers, AI-based technologies have proven complementary to human experts in handling safe and expeditious air traffic.

*Fatma Derya Mentes*

*See also:* Intelligent Transportation.

**Further Reading**

Federal Aviation Administration. 2013. *Aeronautical Information Manual: Official Guide to Basic Flight Information and ATC Procedures.* Washington, DC: FAA. https:// www.faa.gov/air_traffic/publications/.

International Civil Aviation Organization. 2018. "Potential of Artificial Intelligence (AI) in Air Traffic Management (ATM)." In *Thirteenth Air Navigation Conference,* 1–3. Montreal, Canada. https://www.icao.int/Meetings/anconf13/Documents/WP /wp_232_en.pdf.

Nolan, Michael S. 1999. *Fundamentals of Air Traffic Control.* Pacific Grove, CA: Brooks /Cole.

## Alchemy and Artificial Intelligence

The RAND Corporation report *Alchemy and Artificial Intelligence*, written by Massachusetts Institute of Technology (MIT) philosopher Hubert Dreyfus and published as a mimeographed memo in 1965, criticized both the goals and basic assumptions of artificial intelligence researchers. Written while Dreyfus did consulting work for RAND, the report generated a strongly negative reaction in the AI community.

RAND, a nonprofit American global policy think tank, had hired Dreyfus to assess the prospects for artificial intelligence research from a philosopher's point of view. At the time, glowing predictions for the future of AI were common among researchers such as Herbert Simon and Marvin Minsky, who, in the late 1950s, declared that machines capable of doing anything humans could do would exist within decades. For most AI researchers, the goal was not merely to create programs that processed information in such a way that the product or outcome appeared to be the result of intelligent behavior. Rather, they intended to create programs that replicated human thought processes. Artificial intelligence experts viewed human cognitive processes as a model for their programs, and likewise, they believed AI would offer insight into human psychology.

Dreyfus's philosophy was influenced by the work of phenomenologists Maurice Merleau-Ponty, Martin Heidegger, and Jean-Paul Sartre. In his report, Dreyfus argued that the theory and goals of AI were based on a philosophy of human psychology called associationism, which contains a key concept: that thinking occurs in a series of simple, determinate steps. Belief in associationism (which Dreyfus argued is incorrect) led artificial intelligence researchers to think they could use computers to replicate human thought processes. Dreyfus contrasted the features of human thinking (as he interpreted them) with the information processing of computers and the inner workings of specific AI programs.

The gist of his argument was that the information processing techniques of humans and machines are qualitatively different. Computers can only be programmed to handle "unambiguous, completely structured information," which make them incapable of handling the "ill-structured data of daily life," (Dreyfus 1965, 66) and hence incapable of intelligence. On the other hand, Dreyfus argued, many aspects of human intelligence cannot be described by rules or associationist psychology, contrary to the core assumption of AI research.

Dreyfus identified three areas in which human information processing differs from that of computers: the fringe consciousness, insight, and tolerance for ambiguity. The *fringe consciousness* is used, for example, by chess players in deciding what part of the board or pieces to focus on when choosing a move. The human player functions unlike a chess-playing program, in that the human player does not explicitly consider the information or count out potential moves the way the computer does, consciously or unconsciously. Only when the player has used their fringe consciousness to decide on the pieces to focus on do they switch to consciously calculating the effects of potential moves in a way similar to computer processing.

*Insight* into the deep structure of a complex problem allows the (human) problem-solver to develop a series of steps for solving it. Problem-solving programs lack this insight. Rather, the problem-solving strategy must be preliminarily structured as part of the program. *Tolerance for ambiguity* is best seen in the understanding of natural language, in which a word or sentence might have an ambiguous meaning but is correctly understood by the listener. There are an infinite number of cues that might be considered when interpreting ambiguous syntax or semantics, but somehow the human processor is able to choose relevant information from this infinite realm in order to correctly interpret the meaning. A computer, on the other hand, cannot be programmed to search through all possible facts to interpret ambiguous syntax or semantics. Either the number of facts is too large or the rules necessary for interpretation are too complicated.

Dreyfus was criticized by AI researchers for over-simplifying the issues and misunderstanding the capabilities of computers. RAND commissioned a reply to the report by MIT computer scientist Seymour Papert, which he disseminated in 1968 as *The Artificial Intelligence of Hubert L. Dreyfus: A Budget of Fallacies.* Papert also arranged a match between Dreyfus and the chess-playing program Mac Hack, which, to the delight of the artificial intelligence community, Dreyfus lost. Still, some of the criticisms he formulated in this report and in his subsequent books seem to have anticipated intractable problems later acknowledged by AI researchers themselves, including artificial general intelligence (AGI), artificial simulation of analog neurons, and the limitations of symbolic artificial intelligence as a model of human reasoning.

More publicly, artificial intelligence researchers proclaimed Dreyfus's work as irrelevant, claiming that he misunderstood their research. Dreyfus's criticisms of AI, which often included inflammatory rhetoric, had provoked their ire. Even *The New Yorker* magazine published excerpts from the report in its "Talk of the Town" column. Dreyfus later revised and expanded his argument, resulting in his 1972 book *What Computers Can't Do: The Limits of Artificial Intelligence.*

*Juliet Burba*

*See also:* Mac Hack; Simon, Herbert A.; Minsky, Marvin.

## Further Reading

Crevier, Daniel. 1993. *AI: The Tumultuous History of the Search for Artificial Intelligence.* New York: Basic Books.

Dreyfus, Hubert L. 1965. *Alchemy and Artificial Intelligence.* P-3244. Santa Monica, CA: RAND Corporation.

Dreyfus, Hubert L. 1972. *What Computers Can't Do: The Limits of Artificial Intelligence.* New York: Harper and Row.

McCorduck, Pamela. 1979. *Machines Who Think: A Personal Inquiry into the History and Prospects of Artificial Intelligence.* San Francisco: W. H. Freeman.

Papert, Seymour. 1968. *The Artificial Intelligence of Hubert L. Dreyfus: A Budget of Fallacies.* Project MAC, Memo No. 154. Cambridge, MA: Massachusetts Institute of Technology.

## Algorithmic Bias and Error

Bias in algorithmic systems has become one of the most critical concerns surrounding the ethics of artificial intelligence. Algorithmic bias refers to repeatable and systemic errors in a computer system that discriminates against certain groups or individuals. It is important to note that bias in and of itself is not always problematic: bias can be designed into a system in an effort to correct an unfair system or reality. Problems occur when bias leads to an unfair or discriminatory outcome, impacting the lives and opportunities of individuals and communities. Very often, those at risk from algorithmic bias and error are individuals and communities already vulnerable within the society. In this way, algorithmic bias can perpetuate and deepen social inequality, limiting people's access to services and products.

Algorithms are increasingly being used to inform government decision-making, particularly within the criminal justice sector for sentencing and bail, as well as in the management of migration with the use of biometric technologies, such as facial and gait recognition. When a government's algorithms are shown to be biased, citizens can lose trust in both the AI system and its use by institutions, whether it's a government office or private company.

Over the last few years, examples of algorithmic bias have abounded. The targeted advertising of Facebook, premised on algorithms that determine which demographic groups a particular advertisement should be seen by, is a high-profile example. Indeed, one study showed that on Facebook job advertisements for janitors and similar positions are often directed at lower-income groups and minorities, while advertisements for nurses or secretaries are targeted at women (Ali et al. 2019). This is effectively profiling of people in protected classes, such as race, gender, and income bracket, for the purposes of maximizing advertising effectiveness and profitability.

Another well-cited example is the algorithm developed by Amazon to improve efficiency, and supposed objectivity, in recruitment by sorting and ranking résumés. Amazon's algorithm was trained on the data of the historical hiring practices of the company. However, as it was rolled out, it became clear that the algorithm was biased toward women, down-ranking résumés that included the words *women* or *gender*, or if it indicated that the applicant had attended a women's college. As the algorithm was trained on the historical recruitment patterns of Amazon, little could be done to correct its biases. In fact, while the algorithm was clearly biased, what this example shows is how such biases may reflect societal biases and, in this case, the deeply embedded biases against hiring women at Amazon.

Indeed, there are several ways in which bias in an algorithmic system can occur. Broadly speaking, algorithmic bias tends to occur when a group of persons, and their lived realities, are not taken into account in the design of the algorithm. This can occur at various stages of the process of developing an algorithm, from the collection of data that is not representative of all demographic groups to the labeling of data in ways that reproduce discriminatory profiling, to the rollout of an algorithm where the differential impact it may have on a particular group is not taken into account.

Partly in response to significant publicity of algorithmic biases, in recent years there has been a proliferation of policy documents addressing the ethical responsibilities of state and non-state bodies using algorithmic processing—to ensure against unfair bias and other negative effects of algorithmic processing (Jobin et al. 2019). One of the key policies in this space is the European Union's "Ethics Guidelines for Trustworthy AI" published in 2018. The EU document outlines seven principles for the fair and ethical governance of AI and algorithmic processing.

In addition, the European Union has been at the forefront of regulatory responses to algorithmic processing with the promulgation of the General Data Protection Regulation (GDPR), also published in 2018. Under the GDPR, which applies in the first instance to the processing of all personal information within the EU, a company can be fined up to 4 percent of its annual global revenue for using an algorithm that is shown to be biased on the basis of race, gender, or other protected category.

A lingering concern for the regulation of algorithmic processing is the difficulty of ascertaining where a bias occurred and what dataset led to bias. Typically, this is known as the algorithmic black box problem: the deep data processing layers of an algorithm are so complex and numerous they simply cannot be understood by a human. Based on the right to an explanation where, subject to an automated decision under the GDPR, one of the responses has been to ascertain where the bias occurred through counterfactual explanations, different data is inputted into the algorithm to see where the differential outcomes occur (Wachter et al. 2018).

In addition to legal and policy tools for addressing algorithmic bias, technical solutions to the problem included developing synthetic datasets that attempt to fix naturally occurring biases in datasets or offer an unbiased and representative dataset. While such avenues for redress are important, one of the more holistic responses to the problem are that human teams that develop, produce, use, and monitor the impact of algorithms should be much more diverse. Within diverse teams, a combination of lived experiences make it more likely that biases can be detected sooner and addressed.

*Rachel Adams*

*See also:* Biometric Technology; Explainable AI; Gender and AI.

**Further Reading**

Ali, Muhammed, Piotr Sapiezynski, Miranda Bogen, Aleksandra Korolova, Alan Mislove, and Aaron Rieke. 2019. "Discrimination through Optimization: How

Facebook's Ad Delivery Can Lead to Skewed Outcomes." In *Proceedings of the ACM on Human-Computer Interaction*, vol. 3, CSCW, Article 199 (November). New York: Association for Computing Machinery.

European Union. 2018. "General Data Protection Regulation (GDPR)." https://gdpr-info.eu/.

European Union. 2019. "Ethics Guidelines for Trustworthy AI." https://ec.europa.eu/digital-single-market/en/news/ethics-guidelines-trustworthy-ai.

Jobin, Anna, Marcello Ienca, and Effy Vayena. 2019. "The Global Landscape of AI Ethics Guidelines." *Nature Machine Intelligence* 1 (September): 389–99.

Noble, Safiya Umoja. 2018. *Algorithms of Oppression: How Search Engines Reinforce Racism*. New York: New York University Press.

Pasquale, Frank. 2016. *The Black Box Society: The Secret Algorithms that Control Money and Information*. Cambridge, MA: Harvard University Press.

Wachter, Sandra, Brent Mittelstadt, and Chris Russell. 2018. "Counterfactual Explanations Without Opening the Black Box: Automated Decisions and the GDPR." *Harvard Journal of Law & Technology* 31, no. 2 (Spring): 841–87.

Zuboff, Shoshana. 2018. *The Age of Surveillance Capitalism: The Fight for a Human Future at the New Frontier of Power*. London: Profile Books.

## Animal Consciousness

In recent decades, researchers have developed an increasing appreciation of animal and other nonhuman intelligences. Consciousness or sentience, complex forms of cognition, and personhood rights have been argued for in ravens, bower birds, gorillas, elephants, cats, crows, dogs, dolphins, chimpanzees, grey parrots, jackdaws, magpies, beluga whales, octopi, and several other species of animals. The Cambridge Declaration on Consciousness and the separate Nonhuman Rights Project mirror the contemporary struggle against racism, classism, sexism, and ethnocentrism by adding one more prejudice: "speciesism," coined in 1970 by psychologist Richard Ryder and popularized by the philosopher Peter Singer. Animal consciousness, indeed, may pave the way for consideration and appreciation of other types of proposed intelligences, including those that are artificial (traditionally considered, such as animals, to be "mindless automata") and extraterrestrial.

One of the most important questions experts in many fields grapple with today is the knowability of the subjective experience and objective qualities of other types of consciousness. "What is it like to be a bat?" the philosopher Thomas Nagel famously asked, especially as they are capable of echolocation and humans are not. Most selfishly, understanding animal consciousness might open a window to better understanding of human consciousness by way of comparison. Looking to animals also might reveal new perspectives on the mechanisms for evolution of consciousness in human beings, which might in turn help scientists equip robots with similar traits, appreciate their moral status, or sympathize with their behavior. History is littered with examples of animals used as a means for human ends, rather than as ends themselves. Cows produce milk for human consumption. Sheep make wool for clothing. Horses once provided transportation and power for agriculture, and now afford opportunities for entertainment and gambling. The "discovery" of animal consciousness may mean removing the human species from the center of its own mental universe.

The "Cognitive Revolution" of the twentieth century, which seemingly removed the soul as a scientific explanation of mental life, opened the door to studying and making experiments in perception, memory, cognition, and reasoning in animals and also exploring the possibilities for incorporating sophisticated information processing convolutions and integrative capabilities into machines. The possibility of a basic cognitive "software" common to humans, animals, and artificial general intelligences is often discussed from the perspective of newer interdisciplinary fields such as neuroscience, evolutionary psychology, and computer science.

The independent researcher John Lilly was among the first to argue, in his book *Man and Dolphin* (1961), that dolphins are not merely intelligent, but in many ways, they possess qualities and communication skills beyond the human. Other researchers such as Lori Marino and Diana Reiss have since confirmed many of his findings, and rough agreement has been reached that dolphin's self-awareness lies somewhere on the continuum between humans and chimps. Dolphins have been observed to fish cooperatively with human fishermen, and the most famous dolphin in history, Pelorus Jack, faithfully and voluntarily escorted ships through the dangerous rocks and tidal flows of Cook Strait in New Zealand for twenty-four years.

Some animals appear to pass the famous mirror test of self-recognition. These include dolphins and killer whales, chimpanzees and bonobos, magpies, and elephants. The test is usually administered by painting a small mark on an animal, in a place where it cannot see without recourse to the mirror. If the animal touches the mark on their own body after seeing it reflected, they may be said to recognize themselves. Some critics have argued that the mirror-mark test is unfair to some species of animals because it privileges vision over other sense organs.

SETI researchers acknowledge that study of animal consciousness may partially prepare human beings to grapple with the existential ramifications of self-aware extraterrestrial intelligences. Similarly, work with animals has spawned parallel interest in consciousness in artificial intelligences. To cite one direct example: In his autobiography *The Scientist* (1978), John Lilly describes a hypothetical Solid State Intelligence (SSI) that would inevitably arise from the work of human computer scientists and engineers. This SSI would be made of computer parts, produce its own integrations and enhancements, and ultimately engage in self-replication to challenge and defeat humanity. The SSI would protect some human beings in domed "reservations" completely subject to its own maintenance and control. Eventually, the SSI would master the ability to move the planet and explore the galaxy looking for other intelligences like itself.

Self-consciousness in artificial intelligences has been critiqued on many levels. John Searle has argued vigorously that machines lack intentionality, that is, the ability to find meaning in the computations they execute. Inanimate objects are rarely thought of as possessing free will and thus are not conceptually human. Further, they might be thought of as having a "missing-something," for instance, no soul, or no ability to discriminate between right and wrong, or no emotion and creativity.

However, discoveries about animal consciousness have added a dimension to arguments about both animal and robot rights because they make possible an argument that these beings have the capacity to recognize that they are having positive or negative experiences. They also open the door to recognizing strong forms of social cognition such as attachment, communication, and empathy. A long list of chimpanzees, gorillas, orangutans, and bonobos—the famous Washoe, Koko, Chantek, and Kanzi among them—have mastered an astonishing number of gestures in American Sign Language or artificial lexigrams (keyboard symbols of objects or ideas), raising the prospect for true interspecies social exchange. In 1993, an international group of primatologists founded the Great Ape Project, with the avowed aim of conferring basic human rights to life, liberty, and protection from torture upon these animals. These animals should, they said, be accorded nonhuman personhood and brought to the fore in the great mammalian "community of equals." Many prominent marine mammal researchers have leapt to the forefront as activists against the indiscriminate killing of cetaceans by fishermen or their use in captive performances.

The Nonhuman Rights Project was founded by American lawyer Steven Wise at the Center for the Expansion of Fundamental Rights in 2007. The aim of the Nonhuman Rights Project is to seek legal personhood for animals that are now considered property of legal persons. These fundamental personhood rights would include bodily liberty (against imprisonment) and bodily integrity (against laboratory experimentation). The organization claims that there is no standard under common law that precludes animal personhood, and indeed, the law eventually permitted human slaves to become legal persons under the writ of habeas corpus without prior precedent. Habeas corpus writs permit individuals to declare the right to liberty and against unlawful detention.

Since 2013, the Nonhuman Rights Project has sought legal redress for animals in the courts. The first lawsuit was filed in New York State to assert the rights of four captive chimpanzees, and evidence included the affidavit of respected primatologist Jane Goodall. The organization asked that the chimps be released and moved to a reserve operated by the North American Primate Sanctuary Alliance. The group's petitions and appeals were denied. Steven Wise has taken solace in the fact that in one decision, justice agreed that the question of personhood is not biological but one shaped by public policy and social principles.

In 2012, a group of neuroscientists attending the Francis Crick Memorial Conference signed the Cambridge Declaration on Consciousness. The three scientists most directly involved in the writing of the document were David Edelman of the Neurosciences Institute in La Jolla, California; Christof Koch of the California Institute of Technology; and Philip Low of Stanford University. All signatories agreed that scientific techniques had yielded evidence that mammals have brain circuits that appear to correlate with the experience with consciousness, affective states, and emotional behaviors. The group also opined that birds seem to have evolved consciousness convergent with mammals. The group also found REM sleep patterns in zebra finches and analogous effects of hallucinogenic drugs notable as evidence of conscious behavior in animals. Signers of the declaration also observed that invertebrate cephalopods appear to have self-conscious awareness despite lacking a neocortex for higher-order brain functions.

Such views have not been without criticism. Attorney Richard Cupp has argued that humans should continue to bear legal responsibility for the care of animals. He also argues that animal personhood might obstruct the rights and autonomy of humans with cognitive impairments, making them vulnerable to diminished personhood under the law. Cupp also views animals as existing outside human moral community and therefore the social compact that defines personhood rights in the first place. Philosopher and cognitive scientist Daniel Dennett is a prominent critic of animal sentience, believing that consciousness is a "fiction" that can only be constructed out of stories created using human language. Because animals cannot create such stories, they must not be conscious. And because consciousness is a narrative we tell ourselves, scientific disciplines can never capture what it means to be like an animal that is conscious, because science relies on objective facts and universal descriptions, not stories.

*Philip L. Frana*

*See also:* Nonhuman Rights and Personhood; Sloman, Aaron.

**Further Reading**

Dawkins, Marian S. 2012. *Why Animals Matter: Animal Consciousness, Animal Welfare, and Human Well-being.* Oxford, UK: Oxford University Press.

Kaplan, Gisela. 2016. "Commentary on 'So Little Brain, So Much Mind: Intelligence and Behavior in Nonhuman Animals' by Felice Cimatti and Giorgio Vallortigara." *Italian Journal of Cognitive Science* 4, no. 8: 237–52.

Solum, Lawrence B. 1992. "Legal Personhood for Artificial Intelligences." *North Carolina Law Review* 70, no. 4: 1231–87.

Wise, Steven M. 2010. "Legal Personhood and the Nonhuman Rights Project." *Animal Law Review* 17, no. 1: 1–11.

Wise, Steven M. 2013. "Nonhuman Rights to Personhood." *Pace Environmental Law Review* 30, no. 3: 1278–90.

# Asimov, Isaac (c. 1920–1992)

Isaac Asimov was a professor of biochemistry at Boston University and also one of the foremost authors of science fiction. A prolific writer in many genres, Asimov's body of science fiction is hugely influential in not only the genre but also ethical issues surrounding science and technology.

Asimov was born in Russia. Unsure of his exact birth date, he celebrated January 2, 1920, as his birthday. His family immigrated to New York City in 1923. At age sixteen, Asimov attempted to attend Columbia College, Columbia University's undergraduate school, but due to anti-Semitic quotas on the number of Jewish students, he was denied entry and eventually enrolled in Seth Low Junior College, a related undergraduate institution. When Seth Low shut its doors, Asimov transferred to Columbia College but received a Bachelor of Science rather than of arts, which Asimov described as "a gesture of second-class citizenship" (Asimov 1994, n.p.).

Around this time, Asimov became interested in science fiction and began writing letters to science fiction magazines, eventually trying to craft his own short stories. He succeeded in publishing his first short story, "Marooned off Vesta" (1938), in *Amazing Stories*. His early writings put him in the orbit of other science fiction pioneers such as Robert Heinlein. Following graduation, Asimov initially tried—but failed—to enter medical school. Instead, he entered graduate school for chemistry in 1939 at age nineteen. His graduate studies were interrupted by World War II, and at Heinlein's suggestion, Asimov fulfilled his military service by working at the Naval Air Experimental Station in Philadelphia during the war. While stationed there, he wrote short stories that became the basis for *Foundation* (1951), one of his most famous works and the first in a multi-volume series that he would later connect to several of his other stories. In 1948, he received his doctorate from Columbia University.

Asimov's seminal *Robots* series (1950s–1990s) has become a basis for ethical guidelines to assuage human fears of technology run amok. For instance, Asimov's Three Laws of Robotics often are cited as guiding principles for artificial intelligence and robotics. The Three Laws were first introduced in the short story "Runaround" (1942), which was later included in the 1950 collection *I, Robot* (1950):

1.  A robot may not injure a human being or, through inaction, allow a human being to come to harm.
2.  A robot must obey orders given to it by human beings except where such orders would conflict with the First Law.
3.  A robot must protect its own existence as long as such protection does not conflict with the First or Second Law.

In *Robots and Empire* (1985), a "zeroth law" is developed in order for robots to foil a plot to destroy Earth: "A robot may not injure humanity, or through inaction, allow humanity to come to harm." This law supersedes the original Three Laws.

In Asimov's *Robot* series of novels and short stories, characters often must solve a mystery where a robot seemingly has acted in violation of the Three Laws. For instance, in "Runaround," two field specialists for U.S. Robots and Mechanical Men, Inc., realize they are in danger of being trapped on Mercury because their robot "Speedy" has not returned with selenium needed to power a protective shield in an abandoned base, to protect themselves from the sun. They realize that Speedy has malfunctioned because he is caught in a conflict between the Second and Third Laws: Speedy is obeying an order given casually by one of the field specialists, but as the robot approaches the selenium, it is forced to retreat in order to protect itself from a corrosive amount of carbon monoxide near the selenium. The humans must figure out how to use the Three Laws to break Speedy out of a feedback loop caused by the conflict.

Later stories and novels feature more complex deliberations about the application of the Three Laws. In "The Evitable Conflict" (1950), four powerful computers called the Machines run the world's economy, and "robopsychologist" Susan Calvin realizes that they have altered the First Law into a forerunner of Asimov's zeroth law: "the Machines work not for any single human being, but for all human-ity" (Asimov 2004b, 222). Calvin worries that the Machines are moving humanity toward what they believe is "the ultimate good of humanity" (Asimov 2004b, 222) even if humanity doesn't know what that is.

Additionally, "psychohistory," a term introduced in Asimov's *Foundation* series (1940s–1990s), could be described as anticipating the algorithms that pro-vide the foundation for artificial intelligence today. In *Foundation,* the main pro-tagonist Hari Seldon develops psychohistory as a way to make general predictions about the future behavior of very large groups of people, including the fall of civi-lization (here, the Galactic Empire) and the inevitable Dark Ages. But Seldon argues that utilizing psychohistory can reduce the period of anarchy:

> Psychohistory, which can predict the fall, can make statements concerning the suc-ceeding dark ages. The Empire . . . has stood twelve thousand years. The dark ages to come will endure not twelve, but *thirty* thousand years. A Second Empire will rise, but between it and our civilization will be one thousand generations of suffer-ing humanity . . . It is possible . . . to reduce the duration of anarchy to a single mil-lennium, if my group is allowed to act now. (Asimov, 2004a, 30–31)

Psychohistory makes "a prediction which is made by mathematics" (Asimov, 2004a, 30), just as artificial intelligence might make a prediction. In the *Founda-tion* series, Seldon establishes the Foundation, a secret group of individuals keep-ing the collected knowledge of humanity enact and thus being the literal foundation for a potential second Galactic Empire. In later installments of the *Foundation* series, the Foundation finds itself threatened by the Mule, a mutant and thus an aberration that was not foreseen through the predictive analysis of psychohistory. The tension between large-scale theories and individual actions is a key force driv-ing *Foundation*: although Seldon's thousand-year plan relies on macro theories—"the future isn't nebulous. It's been calculated out by Seldon and charted" (Asimov 2004a, 100)—the plan is saved or threatened by individual actions.

Asimov's writings were often predictive, leading some to call his and col-leagues' work to be "future history" or "speculative fiction." Because of the ethi-cal dilemmas explored in Asimov's writings, they are often invoked in legal, political, and policy debates years after they were written. For instance, in 2007, the South Korean Ministry of Commerce, Industry and Energy developed a Robot Ethics Charter with ethical guidelines based on the Three Laws, as the govern-ment speculated that all Korean homes would have a robot by 2020. In 2017, the Artificial Intelligence Committee of the British House of Lords included a set of principles that parallel the Three Laws.

Others have questioned the usefulness of the Three Laws. First, some critics note that robots are often used for military functions, so the Three Laws would reduce these functions, something that Asimov might have been supportive of given his anti-war short stories such as "The Gentle Vultures" (1957). Second, some note that the types of robots and the use of AI today are very different from those in the *Robot* series. The "positronic brain" that powers Asimov's fictional robots is still science fiction today, and beyond current computing power. Third, the Three Laws are obviously fiction, and Asimov's *Robot* series is based on misinterpretations in order to further the ethical questions and for dramatic effect. But because they can be misinterpreted just like other laws, critics argue that the Three Laws cannot serve as a real moral code for regulating AI or robots. Finally, some question to whom these ethical guidelines should apply.

Asimov died in 1992 due to complications from AIDS, which he contracted via a contaminated blood transfusion during a 1983 heart bypass surgery.

*Oliver J. Kim*

*See also:* Beneficial AI, Asilomar Meeting on; Pathetic Fallacy; Robot Ethics.

**Further Reading**

Asimov, Isaac. 1994. *I, Asimov: A Memoir*. New York: Doubleday.
Asimov, Isaac. 2002. *It's Been a Good Life*. Amherst: Prometheus Books.
Asimov, Isaac. 2004a. *The Foundation Novels*. New York: Bantam Dell.
Asimov, Isaac. 2004b. *I, Robot*. New York: Bantam Dell.

## Automated Machine Learning

Machine learning algorithms are designed specifically to detect and characterize complex patterns in large datasets. Automated machine learning (AutoML) promises to bring these analytical tools to everyone interested in the analysis of big data by taking the guesswork out of building tools of convenience. These tools are termed "computational analysis pipelines."

While there is much work to do in automated machine learning, notable early successes suggest that automated machine learning software will be a crucial tool in the computer or data scientists' toolbox. It will be important to tailor these software packages to novice users by allowing them to perform complex machine learning tasks in a user-friendly manner, while at the same time facilitating the integration of domain-specific knowledge and model interpretation and action. These latter goals have largely not been addressed, but will need to be the focus of coming research, before AutoML is ready to solve complex real-world problems.

Automated machine learning is a relatively new area of study that has emerged within the last ten years as a result of the availability of powerful open-source machine learning libraries and plentiful high-performance computing. There are now multiple open-source and commercial AutoML software packages available for use. Many of these packages enable the exploration of machine learning pipelines that can include feature transformation algorithms such as discretization (transforming continuous equations, functions, models, and variables into discrete equations, functions, and so forth for digital computers), feature engineering algorithms such as principal components analysis (a process that removes large dimension "less important" data while retaining a subset of "more important" variables), feature selection algorithms such as ReliefF (a technique that minimizes error), and multiple different machine learning algorithms along with their parameter settings. Stochastic search algorithms used in AutoML have included Bayesian optimization, ensemble methods, and genetic programming. Stochastic search algorithms may be deployed in computational problems with intrinsic random noise or deterministic problems unraveled by injection of randomness. New approaches to remove the "signal from the noise" in datasets, and find insights and make predictions, are being actively developed and evaluated.

One of the challenges of machine learning is that each algorithm looks at data in a different way. That is, each algorithm detects and characterizes different patterns. Some algorithms such as linear support vector machines are good at detecting linear patterns while k-nearest neighbor algorithms can detect nonlinear patterns. The challenge is that scientists do not know when they begin their work which algorithm(s) to use because they do not know what patterns they are looking for in the data. What most people do is choose an algorithm they are familiar with or choose one that seems to work well across a wide range of datasets. Some may choose an algorithm because the models that are generated are easy to interpret. There are lots of different reasons that certain algorithms are chosen for a data analysis. Nonetheless, the chosen algorithm may not be ideal for a given dataset. This challenge is particularly difficult for the novice user who may not understand the strengths and weaknesses of each algorithm.

One approach to this problem is to perform a grid search. Here, multiple machine learning algorithms and parameter settings are applied to a dataset in a systematic way and the results compared to identify a best algorithm. This is a commonly used approach and can yield good results. The challenge with the grid search is that it can be computationally intensive if many algorithms, each with several parameter settings, need to be evaluated. Random forests are classification algorithms built from multiple decision trees that have several commonly used parameters that must be tuned for optimal performance on a given dataset. Parameters are configuration variables that the adopted machine learning technique uses to adjust the data. A common parameter is the maximum number of features that will be permitted in the decision trees that are created and evaluated.

Automated machine learning can help manage the complex, computationally expensive combinatorial explosion in specific analyses that need to be run. A single parameter might have, for example, ten different settings. A second parameter could be the number of decision trees to be included in the forest, perhaps another ten different settings. A third parameter could be the minimum number of samples that will be allowed in the "leaves" of the decision trees, another ten different settings. This example yields 1000 different possible parameter settings assuming the exploration of only three parameters. A data scientist investigating ten different machine learning algorithms, each with 1000 possible parameter settings, would need to run 10,000 specific analyses.

On top of these analyses are so-called hyperparameters, which involve characteristics of the analyses that are set beforehand and thus not learned from the data. They are often specified by the data scientist using various rules of thumbs or values drawn from other problems. Hyperparameter configurations might involve comparisons of several different cross-validation strategies or examine the effect of sample size on results. In a typical example, hundred hyperparameter combinations might need to be evaluated. The combination of machine learning algorithms, parameter settings, and hyperparameter settings could in this way yield a total of one million analyses that the data scientist would need to perform.

Depending on the sample size of the data to be analyzed, the number of features, and the types of machine learning algorithms selected, so many separate analyses could be prohibitive given the computing resources that are available to the user.

An alternative approach is to use a stochastic search to approximate the best combination of machine learning algorithms, parameter settings, and hyperparameter settings. A random number generator is used to sample from all possible combinations until some computational limit is reached. The user manually explores additional parameter and hyperparameter settings around the best method before making a final choice. This has the advantage of being computationally manageable but suffers from the stochastic element where chance might not explore the optimal combinations.

A solution to this is to add a heuristic element—a practical method, guide, or rule—to create a stochastic search algorithm that can adaptively explore algorithms and settings while improving over time. Approaches that employ stochastic searches with heuristics are called automated machine learning because they automate the search for optimal machine learning algorithms and settings. A stochastic search might start by randomly generating a number of machine learning algorithm, parameter setting, and hyperparameter setting combinations and then evaluating each one using cross-validation, a technique for testing the effectiveness of a machine learning model. The best of these is selected, randomly modified, and then evaluated again. This process is repeated until a computational limit or a performance objective is reached. The heuristic algorithm governs this process of stochastic search. The development of optimal search strategies is an active area of research.

The AutoML approach has numerous advantages. First, it can be more computationally efficient than the exhaustive grid search approach. Second, it makes machine learning more approachable because it takes some of the guesswork out of selecting an optimal machine learning algorithm and its many settings for a given dataset. This helps bring machine learning to the novice user. Third, it can yield more reproducible results if generalizability metrics are built into the heuristic that is used. Fourth, it can yield more interpretable results if complexity metrics are built into the heuristic. Fifth, it can yield more actionable results if expert knowledge is built into the heuristic.

Of course, there are some challenges with AutoML approaches. First is the challenge of overfitting—producing an analysis that corresponds too closely to known data but does not fit or predict unseen or new data—due to the evaluation of many different algorithms. The more analytical methods that are applied to a dataset, the higher the chance of learning the noise in the data that leads to a model unlikely to generalize to independent data. This needs to be rigorously addressed with any AutoML method. Second, AutoML can be computationally intensive in its own right. Third, AutoML methods can generate very complex pipelines that include multiple different machine learning methods. This can make interpretation much more difficult than picking a single algorithm for the analysis. Fourth, this field is still in its infancy. Ideal AutoML methods may not have been developed as yet despite some promising early examples.

*Jason H. Moore*

*See also:* Deep Learning.

**Further Reading**

Feurer, Matthias, Aaron Klein, Katharina Eggensperger, Jost Springenberg, Manuel Blum, and Frank Hutter. 2015. "Efficient and Robust Automated Machine Learning." In *Advances in Neural Information Processing Systems*, 28. Montreal, Canada: Neural Information Processing Systems. http://papers.nips.cc/paper/5872 -efficient-and-robust-automated-machine-learning.

Hutter, Frank Hutter, Lars Kotthoff, and Joaquin Vanschoren, eds. 2019. *Automated Machine Learning: Methods, Systems, Challenges.* New York: Springer.

# Automated Multiphasic Health Testing

Automated Multiphasic Health Testing (AMHT) is an early medical computing system for semiautomatically screening large numbers of sick or well individuals in a short amount of time. Public health officer Lester Breslow introduced the AMHT concept in 1948 by combining traditional automated medical questionnaires and mass screening techniques for groups of people being tested for single diseases such as diabetes, tuberculosis, or heart disease. In multiphasic health testing, groups of people are checked for multiple diseases, illnesses, or injuries by combining a number of tests into a single package. AMHT may be linked to routine physicals or periodic health campaigns. Humans are examined under circumstances not unlike those in place for the state inspection of automobiles. In other words, AMHT takes an assembly line approach to preventive medical care.

Automated Multiphasic Health Testing (AMHT) came of age in the 1950s, giving health provider networks a way of quickly testing new applicants. The Kaiser Foundation Health Plan made a Multiphasic Health Checkup available to its subscribers in 1951. Electrical engineer and physician Morris F. Collen headed the program from 1961 to 1979. The "Kaiser Checkup," which relied on an IBM 1440 computer, crunched data from patient interviews, laboratory tests, and clinician findings, looking for undiagnosed conditions and offering recommendations for treatment. At the questionnaire station (one of twenty such stations), patients hand-sorted 200 prepunched cards, upon which were printed questions requiring "yes" or "no" responses. The computer arranged the cards and applied a likelihood ratio test developed by the famous statistician Jerzy Neyman. Kaiser's computer system also recorded electrocardiographic, spirographic, and ballistocardiographic medical data. An entire Kaiser Checkup took approximately two-and-a-half hours. Similar AMHT programs were launched overseas, including BUPA in the United Kingdom and a national program initiated by the Swedish government.

In recent decades, enthusiasm for automated health testing has waned. There are privacy concerns as well as questions about cost. Physicians and computer scientists working with AMHT have also discovered that the body often obscures symptoms. A sick individual may successfully pass through diagnostic machines one day and still drop dead the next. Electronic medical recordkeeping, however, has succeeded where AMHT generally has not. Records may be transmitted, amended, and returned without physical handling or duplication. Patient charts can even be used simultaneously by multiple health professionals. Uniform data entry promises legibility and uniform structure.

Electronic medical records software today allows summary reports to be compiled automatically from the facts collected in individual electronic medical records. These "big data" reports permit the tracking of changes in medical practice and assessment of outcomes over time. Summary reports also allow cross-patient analysis, deep algorithmic study of prognoses by classes of patients, and identification of risk factors before treatment becomes necessary. Mining of medical data and application of deep learning algorithms have led to an explosion of interest in so-called cognitive computing for health care. Two current leaders, IBM's Watson system and Google DeepMind Health, promise revolutions in eye disease and cancer diagnosis and treatment. And IBM has announced the Medical Sieve system, which analyzes radiological scans as well as written records.

*Philip L. Frana*

*See also:* Clinical Decision Support Systems; Computer-Assisted Diagnosis; INTERNIST-I and QMR.

**Further Reading**

Ayers, W. R., H. M. Hochberg, and C. A. Caceres. 1969. "Automated Multiphasic Health Testing." *Public Health Reports* 84, no. 7 (July): 582–84.

Bleich, Howard L. 1994. "The Kaiser Permanente Health Plan, Dr. Morris F. Collen, and Automated Multiphasic Testing." *MD Computing* 11, no. 3 (May–June): 136–39.

Collen, Morris F. 1965. "Multiphasic Screening as a Diagnostic Method in Preventive Medicine." *Methods of Information in Medicine* 4, no. 2 (June): 71–74.

Collen, Morris F. 1988. "History of the Kaiser Permanente Medical Care Program." Interviewed by Sally Smith Hughes. Berkeley: Regional Oral History Office, Bancroft Library, University of California.

Mesko, Bertalan. 2017. "The Role of Artificial Intelligence in Precision Medicine." *Expert Review of Precision Medicine and Drug Development* 2, no. 5 (September): 239–41.

Roberts, N., L. Gitman, L. J. Warshaw, R. A. Bruce, J. Stamler, and C. A. Caceres. 1969. "Conference on Automated Multiphasic Health Screening: Panel Discussion, Morning Session." *Bulletin of the New York Academy of Medicine* 45, no. 12 (December): 1326–37.

## Automatic Film Editing

Automatic film editing is a process by which an algorithm, trained to follow basic rules of cinematography, edits and sequences video in the assembly of completed motion pictures. Automated editing is part of a broader effort to introduce artificial intelligence into movie making, sometimes called intelligent cinematography.

Already by the mid-1960s, the influential filmmaker Alfred Hitchcock could imagine that an IBM computer might one day be capable of turning a written screenplay into a finished film. Hitchcock invented many of the principles of modern moviemaking. One well-known rule of thumb, for instance, is his assertion that, where possible, the size of a person or object in frame should be proportional to their importance in the story at that particular point in time. Other film editing tenets that emerged from long experience by filmmakers are "exit left, enter right," which helps the viewer follow lateral movements of characters on the screen, and the 180- and 30-degree rules for maintaining spatial relationships between subjects and the camera. Such rules over time became codified as heuristics governing shot selection, cutting, and rhythm and pacing. One example is Joseph Mascelli's *Five C's of Cinematography* (1965), which has grown into a vast knowledge base for making decisions about camera angles, continuity, cutting, closeups, and composition.

The first artificial intelligence film editing systems developed from these human-curated rules and human-annotated movie stock footage and clips. An early 1990s system is IDIC, developed by Warren Sack and Marc Davis in the MIT Media Lab. IDIC is designed to solve the real-world problem of film editing using Herbert Simon, J. C. Shaw, and Allen Newell's General Problem Solver, an early artificial intelligence program that was intended to solve any general problem using the same base algorithm. IDIC has been used to generate hypothetical *Star Trek* television trailers assembled from a human-specified story plan centered on a particular plot point.

Several film editing systems rely on idioms, that is, conventional procedures for editing and framing filmed action in specific situations. The idioms themselves will vary based on the style of film, the given context, or the action to be portrayed. In this way, the knowledge of expert editors can be approached in terms of case-based reasoning, using a past editing recipe to solve similar current and future problems. Editing for fight scenes follows common idiomatic pathways, as does ordinary conversations between characters. This is the approach modeled by Li-wei He, Michael F. Cohen, and David H. Salesin's Virtual Cinematographer, which relies on expert knowledge of idioms in the editing of entirely computer-generated video for interactive virtual worlds. The Declarative Camera Control Language (DCCL) developed by He's group formalizes the control of camera positions to follow cinematographic conventions in the editing of CGI animated films.

More recently, researchers have been working with deep learning algorithms and training data pulled from existing collections of recognized films possessing high cinematographic quality, to create proposed best cuts of new films. Many of the newer applications are available on mobile, drone, or handheld equipment. Easy automatic video editing is expected to make the sharing of short and interesting videos, assembled from shots made by amateurs with smartphones, a preferred medium of exchange over future social media. That niche is currently occupied by photography. Automatic film editing is also in use as an editing technique in machinima films made using 3D virtual game engines with virtual actors.

*Philip L. Frana*

*See also:* Workplace Automation.

## Further Reading

Galvane, Quentin, Rémi Ronfard, and Marc Christie. 2015. "Comparing Film-Editing."
    In *Eurographics Workshop on Intelligent Cinematography and Editing*, edited by

William H. Bares, Marc Christie, and Rémi Ronfard, 5–12. Aire-la-Ville, Switzerland: Eurographics Association.

He, Li-wei, Michael F. Cohen, and David H. Salesin. 1996. "The Virtual Cinematographer: A Paradigm for Automatic Real-Time Camera Control and Directing." In *Proceedings of SIGGRAPH '96*, 217–24. New York: Association for Computing Machinery.

Ronfard, Rémi. 2012. "A Review of Film Editing Techniques for Digital Games." In *Workshop on Intelligent Cinematography and Editing.* https://hal.inria.fr/hal -00694444/.

## Autonomous and Semiautonomous Systems

Autonomous and semiautonomous systems are generally distinguished by their reliance on external commands for decision-making. They are related to conditionally autonomous systems and automated systems. *Autonomous* systems are capable of decision-making within a specified domain of activity without human input, whereas *semiautonomous* systems rely upon a human user somewhere "in the loop" for decision-making, behavior regulation, or circumstantial interventions. *Conditionally* autonomous systems function autonomously under certain conditions.

Semiautonomous and autonomous systems (autonomy) are also distinct from automated systems (automation). The former systems include decision-making capability inclusive of assessing contextual inputs, whereas the latter systems' actions are predetermined sequences directly tied to specified inputs. Systems are considered automated when their actions, and alternatives for action, are predetermined in advance as responses to specific inputs. An example of an automated system is an automatic garage door that stops closing when a sensor detects an obstruction in the path of the door. Inputs can be received via not only sensors but also user interaction. An example of a user-initiated automatic system would be an automatic dishwasher or clothes washer where the human user specifies the sequences of events and behaviors through a user interface, and the machine then proceeds to execute the commands according to predetermined mechanical sequences.

In contrast, autonomous systems are those systems wherein the ability to evaluate circumstances and select actions is internal to the system. Like an automated system, the autonomous system still relies upon sensors, cameras, or user input to provide information, but the system's responses can be characterized by more complex decision-making based upon the situated assessment of multiple simultaneous inputs such as user intent, environment, and capability.

In considering real-world examples of systems, *automated*, *semiautonomous*, and *autonomous* are categories that have some overlap depending on the nature of the tasks under consideration and upon the specifics of decision-making. These categories aren't always clearly or precisely delineated. Lastly, the extent to which these categories apply depends upon the scale and level of the activity under consideration.

While the rough distinctions outlined above between automated, semiautonomous, and autonomous systems are generally agreed upon, ambiguity exists where these system categories are present in actual systems. One example of such ambiguity is in the levels of autonomy designated by SAE (formerly the Society of Automotive Engineers) for driverless cars. A single system may be Level 2 semiautonomous, Level 3 conditionally autonomous, or Level 4 autonomous depending on road or weather conditions or upon circumstantial indices such as the presence of road barriers, lane markings, geo-fencing, surrounding vehicles, or speed. Autonomy level may also depend upon how an automotive task is defined. In this way, the classification of a system depends as much upon the technological constitution of the system itself as the circumstances of its functioning or the parameters of the activity focus.

## EXAMPLES

### Autonomous Vehicles

Automated, semiautonomous, conditionally autonomous, and fully autonomous vehicle systems help illustrate the distinctions between these types of systems. Cruise control functionality is an example of an automated technology. The user sets a speed target for the vehicle and the vehicle maintains that speed, adjusting acceleration and deceleration as the terrain requires. In the case, however, of semiautonomous vehicles, a vehicle may be equipped with an adaptive cruise control feature (one that regulates the speed of a vehicle relative to a leading vehicle and to a user's input) coupled with lane keeping assistance, automatic braking, and collision mitigation technology that together make up a semiautonomous system.

Today's commercially available vehicles are considered semiautonomous. Systems are capable of interpreting many potential inputs (surrounding vehicles, lane markings, user input, obstacles, speed limits, etc.) and can regulate longitudinal and latitudinal control to semiautonomously guide the trajectory of the vehicle. Within this system, the human user is still enrolled in decision-making, monitoring, and interventions. Conditional autonomy refers to a system that (under certain conditions) permits a human user to "exit the loop" of control and decision-making. Once a goal is established (e.g., to continue on a path), the vehicle processes emergent inputs and regulates its behavior to achieve the goal without human monitoring or intervention. Behaviors internal to the activity (the activity is defined by the goal and the available means) are regulated and controlled without the participation of the human user. It's important to note that any classification is contingent on the operationalization of the goal and activity. Finally, an autonomous system possesses fewer limitations than conditional autonomy and entails the control of all tasks in an activity. Like conditional autonomy, an autonomous system operates independently of a human user within the activity structure.

### Autonomous Robotics

Examples of autonomous systems can be found across the field of robotics for a variety of purposes. There are a number of reasons that it is desirable to replace or augment humans with autonomous robots, and some of the reasons include safety (for example, spaceflight or planetary surface exploration), undesirable circumstances (monotonous tasks such as domestic chores and strenuous labor such as heavy lifting), or where human action is limited or impossible (search and rescue in confined conditions). As with automotive applications, robotics applications may be considered autonomous within the constraints of a narrowly defined domain or activity space, such as a manufacturing facility assembly line or home. Like autonomous vehicles, the degree of autonomy is conditional upon the specified domain, and in many cases excludes maintenance and repair. However, unlike automated systems, an autonomous robot within such a defined activity structure will act to complete a specified goal through sensing its environment, processing circumstantial inputs, and regulating behavior accordingly without necessitating human intervention. Current examples of autonomous robots span an immense variety of applications and include domestic applications such as autonomous lawn care robots and interplanetary exploration applications such as the MER-A and MER-B Mars rovers.

## Semiautonomous Weapons

Autonomous and semiautonomous weapon systems are currently being developed as part of modern warfare capability. Like the above automotive and robotics examples, the definition of, and distinction between, autonomous and semiautonomous varies substantially on the operationalization of the terms, the context, and the domain of activity. Consider the landmine as an example of an automated weapon with no autonomous capability. It responds with lethal force upon the activation of a sensor and involves neither decision-making capability nor human intervention. In contrast, a semiautonomous system processes inputs and acts accordingly for some set of tasks that constitute the activity of weaponry in conjunction with a human user. Together, the weapons system and the human user are necessary contributors to a single activity. In other words, the human user is "in the loop." These tasks may include identifying a target, aiming, and firing. They may also include navigating toward a target, positioning, and reloading. In a semiautonomous weapon system, these tasks are distributed between the system and the human user. By contrast, an autonomous system would be responsible for the whole set of these tasks without requiring the monitoring, decision-making, or intervention of the human user once the goal was set and the parameters specified. By these criteria, there are currently no fully autonomous weapons systems. However, as noted above, these definitions are technologically as well as socially, legally, and linguistically contingent. Most conspicuously in the case of weapons systems, the definition of semiautonomous and autonomous systems has ethical, moral, and political significance. This is especially true when it comes to determining responsibility, because causal agency and decision-making may be scattered across developers and users. The sources of agency and decision-making may also be opaque as in the case of machine learning algorithms.

## USER-INTERFACE CONSIDERATIONS

Ambiguity in definitions of semiautonomous and autonomous systems mirrors the many challenges in designing optimized user interfaces for these systems. In the case of vehicles, for example, ensuring that the user and the system (as developed by a system's designers) share a common model of the capabilities being automated (and the expected distribution and extent of control) is critical for safe transference of control responsibility. Autonomous systems are theoretically simpler user-interface challenges insofar as once an activity domain is defined, control and responsibility are binary (either the system or the human user is responsible). Here the challenge is reduced to specifying the activity and handing over control.

Semiautonomous systems present more complex challenges for the design of user-interfaces because the definition of an activity domain has no necessary relationship to the composition, organization, and interaction of constituting tasks. Particular tasks (such as a vehicle maintaining lateral position in a lane) may be determined by an engineer's application of specific technological equipment (and the attendant limitations) and thus bear no relationship to the user's mental representation of that task. An illustrative example is an obstacle detection task in which a semiautonomous system relies upon avoiding obstacles to move around an environment. The obstacle detection mechanisms (camera, radar, optical sensors, touch sensors, thermo sensors, mapping, etc.) determine what is or is not considered an obstacle by the machine, and those limitations may be opaque to a user. The resultant ambiguity necessitates that the system communicates to a human user when intervention is necessary and relies upon the system (and the system's designers) to understand and anticipate potential incompatibility between system and user models.

In addition to the issues above, other considerations for designing semiautonomous and autonomous systems (specifically in relation to the ethical and legal dimensions complicated by the distribution of agency across developers and users) include identification and authorization methods and protocols. The problem of identifying and authorizing users for the activation of autonomous technologies is critical where systems, once initiated, no longer rely upon continual monitoring, intermittent decision-making, or intervention.

*Michael Thomas*

*See also:* Autonomy and Complacency; Driverless Cars and Trucks; Lethal Autonomous Weapons Systems.

### Further Reading

Antsaklis, Panos J., Kevin M. Passino, and Shyh Jong Wang. 1991. "An Introduction to Autonomous Control Systems." *IEEE Control Systems* 11, no. 4 (June): 5–13.

Bekey, George A. 2005. *Autonomous Robots: From Biological Inspiration to Implementation and Control.* Cambridge, MA: MIT Press.

Norman, Donald A., Andrew Ortony, and Daniel M. Russell. 2003. "Affect and Machine Design: Lessons for the Development of Autonomous Machines." *IBM Systems Journal* 42, no. 1: 38–44.

Roff, Heather. 2015. "Autonomous or 'Semi' Autonomous Weapons? A Distinction without a Difference?" *Huffington Post*, January 16, 2015. https://www.huffpost.com/entry/autonomous-or-semi-autono_b_6487268.

SAE International. 2014. "Taxonomy and Definitions for Terms Related to On-Road Motor Vehicle Automated Driving Systems." J3016. SAE International Standard. https://www.sae.org/standards/content/j3016_201401/.

## Autonomous Weapons Systems, Ethics of

Autonomous weapons systems (AWS) involve armaments that are programmed to make decisions without continuous input from their programmers. These decisions include, but are not limited to, navigation, target selection, and when to engage with enemy combatants. The imminent nature of this technology has led to many ethical considerations and debates about whether they should be developed and how they should be used. The seeming inevitability of the technology led Human Rights Watch to begin a campaign in 2013: "Stop Killer Robots," which calls for prohibitions on their use globally. This movement is still active today. Other scholars and military strategists point to strategic and resource advantages of AWS that lead to support for their continued development and use. Integral to this debate is a discussion of whether it is desirable or possible to create an international agreement on their development and/or use.

Those scholars who are proponents of further technology development in these areas focus on the positive aspects that a military power can gain from the use of AWS. These systems have the potential to lead to less collateral damage, fewer combat casualties, the ability to avoid unnecessary risk, more efficient military operations, decreased psychological damage to soldiers from combat, as well as bolstering armies that tend to have dwindling human numbers. In other words, they focus on the benefits to the military that will end up using the weapon. These conversations tend to include the basic assumption that the goals of the military are themselves ethically worthy. AWS may lead to fewer civilian casualties as the systems are able to make decisions more quickly than their human counterparts; however, this is not necessarily guaranteed with the technology as the decision-making processes of AWS may lead to increased civilian casualties rather than the reverse. However, if they are able to prevent civilian deaths and the destruction of property more than traditional combat, this means that they are more efficient and therefore desirable. Another way they may increase efficiency is through minimizing resource waste in times of conflict. Transporting people and the supplies needed to sustain them is an inefficient and difficult aspect of war. AWS provide a solution to difficult logistic problems. Drones and other autonomous systems do not need rain gear, food, water, or access to health care, making them less cumbersome and thereby potentially more effective in achieving their goals. In these and other ways, AWS are seen as reducing waste and providing the best possible outcome in a combat scenario.

Conversation about use of AWS in military action is closely connected to Just War Theory. Just War Theory focuses on when it is ethically permissible or required for a military power to engage in war and theorizes about what actions are ethically permitted in times of war. If it is permissible to use an autonomous system in a military attack, it is only permissible to do so if the attack itself is justified. According to this consideration, the how of being killed is less important than the why.

Those who deem AWS as ethically impermissible focus on the inherent risks of such technology. These include scenarios where enemy combatants obtain the weaponry and use it against the military power that deploys them or when there is increased (and uncontrolled) collateral damage, reduced ability of retaliation (against enemy combatant aggressors), and loss of human dignity. A major concern is whether it is compatible with human dignity on being killed by a machine without a human as the ultimate decision-maker. It seems that there is something dehumanizing to be killed by an AWS that was provided with little human input. Another major concern is the risk factor, including the risk to the user of the technology that if there is a situation in which the AWS is shut down (either through a malfunction or an attack by an enemy) it will be confiscated and then used against the owner.

Just War is also a concern for those who condemn the use of AWS. Just War Theory explicitly prohibits the targeting of civilians by military agents; the only legitimate military targets are other military bases or persons. However, the advent of autonomous weaponry may mean that a state, especially one that does not have access to AWS, will not be able to respond to military strikes made by AWS. In an imbalanced situation where one side has access to AWS and another does not, it is necessarily the case that the side lacking the weapons will not have a legitimate military target, meaning that they must either target nonmilitary (civilian) targets or not respond at all. Neither option is ethically or practically viable.

Generally, it is accepted that automated weaponry is imminent, and so another aspect of the ethical consideration is how to regulate its use. This discussion has been particularly prominent due to the United States' prolific use of remote control drones in the Middle East. Some are proponents of an international ban on the technology; although this is generally deemed as naive and therefore implausible, these often focus on the UN prohibition against blinding lasers, which has been agreed upon by 108 states. Rather than focus on establishing a complete ban, others focus on creating an international treaty that regulates the legitimate use of these systems with sanctions and punishments for states that violate these norms. Currently no such agreement exists, and each state must determine for itself how it wants to regulate the use of these systems.

*Laci Hubbard-Mattix*

*See also:* Battlefield AI and Robotics; Campaign to Stop Killer Robots; Lethal Autonomous Weapons Systems; Robot Ethics.

## Further Reading

Arkin, Ronald C. 2010. "The Case for Ethical Autonomy in Unmanned Systems." *Journal of Military Ethics* 9, no. 4: 332–41.

Bhuta, Nehal, Susanne Beck, Robin Geiss, Hin-Yan Liu, and Claus Kress, eds. 2016. *Autonomous Weapons Systems: Law, Ethics, Policy.* Cambridge, UK: Cambridge University Press.

Killmister, Suzy. 2008. "Remote Weaponry: The Ethical Implications." *Journal of Applied Philosophy* 25, no. 2: 121–33.

Leveringhaus, Alex. 2015. "Just Say 'No!' to Lethal Autonomous Robotic Weapons." *Journal of Information, Communication, and Ethics in Society* 13, no. 3–4: 299–313.

Sparrow, Robert. 2016. "Robots and Respect: Assessing the Case Against Autonomous Weapon Systems." *Ethics & International Affairs* 30, no. 1: 93–116.

## Autonomy and Complacency

Machine autonomy and human autonomy and complacency are interlinked concepts. As artificial intelligences are programmed to *learn* from their own experiences and data input, they are arguably becoming more autonomous. Machines that gather more abilities than their human counterparts tend to become more reliant on these machines to both make decisions and respond appropriately to novel scenarios. This reliance on the decision-making processes of AI systems can lead to diminished human autonomy and over-complacency. This complacency can lead to a lack of response to critical errors in the AI's system or its decision-making processes.

Autonomous machines are those that can act in unsupervised environments, adapt to their circumstances and new experiences, learn from past mistakes, and determine the best possible outcomes in each situation without new input from programmers. In other words, these machines learn from their experiences and are in some ways capable of reaching beyond their initial programming. The idea is that it is impossible for programmers to anticipate every scenario that a machine equipped with AI may face congruent with its actions and so it must be able to adapt. However, this is not universally accepted as some argue that the very adaptability of these programs is not beyond their programming, as their programs are built to be adapted. These arguments are exacerbated by the debate about whether any actor, including human beings, can express free-will and act autonomously.

The autonomy of AI programs is not the only aspect of autonomy that is being considered with the advent of the technology. There are also concerns about the impact on human autonomy as well as concerns about complacency regarding the machines. As AI systems become better adapted to anticipating the desires and preferences of the people, they serve the people who benefit from the choice of the machine becoming moot as they no longer have to make choices.

Significant research has been done on the interaction of human workers with automated processes. Studies have found that human beings are likely to miss issues in these processes, especially when these processes become routinized, which leads to an expectation of success rather than an anticipation of failure. This anticipation of achievement leads to the operators or supervisors of the automated processes to trust faulty readouts or decisions of the machines, which can lead to ignored errors and accidents.

*Laci Hubbard-Mattix*

*See also:* Accidents and Risk Assessment; Autonomous and Semiautonomous Systems.

**Further Reading**

André, Quentin, Ziv Carmon, Klaus Wertenbroch, Alia Crum, Frank Douglas, William Goldstein, Joel Huber, Leaf Van Boven, Bernd Weber, and Haiyang Yang. 2018. "Consumer Choice and Autonomy in the Age of Artificial Intelligence and Big Data." *Customer Needs and Solutions* 5, no. 1–2: 28–37.

Bahner, J. Elin, Anke-Dorothea Hüper, and Dietrich Manzey. 2008. "Misuse of Automated Decision Aids: Complacency, Automation Bias, and the Impact of Training Experience." *International Journal of Human-Computer Studies* 66, no. 9: 688–99.

Lawless, W. F., Ranjeev Mittu, Donald Sofge, and Stephen Russell, eds. 2017. *Autonomy and Intelligence: A Threat or Savior?* Cham, Switzerland: Springer.

Parasuraman, Raja, and Dietrich H. Manzey. 2010. "Complacency and Bias in Human Use of Automation: An Attentional Integration." *Human Factors* 52, no. 3: 381–410.

# B

## Battlefield AI and Robotics

Generals on the modern battlefield are witnessing a potential tactical and strategic revolution due to the advancement of artificial intelligence (AI) and robotics and their application to military affairs. Robotic devices, such as unmanned aerial vehicles (UAVs), also known as drones, played a major role in the wars in Afghanistan (2001–) and Iraq (2003–2011), as did other robots. It is conceivable that future wars will be fought without human involvement. Autonomous machines will engage in battle on land, in the air, and under the sea without human control or direction. While this vision still belongs to the realm of science fiction, battlefield AI and robotics raises a variety of practical, ethical, and legal questions that military professionals, technological experts, jurists, and philosophers must grapple with.

What comes first in many people's minds when thinking about the application of AI and robotics to the battlefield is "killer robots," armed machines indiscriminately destroying everything in their path. There are, however, many uses for battlefield AI technology that do not involve killing. The most prominent use of such technology in recent conflicts has been nonviolent in nature. UAVs are most often used for monitoring and reconnaissance. Other robots, such as the PackBot manufactured by iRobot (the same company that produces the vacuum-cleaning Roomba), are used to detect and examine improvised explosive devices (IEDs), thereby aiding in their safe removal. Robotic devices are capable of traversing treacherous ground, such as the caves and mountain crags of Afghanistan, and areas too dangerous for humans, such as under a vehicle suspected of being rigged with an IED. Unmanned Underwater Vehicles (UUVs) are similarly used underwater to detect mines. The ubiquity of IEDs and mines on the modern battlefield make these robotic devices invaluable.

Another potential, not yet realized, life-saving capability of battlefield robotics is in the field of medicine. Robots can safely retrieve wounded soldiers on the battlefield in places unreachable by their human comrades, without putting their own lives at grave risk. Robots can also be used to carry medical equipment and medicines to soldiers on the battlefield and potentially even perform basic first aid and other emergency medical procedures.

It is in the realm of lethal force that AI and robotics have the greatest potential to alter the battlefield—whether on land, sea, or in the air. The Aegis Combat System (ACS) is an example of an automatic system currently deployed on destroyers and other naval combat vessels by numerous navies throughout the world. The system can track incoming threats—be they missiles from the surface or air or mines or torpedoes from the sea—through radar and sonar. The system is integrated with a powerful computer system and has the capability to destroy identified threats with its own munitions. Though Aegis is activated and supervised manually, the system has the capability to act independently, so as to counter threats more quickly than would be possible for humans.

In addition to such partially automated systems such as the ACS and UAVs, the future may see the rise of fully autonomous military robots capable of making decisions and acting of their own accord. The most potentially revolutionary aspect of AI empowered robotics is that of lethal autonomous weapons (LAWs)—more colloquially referred to as "killer robots." Robotic autonomy exists on a sliding scale. At one end of the scale are robots programmed to function automatically, but in response to a given stimulus and only in one way. A mine that detonates automatically when stepped on is an example of this level of autonomy. Also, at the lower end of the spectrum are remotely controlled machines that, while unmanned, are remotely controlled by a human.

Semiautonomous systems are found near the middle of the spectrum. These systems may be able to function independently of a human being, but only in limited ways. An example of such a system is a robot directed to launch, travel to a specified location, and then return at a given time. In this scenario, the machine does not make any "decisions" on its own. Semiautonomous devices may also be programmed to complete part of a mission and then to wait for additional inputs before proceeding to the next level of action. The final stage is full autonomy. Fully autonomous robots are programmed with a goal and can carry out that goal completely on their own. In battlefield scenarios, this may include the ability to employ lethal force without direct human instruction.

Lethally equipped, AI-enhanced, fully autonomous robotic devices have the potential to completely change the modern battlefield. Military ground units comprising both human beings and robots, or only robots with no humans at all, would increase the size of militaries. Small, armed UAVs would not be limited by the need for human operators and would be gathered in large swarms with the potential ability to overwhelm larger, but less mobile, forces. Such technological changes would necessitate similarly revolutionary changes in tactics, strategy, and even the concept of war itself. As this technology becomes more widely available, it will also become cheaper. This could upset the current balance of military power. Even relatively small countries, and perhaps even some nonstate actors, such as terrorist groups, may be able to establish their own robotic forces.

Fully autonomous LAWs raise a host of practical, ethical, and legal questions. Safety is one of the primary practical concerns. A fully autonomous robot equipped with lethal weaponry that malfunctions could pose a serious risk to anyone in its path. Fully autonomous missiles could conceivably, due to some mechanical fault, go off course and kill innocent people. Any kind of machinery is liable to unpredictable technical errors and malfunctions. With lethal robotic devices, such problems pose a serious safety risk to those who deploy them as well as innocent bystanders. Even aside from potential malfunctions, limitations in programming could lead to potentially calamitous mistakes. Programming robots to distinguish between combatants and noncombatants, for example, poses a major difficulty, and it is easy to imagine mistaken identity resulting in inadvertent casualties. The ultimate worry, however, is that robotic AI will advance too rapidly and break away from human control. Like popular science fiction movies and literature, and in fulfilment of the prominent scientist Stephen Hawking's dire prediction that the development of AI could result in the extinction of humanity, sentient robots could turn their weaponry on people.

LAWs raise serious legal dilemmas as well. Human beings are subject to the laws of war. Robots cannot be held liable, criminally, civilly, or in any other way, for potential legal violations. This poses the potential, therefore, of eliminating accountability for war crimes or other abuses of law. Serious questions are relevant here: Can a robot's programmer or engineer be held accountable for the actions of the machine? Could a human who gave the robot its "orders" be held responsible for unpredictable choices or mistakes made on an otherwise self-directed mission? Such issues require thorough consideration prior to the deployment of any fully autonomous lethal machine.

Apart from legal matters of responsibility, a host of ethical considerations also require resolution. The conduct of war requires split-second moral decision-making. Will autonomous robots be able to differentiate between a child and a soldier or recognize the difference between an injured and defenseless soldier and an active combatant? Can a robot be programmed to act mercifully when a situation dictates, or will a robotic military force always be considered a cold, ruthless, and merciless army of extermination? Since warfare is fraught with moral dilemmas, LAWs engaged in war will inevitably be faced with such situations. Experts doubt lethal autonomous robots can ever be depended upon to take the correct action. Moral behavior requires not only rationality—something that might be programmed into robots—but also emotions, empathy, and wisdom. These latter things are much more difficult to write into code.

The legal, ethical, and practical concerns raised by the prospect of ever more advanced AI-powered robotic military technology has led many people to call for an outright ban on research in this area. Others, however, argue that scientific progress cannot be stopped. Instead of banning such research, they say, scientists and society at large should look for pragmatic solutions to those problems. Some claim, for example, that many of the ethical and legal problems can be resolved by maintaining constant human supervision and control over robotic military forces. Others point out that direct supervision is unlikely over the long run, as human cognition will not be capable of matching the speed of computer thinking and robot action. There will be an inexorable tendency toward more and more autonomy as the side that provides its robotic forces with greater autonomy will have an insurmountable advantage over those who try to maintain human control. Fully autonomous forces will win every time, they warn.

Though still in its emergent phase, the introduction of continually more advanced AI and robotic devices to the battlefield has already resulted in tremendous change. Battlefield AI and Robotics have the potential to radically alter the future of war. It remains to be seen if, and how, the technological, practical, legal, and ethical limitations of this technology can be overcome.

*William R. Patterson*

*See also:* Autonomous Weapons Systems, Ethics of; Lethal Autonomous Weapons Systems.

**Further Reading**

Borenstein, Jason. 2008. "The Ethics of Autonomous Military Robots." *Studies in Ethics, Law, and Technology* 2, no. 1: n.p. https://www.degruyter.com/view/journals/selt /2/1/article-selt.2008.2.1.1036.xml.xml.

Morris, Zachary L. 2018. "Developing a Light Infantry-Robotic Company as a System." *Military Review* 98, no. 4 (July–August): 18–29.

Scharre, Paul. 2018. *Army of None: Autonomous Weapons and the Future of War.* New York: W. W. Norton.

Singer, Peter W. 2009. *Wired for War: The Robotics Revolution and Conflict in the 21st Century.* London: Penguin.

Sparrow, Robert. 2007. "Killer Robots," *Journal of Applied Philosophy* 24, no. 1: 62–77.

# Bayesian Inference

Bayesian inference is a way to calculate the probability of the validity of a proposition based on a prior estimate of its probability plus any new and relevant data. Bayes' Theorem, from which Bayesian statistics are drawn, was a popular mathematical approach used in expert systems in the twentieth century. The Bayesian theorem remains useful to artificial intelligence in the twenty-first century and has been applied to problems such as robot locomotion, weather forecasting, jurimetrics (the application of quantitative methods to law), phylogenetics (the evolutionary relationships among organisms), and pattern recognition. It is also useful in solving the famous Monty Hall problem and is often utilized in email spam filters.

The mathematical theorem was developed by the Reverend Thomas Bayes (1702–1761) of England and published posthumously as "An Essay Towards Solving a Problem in the Doctrine of Chances" in the *Philosophical Transactions of the Royal Society of London* in 1763. It is sometimes called Bayes' Theorem of Inverse Probabilities. The first notable discussion of Bayes' Theorem as applied to the field of medical artificial intelligence appeared in a classic article entitled "Reasoning Foundations of Medical Diagnosis," written by George Washington University electrical engineer Robert Ledley and Rochester School of Medicine radiologist Lee Lusted and published by *Science* in 1959. As Lusted later remembered, medical knowledge in the mid-twentieth century was usually presented as symptoms associated with a disease, rather than as diseases associated with a symptom. Bayesian inference led them to consider the idea that medical knowledge could be expressed as the probability of a disease given the patient's symptoms.

Bayesian statistics are conditional, allowing one to determine the chance that a certain disease is present given a certain symptom, but only with prior knowledge of how often the disease and symptom are correlated and how often the symptom is present in the absence of the disease. It is very close to what Alan Turing described as the factor in favor of the hypothesis provided by the evidence. Bayes'

Theorem may also be utilized in resolving disorders involving multiple symptoms in a patient, that is, the symptom-disease complex. Bayesian statistics as applied in computer-aided diagnosis compares the probability of the manifestation of each disease in a population with the probability of manifestation of each symptom given each disease in order to calculate the probability of all possible diseases given each patient's symptom-disease complex. Bayes' Theorem assumes that all induction is statistical.

The first practical demonstration of the theorem in generating the posterior probabilities of particular diseases came in 1960. In that year, University of Utah cardiologist Homer Warner, Jr. took advantage of his access to a Burroughs 205 digital computer to diagnose well-defined congenital heart diseases at Salt Lake's Latter-Day Saints Hospital by using Bayesian statistics. Warner and his staff applied the theorem to determine the probabilities by which an undiagnosed patient with definable symptoms, signs, or laboratory results might fit into previously established disease categories. The computer program could be used over and over as new information presented itself, establishing or ranking diagnoses by serial observation. Warner found that in applying Bayesian conditional-probability algorithms to a symptom-disease matrix of thirty-five coronary conditions, the Burroughs machine performed just as well as any trained cardiologist. Other enthusiastic early adopters of Bayesian estimation included John Overall, Clyde Williams, and Lawrence Fitzgerald for thyroid disorders; Charles Nugent for Cushing's disease; Gwilym Lodwick for primary bone tumors; Martin Lipkin for hematological diseases; and Tim de Dombal for acute abdominal pain. The Bayesian model has been extended and modified many times in the last half century to account or correct for sequential diagnosis and conditional independence and to weight various factors.

Bayesian computer-aided diagnosis is criticized for poor prediction of rare diseases, inadequate discrimination between diseases with similar symptom-complexes, inability to quantify qualitative evidence, troubling conditional dependence between evidence and hypotheses, and the overwhelming amount of manual labor required to maintain the requisite joint probability distribution tables. Bayesian diagnostic assistants have also been critiqued for their shortcomings outside of the populations for which they were designed. A nadir in use of Bayesian statistics in differential diagnosis was reached in the mid-1970s when rule-based decision support algorithms became more popular.

Bayesian methods recovered in the 1980s and are widely used today in the field of machine learning. Artificial intelligence researchers have extracted rigorous methods for supervised learning, hidden Markov models, and mixed methods for unsupervised learning from the idea of Bayesian inference. In a practical context, Bayesian inference has been used controversially in artificial intelligence algorithms that attempt to determine the conditional probability of a crime being committed, to screen welfare recipients for drug use, and to detect potential mass shooters and terrorists. The approach has again faced scrutiny, particularly when the screening involves rare or extreme events, where the AI algorithm can behave indiscriminately and identify too many individuals as at-risk of engaging in the undesirable behavior. Bayesian inference has also been introduced into the court-room in the United Kingdom. In *Regina* v. *Adams* (1996), jurors were offered the Bayesian approach by the defense team to help jurors form an unbiased mecha-nism for combining introduced evidence, which involved a DNA profile and dif-fering match probability calculations and constructing a personal threshold for forming a judgment about convicting the accused "beyond a reasonable doubt."

Bayes' theorem had already been "rediscovered" several times before its 1950s' revival under Ledley, Lusted, and Warner. The circle of historic luminaries who perceived value in the Bayesian approach to probability included Pierre-Simon Laplace, the Marquis de Condorcet, and George Boole. The Monty Hall problem, named for the host of the classic game show *Let's Make a Deal*, involves a contes-tant deciding whether to stick with the door they have picked or switch to another unopened door after Monty Hall (with full knowledge of the location of the prize) opens a door to reveal a goat. Rather counterintuitively, the chances of winning under conditional probability are twice as large by switching doors.

*Philip L. Frana*

*See also:* Computational Neuroscience; Computer-Assisted Diagnosis.

**Further Reading**

Ashley, Kevin D., and Stefanie Brüninghaus. 2006. "Computer Models for Legal Predic-tion." *Jurimetrics* 46, no. 3 (Spring): 309–52.

Barnett, G. Octo. 1968. "Computers in Patient Care." *New England Journal of Medicine* 279 (December): 1321–27.

Bayes, Thomas. 1763. "An Essay Towards Solving a Problem in the Doctrine of Chances." *Philosophical Transactions* 53 (December): 370–418.

Donnelly, Peter. 2005. "Appealing Statistics." *Significance* 2, no. 1 (February): 46–48.

Fox, John, D. Barber, and K. D. Bardhan. 1980. "Alternatives to Bayes: A Quantitative Comparison with Rule-Based Diagnosis." *Methods of Information in Medicine* 19, no. 4 (October): 210–15.

Ledley, Robert S., and Lee B. Lusted. 1959. "Reasoning Foundations of Medical Diagno-sis." *Science* 130, no. 3366 (July): 9–21.

Lusted, Lee B. 1991. "A Clearing 'Haze': A View from My Window." *Medical Decision Making* 11, no. 2 (April–June): 76–87.

Warner, Homer R., Jr., A. F. Toronto, and L. G. Veasey. 1964. "Experience with Bayes' Theorem for Computer Diagnosis of Congenital Heart Disease." *Annals of the New York Academy of Sciences* 115: 558–67.

# Beneficial AI, Asilomar Meeting on

Social concerns surrounding artificial intelligence and harm to humans have most famously been represented by Isaac Asimov's Three Laws of Robotics. Asimov's laws maintain that: "A robot may not injure a human being or, through inaction, allow a human being to come to harm; A robot must obey the orders given it by human beings except where such orders would conflict with the First Law; A robot must protect its own existence as long as such protection does not conflict with the First or Second Law" (Asimov 1950, 40). Asimov, in later writings, added a Fourth Law or Zeroth Law, commonly paraphrased as "A robot may not harm humanity, or, by inaction, allow humanity to come to harm" and described in detail by the robot character Daneel Olivaw in *Robots and Empire* (Asimov 1985, chapter 18).

Asimov's zeroth law subsequently provoked discussion as to how harm to humanity should be determined. The 2017 Asilomar Conference on Beneficial AI took on this question, moving beyond the Three Laws and the Zeroth Law and establishing twenty-three principles to safeguard humanity with respect to the future of AI. The Future of Life Institute, sponsor of the conference, hosts the principles on their website and has gathered 3,814 signatures supporting the principles from AI researchers and other interdisciplinary supporters. The principles fall into three main categories: research questions, ethics and values, and longer-term concerns.

Those principles related to research aim to ensure that the goals of artificial intelligence remain beneficial to humans. They are intended to guide financial investments in AI research. To achieve beneficial AI, Asilomar signatories contend that research agendas should support and maintain openness and dialogue between AI researchers, policymakers, and developers. Researchers involved in the development of artificial intelligence systems should work together to prioritize safety.

Proposed principles related to ethics and values are meant to reduce harm and encourage direct human control over artificial intelligence systems. Parties to the Asilomar principles ascribe to the belief that AI should reflect the human values of individual rights, freedoms, and acceptance of diversity. In particular, artificial intelligences should respect human liberty and privacy and be used solely to empower and enrich humanity. AI must align with the social and civic standards of humans. The Asilomar signatories maintain that designers of AI need to be held responsible for their work. One noteworthy principle addresses the possibility of an arms race of autonomous weapons.

The creators of the Asilomar principles, noting the high stakes involved, included principles covering longer term issues. They urged caution, careful planning, and human oversight. Superintelligences must be developed for the larger good of humanity, and not only to advance the goals of one company or nation. Together, the twenty-three principles of the Asilomar Conference have sparked ongoing conversations on the need for beneficial AI and specific safeguards concerning the future of AI and humanity.

*Diane M. Rodgers*

*See also:* Accidents and Risk Assessment; Asimov, Isaac; Autonomous Weapons Systems, Ethics of; Campaign to Stop Killer Robots; Robot Ethics.

**Further Reading**

Asilomar AI Principles. 2017. https://futureoflife.org/ai-principles/.

Asimov, Isaac. 1950. "Runaround." In *I, Robot*, 30–47. New York: Doubleday.

Asimov, Isaac. 1985. *Robots and Empire.* New York: Doubleday.

Sarangi, Saswat, and Pankaj Sharma. 2019. *Artificial Intelligence: Evolution, Ethics, and Public Policy.* Abingdon, UK: Routledge.

# Berger-Wolf, Tanya (1972–)

Tanya Berger-Wolf is a professor in the Department of Computer Science at the University of Illinois at Chicago (UIC). She is known for her contributions to computational ecology and biology, data science and network analysis, and artificial intelligence for social good. She is the leading researcher in the field of computational population biology, which uses artificial intelligence algorithms, computational methods, the social sciences, and data collection to answer question about plants, animals, and humans.

Berger-Wolf leads interdisciplinary field courses at the Mpala Research Centre in Kenya with engineering students from UIC and biology students from Princeton University. She works in Africa for its rich genetic diversity and because it possesses endangered species that are indicators of the health of life on the planet generally. Her group wants to know what the impact of environment on the behavior of social animals is, as well as what puts a given species at risk.

She is cofounder and director of Wildbook, a nonprofit that creates wildlife conservation software. Berger-Wolf's work for Wildbook has included a crowd-sourced project to take as many photographs of Grevy's zebras as possible in order to accomplish a full census of the rare animals. Analysis of the photos with artificial intelligence algorithms allows the group to identify every individual Grevy's zebra by its unique pattern of stripes—a natural bar code or fingerprint. The Wildbook software identifies the animals from hundreds of thousands of pictures using convolutional neural networks and matching algorithms. The census estimates are used to target and invest resources in the protection and survival of the zebras.

The Wildbook deep learning software can be used to identify individual members of any striped, spotted, notched, or wrinkled species. Giraffe Spotter is Wildbook software for giraffe populations. Wildbook crowdsources citizen-scientist's reports of giraffe encounters through its website, which includes gallery images from handheld cameras and camera traps. Wildbook's catalogue of individual whale sharks uses an intelligent agent that extracts still images of tail flukes from uploaded YouTube videos. The whale shark census yielded evidence that led the International Union for Conservation of Nature to change the status of the animals from "vulnerable" to "endangered" on the IUCN Red List of Threatened Species. Wildbook is also using the software to inspect videos of hawksbill and green sea turtles.

Berger-Wolf is also director of tech for conservation nonprofit Wild Me. The nonprofit uses machine vision artificial intelligence algorithms to identify individual animals in the wild. Wild Me records information about animal locations, migration patterns, and social groups. The goal is to develop a comprehensive understanding of global diversity that can inform conservation policy. Wild Me is a partner of Microsoft's AI for Earth program.

Berger-Wolf was born in 1972 in Vilnius, Lithuania. She attended high school in St. Petersburg, Russia, and completed her bachelor's degree at Hebrew University in Jerusalem. She earned her doctorate from the Department of Computer Science at the University of Illinois at Urbana-Champaign and pursued postdoctoral work at the University of New Mexico and Rutgers University. She is the recipient of the National Science Foundation CAREER Award, the Association for Women in Science Chicago Innovator Award, and Mentor of the Year at UIC.

*Philip L. Frana*

*See also:* Deep Learning.

## Further Reading

Berger-Wolf, Tanya Y., Daniel I. Rubenstein, Charles V. Stewart, Jason A. Holmberg, Jason Parham, and Sreejith Menon. 2017. "Wildbook: Crowdsourcing, Computer Vision, and Data Science for Conservation." Chicago, IL: Bloomberg Data for Good Exchange Conference. https://arxiv.org/pdf/1710.08880.pdf.

Casselman, Anne. 2018. "How Artificial Intelligence Is Changing Wildlife Research." *National Geographic*, November. https://www.nationalgeographic.com/animals /2018/11/artificial-intelligence-counts-wild-animals/.

Snow, Jackie. 2018. "The World's Animals Are Getting Their Very Own Facebook." *Fast Company*, June 22, 2018. https://www.fastcompany.com/40585495/the-worlds -animals-are-getting-their-very-own-facebook.

# Berserkers

Berserkers are a fictional type of intelligent killer machines first introduced by science fiction and fantasy author Fred Saberhagen (1930–2007) in a 1962 short story, "Without a Thought." Berserkers subsequently appeared as common antagonists in many more novels and stories by Saberhagen.

Berserkers are an ancient race of sentient, self-replicating, space-faring machines programmed to destroy all life. They were created in a long-forgotten interstellar war between two alien races, as an ultimate doomsday weapon (i.e., one intended as a threat or deterrent more than actual use). The details of how they were unleashed in the first place are lost to time, as the Berserkers apparently wiped out their creators along with their enemies and have been marauding the Milky Way galaxy ever since. They range in size from human-scale units to heavily armored planetoids (cf. Death Star) with a variety of weapons powerful enough to sterilize planets.

The Berserkers prioritize destruction of any intelligent life that fights back, such as humanity. They build factories to replicate and improve themselves, while never changing their central directive of eradicating life. The extent to which they undergo evolution is unclear; some individual units eventually deviate into questioning or even altering their goals, and others develop strategic genius (e.g., *Brother Assassin*, "Mr. Jester," *Rogue Berserker*, *Shiva in Steel*). While the Berserkers' ultimate goal of destroying all life is clear, their tactical operations are unpredictable due to randomness from a radioactive decay component in their cores. Their name is thus based on the Berserkers of Norse legend, fierce human warriors who fought with a wild frenzy.

Berserkers illustrate a worst-case scenario for artificial intelligence: rampant and impassive killing machines that think, learn, and reproduce. They show the perilous hubris of creating AI so advanced as to surpass its creators' comprehension and control and equipping such AI with powerful weapons, destructive intent, and unchecked self-replication. If Berserkers are ever created and unleashed even once, they can pose an endless threat to living beings across vast stretches of space and time. Once unbottled, they are all but impossible to eradicate. This is due not only to their advanced defenses and weapons but also to their far-flung distribution, ability to repair and replicate, autonomous operation (i.e., without any centralized control), capacity to learn and adapt, and infinite patience to lie hiding in wait. In Saberhagen's stories, the discovery of Berserkers is so terrifying that human civilizations become extremely wary of developing their own AI, for fear that it too may turn on its creators. However, some clever humans discover an intriguing counter weapon to Berserkers: Qwib-Qwibs, self-replicating machines programmed to destroy all Berserkers rather than all life ("Itself Surprised" by Roger Zelazny). Cyborgs are another anti-Berserker tactic used by humans, pushing the boundary of what counts as organic intelligence (*Berserker Man, Berserker Prime, Berserker Kill*).

Berserkers also illustrate the potential inscrutability and otherness of artificial intelligence. Even though some communication with Berserkers is possible, their vast minds are largely incomprehensible to the intelligent organic lifeforms fleeing from or fighting them, and they prove difficult to study due to their tendency to self-destruct if captured. What can be understood of their thinking indicates that they see life as a scourge, a disease of matter that must be extinguished. In turn, the Berserkers do not fully understand organic intelligence, and despite many attempts, they are never able to successfully imitate organic life. They do, however, sometimes recruit human defectors (which they call "goodlife") to serve the cause of death and help the Berserkers fight "badlife" (i.e., any life that resists extermination). Nevertheless, the ways that Berserkers and humans think are almost completely incompatible, thwarting efforts toward mutual understanding between life and nonlife. Much of the conflict in the stories hinges on apparent differences between human and machine intelligence (e.g., artistic appreciation, empathy for animals, a sense of humor, a tendency to make mistakes, the use of acronyms for mnemonics, and even fake encyclopedia entries made to detect plagiarism). Berserkers are even sometimes foiled by underestimating nonintelligent life such as plants and mantis shrimp ("Pressure" and "Smasher").

In reality, the idea of Berserkers can be seen as a special case of the von Neumann probe, an idea conceived of by mathematician and physicist John von Neumann (1903–1957): self-replicating space-faring robots that could be dispersed to efficiently explore a galaxy. The Turing Test, proposed by mathematician and computer scientist Alan Turing (1912–1954), is also explored and upended in the Berserker stories. In "Inhuman Error," human castaways compete with a Berserker to convince a rescue team they are human, and in "Without a Thought," a Berserker attempts to determine whether or not its opponent in a game is human. Berserkers also offer a grim explanation for the Fermi paradox—the idea that if advanced alien civilizations exist we should have heard from them by now. It could be that Earth has not been contacted by alien civilizations because they have been destroyed by Berserker-like machines or are hiding from them.

The concept of Berserkers, or something like them, has appeared across numerous works of science fiction in addition to Saberhagen's (e.g., works by Greg Bear,

Gregory Benford, David Brin, Ann Leckie, and Martha Wells; the *Terminator* series of movies; and the Mass Effect series of video games). These examples all show how the potential for existential threats from AI can be tested in the laboratory of fiction.

*Jason R. Finley and Joan Spicci Saberhagen*

*See also:* de Garis, Hugo; Superintelligence; *The Terminator.*

**Further Reading**

Saberhagen, Fred. 2015a. *Berserkers: The Early Tales.* Albuquerque: JSS Literary Productions.

Saberhagen, Fred. 2015b. *Berserkers: The Later Tales.* Albuquerque: JSS Literary Productions.

Saberhagen's Worlds of SF and Fantasy. http://www.berserker.com.

The TAJ: Official Fan site of Fred Saberhagen's Berserker® Universe. http://www.berserkerfan.org.

# Biometric Privacy and Security

Biometrics, a term that comes from the Greek roots *bio* (meaning life) and *metrikos* (meaning to measure), involves statistical or mathematical methods to analyze data in the biological sciences. In recent years, the term has frequently been applied in a more specific, high-technology context to refer to the science of identification of individuals based on biological or behavioral characteristics and the artificial intelligence tools used to these ends.

The systematic measurement of human physical attributes or behaviors for purposes of later identification has been underway for centuries. The earliest documented use of biometrics is found in Portuguese historian Joao de Barros's (1496–1570) writings. De Barros documented how Chinese merchants used ink to stamp and record the hands and footprints of children. In the late nineteenth century, biometric techniques were introduced in criminal justice settings. In Paris, police clerk Alphonse Bertillon (1853–1914) began taking body measurements (the circumference of the head, length of fingers, etc.) of those in custody to keep track of repeat offenders, in particular individuals who frequently used aliases or changed aspects of their appearance to avoid recognition. His system became known as Bertillonage. It fell out of favor after the 1890s, when it became evident that many people had similar measurements. In 1901, Scotland Yard's Richard Edward Henry (1850–1931) developed a far more effective biometric system involving fingerprinting. He measured and classified loops, whorls, and arches and subdivisions of these elements on the tips of peoples' fingers and thumbs.

Fingerprinting remains one of the most common biometric identifiers used by law enforcement agencies worldwide. Fingerprinting systems are evolving alongside networking technology, taking advantage of massive national and international databases and computer matching.

The Federal Bureau of Investigation partnered with the National Bureau of Standards to automate fingerprint identification in the 1960s and 1970s. This involved scanning paper fingerprint cards already on file, as well as developing minutiae feature extraction techniques and automated classifiers for comparing the computerized fingerprint data. The scanned images of fingerprints were not saved in digital form along with the classification data and minutiae because of the high cost of electronic storage. The FBI operationalized the M40 fingerprint matching algorithm in 1980. A full Integrated Automated Fingerprint Identification System (IAFIS) became active in 1999. Palm print, iris, and facial identification was stored in the FBI's Next Generation Identification system, an extension of IAFIS, in 2014.

While biometric technology is frequently perceived as a tool to increase security at the expense of privacy, in some instances it can be used to help maintain privacy. In hospitals, many types of health-care workers need to have access to a common database of patient records. Preventing unauthorized users from accessing this sensitive information is also extremely important under the Health Insurance Portability and Accountability Act (HIPAA). The Mayo Clinic in Florida, for instance, was an early leader in biometric access to medical records. The clinic began using digital fingerprinting technology to control access to patient records in 1997.

Today big data and artificial intelligence recognition software combine to rapidly identify or authenticate individuals on the basis of voice analysis, face or iris recognition, hand geometry, keystroke dynamics, gait, DNA, and even body odor. DNA fingerprinting has grown to the point where its reliability is generally accepted by courts. Criminals have been convicted based on DNA results, even in the presence of little or no other evidence, and wrongfully convicted incarcerated individuals have been exonerated. While biometrics is widely used by enforcement agencies and the courts, as well as within other government agencies, it has also come under strong public scrutiny for invading individual rights to privacy.

Research in biometric artificial intelligence software at universities, government agencies, and private corporations has grown alongside real and perceived threats of criminal and terrorist activity. In 1999, National Bank United installed iris recognition authentication systems on an experimental basis on three ATMs in Texas using technology created by biometric specialists Visionics and Keyware Technologies. The next year, Visage Corporation debuted the FaceFINDER System, an automated facial recognition technology, at Super Bowl XXXV in Tampa, Florida. The system scanned spectators' faces as they entered the stadium and compared them against a database of 1,700 known criminals and terrorists. Law enforcement officials claimed identification of a small number of criminals, but those identifications led to no major arrests or convictions. At that time, indiscriminate use of automated facial recognition generated widespread controversy. The game was even referred to as the Snooper Bowl.

A U.S. public policy debate on the use of biometric solutions for airport security gained additional traction in the wake of the terrorist attacks of September 11, 2001. Polls after 9/11 showed Americans willing to give up substantial aspects of personal privacy for greater security. In other countries, such as The Netherlands, biometric technologies were already in widespread use. The Privium program for iris scan verification of passengers at Schiphol Airport has been in place since

2001. The U.S. Transportation Security Administration (TSA) began testing biometric tools for identity verification purposes in 2015. In 2019, Delta Air Lines, in partnership with U.S. Customs and Border Protection, offered optional facial recognition boarding to passengers at the Maynard Jackson International Terminal in Atlanta. The system allows passengers to pick up boarding passes, self-check luggage, and negotiate TSA checkpoints and gates without interruption. In initial rollout, only 2 percent of passengers opted out.

Financial institutions are now beginning to adopt biometric authentication systems in regular commercial transactions. They are already in widespread use to protect access to personal smart phones. Intelligent security will become even more important as smart home devices connected to the internet demand support for secure financial transactions. Opinions on biometrics often vary with changing situations and environments. Individuals who may favor use of facial recognition technology at airports to make air travel more secure might oppose digital fingerprinting at their bank. Some people perceive private company use of biometric technology as dehumanizing, treating and tracking them in real time as if they were products rather than people.

At the local level, community policing is often cited as a successful way to build relationships between law enforcement officers and the neighborhoods they patrol. But for some critics, biometric surveillance redirects the focus away from community relationship building and on to socio-technical control by the state. Context, however, remains crucial. Use of biometrics in corporations can be perceived as an equalizer, as it places white-collar employees under the same sort of scrutiny long felt by blue-collar laborers. Researchers are beginning to develop video analytics AI software and smart sensors for use in cloud security systems. These systems can identify known people, objects, voices, and movements in real-time surveillance of workplaces, public areas, and homes. They can also be trained to alert users to the presence of unknown people.

Artificial intelligence algorithms used in the creation of biometric systems are now being used to defeat them. Generative adversarial networks (GANs), for instance, simulate human users of network technology and applications. GANs have been used to create imaginary people's faces from sets of real biometric training data. GANs are often composed of a creator system, which makes each new image, and a critic system that compares the artificial face against the original photograph in an iterative process. The startup Icons8 claimed in 2020 that it could create a million fake headshots from only seventy human models in a single day. The company sells the headshots created with their proprietary StyleGAN technology as stock photos. Clients have included a university, a dating app, and a human resources firm. Rosebud AI creates similar GANs generated photos and sells them to online shopping sites and small businesses that cannot afford to hire expensive models and photographers.

Deepfake technology, involving machine learning techniques to create realistic but counterfeit videos, has been used to perpetrate hoaxes and misrepresentations, generate fake news clips, and commit financial fraud. Facebook accounts with deepfake profile photos have been used to amplify social media political campaigns. Smart phones with facial recognition locks are susceptible to deepfake hacking. Deepfake technology also has legitimate applications. Films have used such technology for actors in flashbacks or other similar scenes to make actors appear younger. Films such as *Rogue One: A Star Wars Story* (2016) even used digital technology to include the late Peter Cushing (1913–1994), in which he played the same character from the original, 1977 *Star Wars* film.

Recreational users have access to face-swapping through a variety of software applications. FaceApp allows users to upload a selfie and change hair and facial expression. The program can also simulate aging of a person's features. Zao is a deepfake application that takes a single photo and swaps it with the faces of film and television actors in hundreds of clips. Deepfake algorithms are now in use to detect the very videos created by the deepfakes.

*Philip L. Frana*

*See also:* Biometric Technology.

**Further Reading**

Goodfellow, Ian J., Jean Pouget-Abadie, Mehdi Mirza, Bing Xu, David Warde-Farley, Sherjil Ozair, Aaron Courville, and Yoshua Bengio. 2014. "Generative Adversarial Nets." *NIPS '14: Proceedings of the 27th International Conference on Neural Information Processing Systems* 2 (December): 2672–80.

Hopkins, Richard. 1999. "An Introduction to Biometrics and Large-Scale Civilian Identification." *International Review of Law, Computers & Technology* 13, no. 3: 337–63.

Jain, Anil K., Ruud Bolle, and Sharath Pankanti. 1999. *Biometrics: Personal Identification in Networked Society.* Boston: Kluwer Academic Publishers.

Januškevič, Svetlana N., Patrick S.-P. Wang, Marina L. Gavrilova, Sargur N. Srihari, and Mark S. Nixon. 2007. *Image Pattern Recognition: Synthesis and Analysis in Biometrics.* Singapore: World Scientific.

Nanavati, Samir, Michael Thieme, and Raj Nanavati. 2002. *Biometrics: Identity Verification in a Networked World.* New York: Wiley.

Reichert, Ramón, Mathias Fuchs, Pablo Abend, Annika Richterich, and Karin Wenz, eds. 2018. *Rethinking AI: Neural Networks, Biometrics and the New Artificial Intelligence.* Bielefeld, Germany: Transcript-Verlag.

Woodward, John D., Jr., Nicholas M. Orlans, and Peter T. Higgins. 2001. *Biometrics: Identity Assurance in the Information Age.* New York: McGraw-Hill.

# Biometric Technology

A biometric involves the measurement of some characteristic of a human being. It can be physiological, such as in fingerprint or facial recognition, or it can be behavioral, as in keystroke pattern dynamics or walking stride length. The White House National Science and Technology Council's Subcommittee on Biometrics defines biometric characteristics as "measurable biological (anatomical and physiological) and behavioral characteristics that can be used for automated recognition" (White House, National Science and Technology Council 2006, 4). The International Biometrics and Identification Association (IBIA) defines biometric technologies as "technologies that automatically confirm the identity of people by comparing patterns of physical or behavioral characteristics in real time against enrolled computer records of those patterns" (International Biometrics and Identification Association 2019).

Many types of biometric technologies are in use or in active development. While at one time fingerprints were primarily collected in criminal databases, they are now used to access personal smartphones, pay for goods and services, and verify identities for various online accounts and physical facilities. Fingerprint recognition is the most well-known biometric technology. Fingerprint image collections can be amassed using ultrasound, thermal, optical, or capacitive sensors. AI software applications generally rely on either minutia-based matching or pattern matching in searching for matches. Sensors create images of human veins by lighting up the palm, and vascular pattern recognition is also now possible.

Other common biometrics are derived from characteristics of the face, iris, or voice. Facial recognition AI technology may be used for identification, verification, detection, and characterization of individuals. Detection and characterization processes do not usually involve determining the identity of an individual. Current systems are capable of high accuracy rates, but facial recognition raises privacy concerns because a face can be collected passively, that is, without the subject's knowledge. Iris recognition uses near infrared light to isolate the unique structural patterns of the iris. Retina technology uses a bright light to examine the retinal blood vessels. Recognition is determined by comparing the eyeball being scanned and the stored image. Voice recognition is a more sophisticated system than mere voice activation, which recognizes the content of speech. Voice recognition must be able to identify each unique individual user. The technology to date is not sufficiently accurate for reliable identification in many contexts.

Biometric technology has long been available for security and law enforcement purposes. Today, however, these systems are also increasingly being used as a verification method in the private sector for authentication that previously required a password. Apple's iPhone fingerprint scanner, introduced in 2013, brought new public awareness. A shift to facial recognition access in the company's recent models further normalizes the concept. Biometric technology continues to be adopted in a wide variety of sectors including financial services, transportation, health care, facility access, and voting.

*Brenda Leong*

*See also:* Biometric Privacy and Security.

**Further Reading**

International Biometrics and Identity Association. 2019. "The Technologies." https://www.ibia.org/biometrics/technologies/.

White House. National Science and Technology Council. 2006. *Privacy and Biometrics: Building a Conceptual Foundation*. Washington, DC: National Science and Technology Council. Committee on Technology. Committee on Homeland and National Security. Subcommittee on Biometrics.

## Blade Runner (1982)

Originally published in 1968, Philip Dick's *Do Androids Dream of Electric Sheep?* takes place in postindustrial San Francisco, AD 2020. In 1982, the novel was retitled as *Blade Runner* in a film adaptation that relocates the story to Los Angeles,

AD 2019. While there are significant differences between the texts, both tell the story of bounty hunter Rick Deckard who is tasked with retiring (or killing) escaped replicants/androids (six in the novel, four in the film). The backdrop to both stories is a future where cities have become overpopulated and highly polluted. Nonhuman natural life has largely become extinct (through radiation sickness) and replaced with synthetic and manufactured life. In this future, natural life has become a valuable commodity.

Against this setting, replicants are designed to fulfill a range of industrial functions, most notably as labor for off-world colonies. The replicants can be identified as an exploited group, produced to serve human masters. They are discarded when no longer useful and retired when they rebel against their conditions. Blade runners are specialized law enforcement officers charged with capturing and destroying these rogue replicants. Blade runner Rick Deckard comes out of retirement to hunt down the advanced Nexus-6 replicant models. These replicants have rebelled against the slave-like conditions on Mars and have escaped to Earth.

The handling of artificial intelligence in both texts serves as an implicit critique of capitalism. In the novel, the Rosen Association, and in the film the Tyrell Corporation manufacture replicants in order to create a more docile workforce, thereby suggesting capitalism turns humans into machines. These crass commercial imperatives are underscored by Eldon Rosen (who is renamed Tyrell in the film): "We produced what the colonists wanted. ... We followed the time-honored principle underlying every commercial venture. If our firm hadn't made these progressively more human types, other firms would have."

In the film, there are two categories of replicants: those who are programmed not to know they are replicants, who are replete with implanted memories (like Rachael Tyrell), and those who know they androids and live by that knowledge (the Nexus-6 fugitives). The film version of Rachael is a new Nexus-7 model, implanted with the memories of Eldon Tyrell's niece, Lilith. Deckard is tasked with killing her but falls in love with her instead. The end of the film sees the two fleeing the city together.

The novel treats the character of Rachael differently. Deckard attempts to enlist the help of Rachael to assist him in tracking down the fugitive androids. Rachael agrees to meet Deckard in a hotel in an attempt to get him to abandon the case. During their meeting, Rachael reveals one of the fugitive androids (Pris Stratton) is an exact duplicate of her (making Rachael a Nexus-6 model in the novel). Eventually, Deckard and Rachael have sex and profess their love for one another. However, it is revealed that Rachael has slept with other blade runners. Indeed, she is programmed to do so in order to prevent them from completing their missions. Deckard threatens to kill Rachael but does not follow through, choosing to leave the hotel instead.

In the novel and the film, the replicants are undetectable. They appear to be completely human, even under a microscope. The only way to identify them is through the administration of the Voigt-Kampff test, which distinguishes humans from androids based on emotional responses to various questions. The test is administered with the assistance of a machine that measures blush response, heart rate, and eye movement in response to questions dealing with empathy. Deckard's status as a human or a replicant is not immediately known. Rachael even asks him if he has taken the Voigt-Kampff test. Deckard's status remains ambiguous in the film. Though the viewer may make their own decision, director Ridley Scott has suggested that Deckard is, indeed, a replicant. Toward the end of the novel, Deckard takes and passes the test, but begins questioning the efficacy of blade running.

The book, more than the film, grapples with questions of what it means to be human in the face of technological advances. The book shows the fragility of the human experience and how it might easily be damaged by the very technology designed to serve it. Penfield mood organs, for instance, are devices that individuals can use to regulate their emotions. All that is required is that a person locate an emotion in a manual, dial the appropriate number, and then feel whatever they want. The use of the device and its creation of artificial feelings suggests that humans can become robotic, a point relayed by Deckard's wife Iran:

> My first reaction consisted of being grateful that we could afford a Penfield mood organ. But then I realized how unhealthy it was, sensing the absence of life, not just in this building but everywhere, and not reacting – do you see? I guess you don't. But that used to be considered a sign of mental illness; they called it 'absence of appropriate affect.'

The point Dick makes is that the mood organ prevents people from experiencing the appropriate emotional qualities of life, the very thing the Voigt-Kampff test suggests replicants cannot do.

Philip Dick was particularly noted for his more nebulous and, perhaps, even pessimistic view of artificial intelligence. His robots and androids are decidedly ambiguous. They want to simulate people, but they lack feelings and empathy. This ambiguity strongly informs *Do Androids Dream of Electric Sheep* and evinces itself on-screen in *Blade Runner*.

*Todd K. Platts*

*See also:* Nonhuman Rights and Personhood; Pathetic Fallacy; Turing Test.

### Further Reading

Brammer, Rebekah. 2018. "Welcome to the Machine: Artificial Intelligence on Screen." *Screen Education* 90 (September): 38–45.

Fitting, Peter. 1987. "Futurecop: The Neutralization of Revolt in *Blade Runner*." *Science Fiction Studies* 14, no. 3: 340–54.

Sammon, Paul S. 2017. *Future Noir: The Making of Blade Runner*. New York: Dey Street Books.

Wheale, Nigel. 1991. "Recognising a 'Human-Thing': Cyborgs, Robots, and Replicants in Philip K. Dick's *Do Androids Dream of Electric Sheep?* and Ridley Scott's *Blade Runner*." *Critical Survey* 3, no. 3: 297–304.

## Blue Brain Project

The brain, one of the most complex physical systems known with 100 billion neuron cells, is an organ that necessitates continuous effort to understand and interpret. Similarly, the implementation of digital reconstruction models of the brain and its functioning requires massive and sustained computational power. The Swiss brain research initiative, sponsored by the École Polytechnique Fédérale de Lausanne (EPFL), began in 2005 with the formation of the Blue Brain Project (BBP). The founding director of the Blue Brain Project is Henry Markram.

The Blue Brain Project has set the goal of simulating several mammalian brains and "ultimately, to study the steps involved in the emergence of biological intelligence" (Markram 2006, 153). These simulations were initially supported by the enormous computing power of IBM's BlueGene/L, the top world supercomputer system from November 2004 to November 2007. BlueGene/L was replaced by a BlueGene/P in 2009. The need for even more computational power led to BlueGene/P being replaced in 2014 by BlueGene/Q. In 2018, the BBP selected Hewlett-Packard to create a supercomputer (dubbed Blue Brain 5), which is to be exclusively dedicated to neuroscience simulation.

The implementation of supercomputer-based simulations has shifted neuroscience research from the actual lab into a virtual one. The achievement of digital reconstructions of the brain in the Blue Brain Project allows experiments, through controlled research flow and protocol, to be performed in an "in silico" environment, a Latin pseudo-word referring to simulation of biological systems on computational devices. The potential to convert the analog brain into a digital copy on supercomputers is suggestive of a paradigm shift in brain research. One key assumption is that the digital or artificial copy will behave like an analog or real brain. The software running on Blue Gene hardware, a simulation environment dubbed NEURON that models the neurons, was developed by Michael Hines, John W. Moore, and Ted Carnevale.

Considering the burgeoning budgets, expensive technology, and many interdisciplinary scientists involved, the Blue Brain Project may be considered a typical example of what, after World War II (1939–1945), was called Big Science. In addition, the research approach to the brain through simulation and digital imaging procedures leads to problems such as the management of all of the data produced. Blue Brain became an inaugural member of the Human Brain Project (HBP) consortium and submitted a proposal to the European Commission's Future & Emerging Technologies (FET) Flagship Programme. This application was accepted by the European Union in 2013, and the Blue Brain Project is now a partner in an even broader effort to study and conduct brain simulation.

*Konstantinos Sakalis*

See also: General and Narrow AI; Human Brain Project; SyNAPSE.

**Further Reading**

Djurfeldt, Mikael, Mikael Lundqvist, Christopher Johansson, Martin Rehn, Örjan Ekeberg, Anders Lansner. 2008. "Brain-Scale Simulation of the Neocortex on the IBM Blue Gene/L Supercomputer." *IBM Journal of Research and Development* 52, no. 1–2: 31–41.

Markram, Henry. 2006. "The Blue Brain Project." *Nature Reviews Neuroscience* 7, no. 2: 153–60.

Markram, Henry, et al. 2015. "Reconstruction and Simulation of Neocortical Microcircuitry." *Cell* 63, no. 2: 456–92.

## Bostrom, Nick (1973–)

Nick Bostrom is a philosopher at Oxford University with an interdisciplinary academic background in physics and computational neuroscience. He is a founding director of the Future of Humanity Institute and cofounder of the World Transhumanist Association. He has written or edited a number of books, including *Anthropic Bias* (2002), *Human Enhancement* (2009), *Superintelligence: Paths, Dangers, Strategies* (2014), and *Global Catastrophic Risks* (2014).

Bostrom was born in Helsingborg, Sweden, in 1973. Although he chafed against formal schooling, he loved learning. He especially enjoyed subjects in science, literature, art, and anthropology. Bostrom completed a bachelor's degree in philosophy, math, logic, and artificial intelligence at the University of Gothenburg and master's degrees in philosophy and physics from Stockholm University and computational neuroscience from King's College London. He was awarded his doctorate in philosophy from the London School of Economics. Bostrom is a regular consultant or contributor to the European Commission, U.S. President's Council on Bioethics, the Central Intelligence Agency, and the Centre for the Study of Existential Risk at Cambridge University.

Bostrom is known for his intellectual contributions to many fields and has proposed or written extensively about several well-known philosophical arguments and conjectures, including those on the simulation hypothesis, existential risk, the future of machine intelligence, and transhumanism. The so-called "Simulation Argument" is an extension of Bostrom's interests in the future of technology, as well as his observations on the mathematics of the anthropic bias. The argument consists of three propositions. The first proposition is that almost all civilizations that reach human level of sophistication ultimately become extinct before reaching technological maturity. The second proposition is that most civilizations eventually create "ancestor simulations" of sentient people, but then lose interest in them. The third proposition is that humanity currently lives in a simulation (the "simulation hypothesis"). Only one of the three propositions, he asserts, must be true.

If the first hypothesis would be not true, then some percentage of civilizations at the currant stage of human society would eventually reach ultimate technological maturity. If the second proposition would be not true, some fraction of civilizations would be interested in continuing to run ancestor simulations. Researchers among these civilizations might be running gigantic numbers of these simulations. In that case, there would be many multiples more simulated people living in simulated universes than real people living in real universes. It is most likely, therefore, that humanity lives in one of the simulated universes. The third possibility would not be untrue if the second proposition was true.

It is even possible, Bostrom asserts, that a civilization inside a simulation might be running its own simulations. Simulations could be living inside simulated universes, inside of their own simulated universes, in the manner of an infinite regress. It is also possible that all civilizations will go extinct, perhaps when a particular technology is discovered, which represents an existential risk beyond all ability to control.

Bostrom's argument assumes that the truth of the external world is not somehow hidden from humanity, an argument that extends back to Plato's belief in the reality of universals (the "Forms") and the ability of human senses to perceive only particular instances of universals. His argument also assumes that the capabilities of computers to simulate things now will only grow in power and sophistication. Bostrom points to computer games and literature as current examples of natural human enchantment with simulated reality.

The Simulation Argument is often confused with only the third proposition, the narrow hypothesis that humanity lives in a simulation. Bostrom thinks there is a less than 50 percent chance that humans live in some sort of artificial matrix. He also believes that it is very unlikely that if humanity were living in one, society would observe "glitches" that betrayed the presence of the simulation, because they have complete control over the running of the simulation. Conversely, the makers of the simulation would also let humans know that they are living in a simulation.

Existential risks are those that catastrophically threaten the survival of all humankind. Bostrom believes that the greatest existential risks come from humans themselves rather than natural hazards (e.g., asteroids, earthquakes, and epidemic disease). Artificial risks such as synthetic biology, molecular nanotechnology, or artificial intelligence, he believes, are far more dangerous.

Bostrom distinguishes between local, global, and existential risks. Local risks might involve the loss of a priceless work of art or a car crash. Global risks could involve destruction wrought by a military dictator or the eruption of a supervolcano. Existential risks are different in scope and severity. They are pangenerational and permanent. Reducing the risk of existential risks is, in his view, the most important thing that human beings can do, because of the numbers of lives potentially saved; working against existential risk is also one of the most neglected activities of humanity.

He also defines a number of classes of existential risk. These include human extinction, that is, the extinguishment of the species before it reaches technological maturity; permanent stagnation, or the plateauing of human technological achievement; flawed realization, where humanity fails to use advanced technology for an ultimately worthwhile purpose; and subsequent ruination, where society reaches technological maturity, but then something goes wrong. Bostrom speculates that while humanity has not used human creativity to make a technology that unleashes existentially destructive power, it is conceivable that it might do so in the future. Human society has also not invented a technology so terrible in its consequences that humanity could collectively disremember it. The goal would be to get on to a safe technological course that involves global coordination and is sustainable.

Bostrom uses the metaphor of changed brain complexity in the evolution of humans from apes, which took only a few hundred thousand generations, to argue for the possibility of machine superintelligence. Machine learning (that is, using algorithms that themselves learn) allows artificial systems that are not limited to one domain. He also notes that computers operate at much higher processing speeds than human neurons.

Eventually, Bostrom says, humans will depend on superintelligent machines in the same way that chimpanzees now rely on humans for their ultimate survival—even in the wild. There is also a potential for superintelligent machines to unleash havoc, or even an extinction level event, by creating a potent optimizing process that has a poorly specified objective. A superintelligence might even anticipate a human reaction by subverting humanity to the programmed goal.

Bostrom acknowledges that there are some algorithmic tricks practiced by humans that are not yet understood by computer scientists. He suggests that it is important for artificial intelligences to learn human values as they are engaging in machine learning. On this score, Bostrom is borrowing from artificial intelligence theorist Eliezer Yudkowsky's vision of "coherent extrapolated volition"—also known as "friendly AI," similar in kind to that already available in human goodwill, civil society, and institutions. A superintelligence should want to grant humankind happiness and fulfillment and might even make the hard decisions that benefit the whole community over the individual. In "An Open Letter on Artificial Intelligence" posted on the Future of Life Institute website in 2015, Bostrom joined with Stephen Hawking, Elon Musk, Max Tegmark, and many other top AI researchers to call for research on artificial intelligence that maximizes the benefits to humanity while minimizing "potential pitfalls."

Transhumanism is a theory or belief in the extension and enhancement of the physical, sensory, and cognitive capacities of the human species by technological means. Bostrom established the World Transhumanist Association, now known as Humanity+, in 1998 with fellow philosopher David Pearce to address some of the social challenges to the acceptance and use of emerging transhumanist technologies by all classes in society. Bostrom has said that he is not interested in defending the technologies, but rather in using advanced technologies to confront real risks and promote better lives.

Bostrom is especially interested in the ethics of human enhancement and the future of fundamental technological changes in human nature. He asserts that transhumanist notions are found across time and cultures, expressed in ancient quests such as the Epic of Gilgamesh, as well as in the historical searches for the Fountain of Youth and Elixir of Immortality. The transhumanist project, then, might be considered quite old, with contemporary manifestations in scientific and technological fields such as artificial intelligence and genome editing.

Bostrom takes an activist view against emerging powerful transhumanist tools. He hopes that policymakers can act with foresight and take command of the sequencing of technological developments in order to mitigate the risk of potential applications and reduce the odds of human extinction. He asserts that all people should have the opportunity to become transhuman, or even posthuman (have capacities beyond human nature and intelligence). Preconditions to success, for Bostrom, would include a worldwide commitment to global security and continued technological progress, as well as wide access to the benefits of technologies (cryonics, mind uploading, anti-aging drugs, life extension regimens), holding out the most hope for transhumanist change in our lifetime. While cautious, Bostrom rejects ordinary humility, noting that humans have a long-shared experience of managing potentially devastating risks. He is an active proponent of "individual choice" in such matters, and also of "morphological freedom" to change or reengineer one's body to meet particular wants and needs.

*Philip L. Frana*

*See also:* Superintelligence; Technological Singularity.

**Further Reading**

Bostrom, Nick. 2003. "Are You Living in a Computer Simulation?" *Philosophical Quarterly* 53, no. 211: 243–55.

Bostrom, Nick. 2005. "A History of Transhumanist Thought." *Journal of Evolution and Technology* 14, no. 1: 1–25.

Bostrom, Nick, ed. 2008. *Global Catastrophic Risks.* Oxford, UK: Oxford University Press.

Bostrom, Nick. 2014. *Superintelligence: Paths, Dangers, Strategies.* Oxford, UK: Oxford University Press.

Savulescu, Julian, and Nick Bostrom, eds. 2009. *Human Enhancement.* Oxford, UK: Oxford University Press.

# Brooks, Rodney (1954–)

Rodney Brooks is a computer science researcher, entrepreneur, and business and policy advisor. He is an authority in computer vision, artificial intelligence, robotics, and artificial life. Brooks is famous for his work on behavior-based robotics and artificial intelligence. His iRobot Roomba autonomous robotic vacuum cleaners are among the most ubiquitous domestic robots in use in the United States.

Brooks is influential for his advocacy of a bottom-up approach to computer science and robotics, an epiphany he had while on a long, uninterrupted visit to his wife's family's home in Thailand. Brooks argues that situatedness, embodiment, and perception are as important to modeling the dynamic behaviors of intelligent creatures as cognition in the brain. This approach is now known as action-based robotics or behavior-based artificial intelligence. Brooks' approach to intelligence without explicitly designed reasoning may be contrasted with the symbolic reasoning and representation approach common to the first several decades of artificial intelligence research.

Brooks noted that much of the early progress in robotics and artificial intelligence had been predicated on the formal framework and logical operators of the universal computational architecture created by Alan Turing and John von Neumann. He believed that these artificial systems had diverged widely from the actual biological systems they were intended to represent. Living organisms relied on low-speed, massively parallel processing and adaptive engagement with their environments. These were not, in his view, features of classical computing architecture, but rather aspects of what Brooks began, in the mid-1980s, to refer to as subsumption architecture.

For Brooks, behavior-based robots are situated in real environments in the world, and they learn successful actions from that world. They must be embodied so that they can relate to the world and receive immediate feedback from their sensory inputs. The origin of intelligence ordinarily arises from specific situations, signal transformations, and real-time physical interactions. Sometimes the source of intelligence is hard to pinpoint functionally, as it emerges instead from multiple direct and indirect interactions between various robot components and the environment.

Brooks created several noteworthy mobile robots based on the subsumption architecture as a professor in the Massachusetts Institute of Technology Artificial Intelligence Laboratory. The first of these behavior-based robots was called Allen, which was equipped with sonar range and motion sensors. The robot possessed three layers of control. The first primitive layer gave it the ability to avoid static and dynamic obstacles. The second implanted a random walk algorithm that provided the robot a reflex to occasionally change direction. The third behavioral layer monitored distant places that might be goals, and it suppressed the other two control layers. Another robot named Herbert relied on a distributed arrangement of 8-bit microprocessors and 30 infrared proximity sensors to avoid obstacles, follow walls, and collect empty soda cans placed around various offices. Genghis was a six-legged robot with four onboard microprocessors, 22 sensors, and 12 servo motors, who could walk in rough terrain. Genghis could stand up, balance and stabilize itself, climb stairs, and follow people.

With Anita Flynn, an MIT Mobile Robotics Group research scientist, Brooks began imagining scenarios in which behavior-based robots could help explore the surface of other planets. In their 1989 essay "Fast, Cheap, and Out of Control" published by the British Interplanetary Society, the two roboticists argued that space organizations such as the Jet Propulsion Laboratory should reconsider plans for expensive, large, and slow-moving mission rovers, and instead contemplate using larger sets of small mission rovers in order to reduce cost and obviate risk. Brooks and Flynn concluded that their autonomous robot technology could be built and tested quickly by space agencies and serve reliably on other planets even when beyond the immediate control of human operators. The Sojourner rover, while relatively large and singular in design, possessed some behavior-based autonomous robot capabilities when it landed on Mars in 1997.

Also in the 1990s, Brooks and a new Humanoid Robotics Group in the MIT Artificial Intelligence Laboratory began building a humanoid robot called Cog. Cog had a double meaning, referring to the teeth on gears and the word "cognitive." Cog had several goals, many of them intended to encourage social communication between the robot and a human being. Cog as fabricated possessed a humanlike face, with many degrees of motor freedom in the head, trunk, arms, and legs. Cog possessed sensors capable of sight, hearing, touch, and vocalization. Group researcher Cynthia Breazeal, who had designed the mechanics and control system for Cog, capitalized on the lessons learned from human interaction with the robot to create a new robot in the lab called Kismet. Kismet is an affective robot who can recognize, interpret, and reproduce human emotions. Cog and Kismet are important milestones in the history of artificial emotional intelligence.

In recent decades, Rodney Brooks has pursued commercial and military application of his robotics research as cofounder and chief technology officer of iRobot Corporation. In 1998, the company received a Defense Advanced Research Projects Agency (DARPA) grant to develop PackBot, a robot commonly used to detect and defuse improvised explosive devices in Iraq and Afghanistan. PackBot has also seen service at the site of the World Trade Center following the terrorist attacks of September 11, 2001, and in the initial assessment of the tsunami- and earthquake-damaged Fukushima Daiichi nuclear power plant in Japan. In 2000, Hasbro marketed a toy robot designed by Brooks and others at iRobot. The result, My Real Baby, is a lifelike doll capable of crying, fussing, sleeping, giggling, and expressing hunger.

The iRobot Corporation is also the creator of the Roomba cleaning robot. Roomba, introduced in 2002, is disc-shaped with roller wheels and various brushes, filters, and a squeegee vacuum. Like other behavior-based robots developed by Brooks, the Roomba detects obstacles with sensors and avoids dangers such as falling down stairs. Newer models follow infrared beams and photocell sensors for self-charging and to map out rooms. By 2019 iRobot had sold more than 25 million robots worldwide.

Brooks is also cofounder and chief technology officer of Rethink Robotics. The company, founded as Heartland Robotics in 2008, develops relatively inexpensive industrial robots. Rethink's first robot was Baxter, which is capable of simple repetitive tasks such as loading, unloading, assembling, and sorting. Baxter possesses an animated human face drawn on a digital screen mounted at its top. Baxter has embedded sensors and cameras that help it recognize and avoid collisions when people are near, an important safety feature. Baxter can be used in ordinary industrial environments without a security cage. The robot can be programmed quickly by unskilled workers, who simply move its arms around in the expected way to direct its movements. Baxter stores these motions in memory and adapts them to specific tasks. Fine movements can be inputted using the controls on its arms. Rethink's Sawyer collaborative robot is a smaller version of Baxter, which is marketed for use in completing dangerous or monotonous industrial tasks, often in confined spaces.

Brooks has often expressed the view that the hard problems of consciousness continue to elude scientists. Artificial intelligence and artificial life researchers, he says, have missed something important about living systems that keeps the chasm between the nonliving and living worlds large. This remains true even though all of the living features of our world are built up from nonliving atoms. Brooks suggests that perhaps some of the parameters used by AI and ALife researchers are wrong or that current models lack enough complexity. Or it may be that researchers continue to lack sufficient raw computing power. But Brooks suggests that it may be that there is something—an ingredient or a property— about biological life and subjective experience that is currently undetectable or hidden from scientific view.

Brooks studied pure mathematics at Flinders University in Adelaide, South Australia. He completed his PhD under American computer scientist and cognitive scientist John McCarthy at Stanford University. His doctoral thesis was expanded and published as *Model-Based Computer Vision* (1984). He served as Director of the MIT Artificial Intelligence Laboratory (renamed Computer Science & Artificial Intelligence Laboratory (CSAIL) in 2003) from 1997 to 2007. Brooks is the recipient of numerous honors and awards for artificial intelligence and robotics. He is a Fellow of the American Academy of Arts & Sciences and a Fellow of the Association for Computing Machinery. Brooks is the winner of the prestigious IEEE Robotics and Automation Award and the Joseph F. Engelberger Robotics Award for Leadership. He is currently deputy board chairman of the advisory board of the Toyota Research Institute.

*Philip L. Frana*

*See also:* Embodiment, AI and; Tilden, Mark.

**Further Reading**

Brooks, Rodney A. 1984. *Model-Based Computer Vision.* Ann Arbor, MI: UMI Research Press.

Brooks, Rodney A. 1990. "Elephants Don't Play Chess." *Robotics and Autonomous Systems* 6, no. 1–2 (June): 3–15.

Brooks, Rodney A. 1991. "Intelligence without Reason." AI Memo No. 1293. Cambridge, MA: MIT Artificial Intelligence Laboratory.

Brooks, Rodney A. 1999. *Cambrian Intelligence: The Early History of the New AI.* Cambridge, MA: MIT Press.

Brooks, Rodney A. 2002. *Flesh and Machines: How Robots Will Change Us.* New York: Pantheon.

Brooks, Rodney A., and Anita M. Flynn. 1989. "Fast, Cheap, and Out of Control." *Journal of the British Interplanetary Society* 42 (December): 478–85.

# Brynjolfsson, Erik (1962–)

Erik Brynjolfsson is director of the Massachusetts Institute of Technology Initiative on the Digital Economy. He is also Schussel Family Professor at the MIT Sloan School and Research Associate at the National Bureau of Economic Research (NBER). Brynjolfsson's research and writing is in the area of information technology productivity and its relation to labor and innovation.

Brynjolfsson's work has long been at the center of discussions about the impacts of technology on economic relations. His early work highlighted the relationship between information technology and productivity, especially the so-called productivity paradox. Specifically, Brynjolfsson found "broad negative correlations with economywide productivity and information worker productivity" (Brynjolfsson 1993, 67). He suggested that the paradox might be explained by mismeasurement of impact, a lag between initial cost and eventual benefits, private benefits accruing at the expense of aggregate benefit, or outright mismanagement.

However, numerous empirical studies by Brynjolfsson and collaborators also show that information technology spending has significantly enhanced productivity—at least since 1991. Brynjolfsson has shown that information technology, and more specifically electronic communication networks, increases multitasking. Multitasking in turn improves productivity, the development of knowledge networks, and worker performance. The relationship between IT and productivity thus represents a "virtuous cycle" more than a simple causal link: as performance increases, users are encouraged to adopt knowledge networks that improve productivity and operational performance.

The productivity paradox has attracted renewed interest in the age of artificial intelligence. The struggle between human and artificial labor poses a brand-new set of challenges for the digital economy. Brynjolfsson writes about the phenomenon of frictionless commerce, a feature produced by online activities such as instant price comparison by smart shopbots. Retailers such as Amazon have understood the way online markets work in the age of AI and have remade their supply chains and distribution strategies. This reshaping of online commerce has transformed the concept of efficiency itself. In the brick-and-mortar economy, price and quality comparisons may be done by secret human shoppers. This process can be slow and costly. By contrast, the costs of acquiring some kinds of online information are now effectively zero, since consumers (and web-scraping bots) can easily surf from one website to another.

In the best-selling book *Race Against the Machine* (2011), Brynjolfsson and coauthor Andrew McAfee tackle the influence of technology on employment, the economy, and productivity growth. They are especially interested in the process of creative destruction, a theory popularized by economist Joseph Schumpeter in *Capitalism, Socialism, and Democracy* (1942). Brynjolfsson and McAfee show that while technology is a positive asset for the economy as a whole, it may not automatically benefit everyone in society. In fact, the benefits that accrue from technological innovations may be unequal, favoring the small groups of innovators and investors that dominate digital markets. Brynjolfsson and McAfee's main conclusion is that humans should not compete against machines, but instead partner *with* machines. Innovation is enhanced, and human capital is improved, when people learn skills to participate in the new age of smart machines.

In *The Second Machine Age* (2014), Brynjolfsson and McAfee shed more light on this subject by surveying the role of data in the digital economy and the growing importance of artificial intelligence. The authors note that data-driven intelligent machines are a central feature of internet commerce. Artificial intelligence opens the door to all kinds of new services and features. They argue that these transformations not only affect productivity indexes but also reshape our very sense of what it means to engage in capitalist enterprise. Brynjolfsson and McAfee have much to say about the destabilizing effects of a growing chasm between internet moguls and ordinary people. Of particular concern to the authors is unemployment caused by artificial intelligence and smart machines. In *Second Machine Age*, Brynjolfsson and McAfee reiterate their view that there should not be a race against technology, but meaningful coexistence with it in order to build a better global economy and society.

In *Machine, Platform, Crowd* (2017), Brynjolfsson and McAfee explain that in the future the human mind must learn to coexist with smart machines. The great challenge is to shape how society will use technology and how the positive attributes of data-driven innovation and artificial intelligence can be nourished while the negative aspects are pruned away. Brynjolfsson and McAfee imagine a future where labor is not just suppressed by efficient machines and the disruptive effects of platforms, but where new matchmaking businesses governing intricate economic structures and large enthusiastic online crowds, and copious amounts of human knowledge and expertise are used to strengthen supply chains and economic processes. Brynjolfsson and McAfee assert that machines, platforms, and the crowd can be used in many ways, whether for the concentration of power or to distribute decision-making and prosperity. They conclude that people do not need to remain passively dependent on past technological trends but can reshape technology in order to make it more productive and socially beneficial.

Brynjolfsson remains engaged in research on artificial intelligence and the digital economy, with current interests in productivity, inequality, labor, and welfare. He holds degrees in Applied Mathematics and Decision Sciences from Harvard University. He completed his doctoral work in Managerial Economics at the MIT Sloan School in 1991. The title of his dissertation was "Information Technology and the Re-organization of Work: Theory and Evidence."

*Angelo Gamba Prata de Carvalho*

*See also:* Ford, Martin; Workplace Automation.

**Further Reading**

Aral, Sinan, Erik Brynjolfsson, and Marshall Van Alstyne. 2012. "Information, Technology, and Information Worker Productivity." *Information Systems Research* 23, no. 3, pt. 2 (September): 849–67.

Brynjolfsson, Erik. 1993. "The Productivity Paradox of Information Technology." *Communications of the ACM* 36, no. 12 (December): 67–77.

Brynjolfsson, Erik, Yu Hu, and Duncan Simester. 2011. "Goodbye Pareto Principle, Hello Long Tail: The Effect of Search Costs on the Concentration of Product Sales." *Management Science* 57, no. 8 (August): 1373–86.

Brynjolfsson, Erik, and Andrew McAfee. 2012. *Race Against the Machine: How the Digital Revolution Is Accelerating Innovation, Driving Productivity, and Irreversibly Transforming Employment and the Economy.* Lexington, MA: Digital Frontier Press.

Brynjolfsson, Erik, and Andrew McAfee. 2016. *The Second Machine Age: Work, Progress, and Prosperity in a Time of Brilliant Technologies.* New York: W. W. Norton.

Brynjolfsson, Erik, and Adam Saunders. 2013. *Wired for Innovation: How Information Technology Is Reshaping the Economy.* Cambridge, MA: MIT Press.

McAfee, Andrew, and Erik Brynjolfsson. 2017. *Machine, Platform, Crowd: Harnessing Our Digital Future.* New York: W. W. Norton.

# C

## Calo, Ryan (1977–)

Michael Ryan Calo is a thought leader on the legal and policy implications of artificial intelligence and robotics. Calo helped to create a community of legal scholars focused on robotics and AI; he anticipated the threat AI could pose to consumer privacy and autonomy, and he wrote an early and widely disseminated primer on AI law and policy. In addition to these and other contributions, Calo has forged methodological and practice innovations for early stage tech policy work, demonstrating the importance and efficacy of legal scholars working side by side with technologists and designers to anticipate futures and meaningful policy responses.

Calo grew up in Syracuse, New York, and Florence, Italy. He first encountered robots at a young age, when his parents gave him a wonderful remote-controlled base attached to an inflatable robot. Calo studied philosophy at Dartmouth as an undergraduate under ethics of computing pioneer James Moor, among others. Calo went on to receive his law degree from the University of Michigan in 2005. After law school, a federal appellate clerkship, and two years in private practice, he became a fellow and then research director at Stanford's Center for Internet and Society (CIS). It was at Stanford that Calo first helped to bring robotics law and policy into the mainstream, including founding the Legal Aspects of Autonomous Driving initiative with Sven Beiker at the Center for Automotive Research at Stanford (CARS). Along the way, Calo crossed paths with Canadian law professor and philosopher of technology, Ian Kerr, and cyberlaw pioneer Michael Froomkin. Together Froomkin, Kerr, and Calo cofounded the We Robot conference in 2012. Calo credits Kerr with challenging him to pursue robotics and AI as an area of study.

Today, Calo codirects the Tech Policy Lab at the University of Washington, an interdisciplinary research unit spanning computer science, information science, and law. In this role, he and his codirectors Batya Friedman and Tadayoshi Kohno set the research and practice agenda for the Lab. Calo also cofounded the University of Washington Center for an Informed Public, devoted to the study and resistance of digital and analog misinformation. Calo has written extensively on the legal and policy aspects of robotics and AI. Key contributions include updating the behavioral economic theory of market manipulation in light of artificial intelligence and digital media, urging a social systems approach to the study of AI's impacts, anticipating the privacy harms of robotics and AI, and rigorously examining how the affordances of robotics and AI challenge the American legal system.

Calo is the Lane Powell and D. Wayne Gittinger Associate Professor at the University of Washington School of Law. In 2016, Calo and the Tech Policy Lab hosted the inaugural Obama White House workshop on artificial intelligence policy. He regularly testifies on national and international concerns related to AI and robotics, including in 2013 before the U.S. Senate Judiciary Committee on the domestic use of drones and in 2016 before the German Bundestag (Parliament) about robotics and artificial intelligence. He serves on numerous advisory boards for organizations such as AI Now, Responsible Robotics, and the University of California People and Robots Initiative and on numerous conference program committees such as FAT* and Privacy Law Scholars Conference, where much of the contemporary conversation around AI and its social impacts takes place.

*Batya Friedman*

*See also:* Accidents and Risk Assessment; Product Liability and AI.

**Further Reading**

Calo, Ryan. 2011. "Peeping Hals." *Artificial Intelligence* 175, no. 5–6 (April): 940–41.

Calo, Ryan. 2014. "Digital Market Manipulation." *George Washington Law Review* 82, no. 4 (August): 995–1051.

Calo, Ryan. 2015. "Robotics and the Lessons of Cyberlaw." *California Law Review* 103, no. 3: 513–63.

Calo, Ryan. 2017. "Artificial Intelligence Policy: A Primer and Roadmap." *University of California, Davis Law Review* 51: 399–435.

Crawford, Kate, and Ryan Calo. 2016. "There Is a Blind Spot in AI Research." *Nature* 538 (October): 311–13.

# Campaign to Stop Killer Robots

The Campaign to Stop Killer Robots is a movement dedicated to preventing the development and use of lethal autonomous weapon systems (LAWS). The core concern of the campaign is that armed robots, evaluating and enacting life or death judgments, undermine legal and ethical constraints on violence in human conflicts. Advocates for the development of LAWS characterize these technologies as consistent with existing weapons and regulations, such as in the case of cruise missiles that are programmed and launched by humans to seek and destroy a designated target. Advocates also argue that robots are entirely dependent on humans, constrained by their design, following the behaviors given to them, and that with proper oversight they can save lives by replacing humans in dangerous scenarios. The Campaign to Stop Killer Robots rejects responsible use as a credible scenario, raising the concern that LAWS development could lead to a new arms race. The campaign emphasizes the risk of loss of human control over the use of deadly force in cases where armed robots may designate and eliminate a threat before possible human intervention.

The campaign was formally launched in London, England, on April 22, 2013, and is hosted and coordinated by Human Rights Watch, an international nongovernmental organization (NGO) that promotes universal human rights and investigates abuses of those rights. The Campaign to Stop Killer Robots is composed of many member organizations, such as the International Committee for Robot Arms Control and Amnesty International. Leadership of the campaign consists of a steering committee and a global coordinator. As of 2018, the steering committee is composed of eleven NGOs. The global coordinator of the campaign is Mary Wareham, who previously led international efforts to regulate land mines and cluster munitions.

As with campaigns against land mines and cluster munitions, efforts to prohibit weaponized robots focus on their potential to cause unnecessary suffering and their risk of indiscriminate harm to civilians. The prohibition of weapons on an international scale is coordinated through the United Nations Convention on Certain Conventional Weapons (CCW), which first came into effect in 1983. The Campaign to Stop Killer Robots advocates for the inclusion of LAWS in the CCW, as the CCW has not yet agreed on a ban of weaponized robots and as the CCW does not include any mechanism for enforcing agreed upon prohibitions,

The Campaign to Stop Killer Robots also supports the creation of additional preemptive bans that could be enacted through new international treaties. In addition to lobbying governing bodies for bans by treaty and convention, the Campaign to Stop Killer Robots provides resources for educating and organizing the public, including multimedia databases, campaign reports, and a mailing list. The Campaign also pursues the cooperation of technology companies, seeking their voluntary refusal to engage in the development of LAWS. The Campaign has an active social media presence through the @BanKillerRobots handle, where it tracks and shares the names of corporations that pledge not to participate in the design or distribution of intelligent weapons.

*Jacob Aaron Boss*

*See also:* Autonomous Weapons Systems, Ethics of; Battlefield AI and Robotics; Lethal Autonomous Weapons Systems.

**Further Reading**

Baum, Seth. 2015. "Stopping Killer Robots and Other Future Threats." *Bulletin of the Atomic Scientists*, February 22, 2015. https://thebulletin.org/2015/02/stopping -killer-robots-and-other-future-threats/.

Campaign to Stop Killer Robots. 2020. https://www.stopkillerrobots.org/.

Carpenter, Charli. 2016. "Rethinking the Political / -Science- / Fiction Nexus: Global Policy Making and the Campaign to Stop Killer Robots." *Perspectives on Politics* 14, no. 1 (March): 53–69.

Docherty, Bonnie. 2012. *Losing Humanity: The Case Against Killer Robots*. New York: Human Rights Watch.

Garcia, Denise. 2015. "Killer Robots: Why the US Should Lead the Ban." *Global Policy* 6, no. 1 (February): 57–63.

# Caregiver Robots

Caregiver robots are personal support robots designed to assist people who, for a variety of reasons, may require assistive technology for long-term care, disability, or supervision. Although not in widespread use, caregiver robots are considered important in nations with growing elderly populations or in circumstances where large numbers of people might be stricken simultaneously with a debilitating illness. Reactions to caregiver robots have ranged from horror to relief. Some ethicists have complained that robotics researchers misunderstand or underappreciate the role of compassionate caregivers as they work to remove the toil from caregiving rituals. Most caregiver robots are personal robots for home use, though some are in use in institutional settings such as hospitals, nursing homes, and schools. Some are care robots for the elderly. Others are designed for childcare duties—so-called robot nannies. Many of them are described as social robots.

Interest in caregiver robots has grown in proportion with aging populations around the world. Japan has one of the highest proportions of elderly citizens and is also a world leader in the development of caregiver robots. The United Nations predicts that one-third of the island nation's people will be sixty-five or above by 2050, outstripping the available natural supply of nursing care workers. In 2013, the country's Ministry of Health, Labor and Welfare began a pilot demonstration project to introduce bionic nursing robots into eldercare facilities. In the United States, the number of eligible retirees will double by 2050, and the number over the age of 85 will triple. Around the world in that same year, there will be 1.5 billion people over the age of sixty-five (United Nations 2019).

Interest in caregiver robot technology is growing for a variety of reasons. Physical demands associated with care of the elderly, the infirm, and children are often cited as one impetus for the development of assistive robots. The caregiver role can be arduous, particularly when the client is suffering from a serious or long-term disease such as Alzheimer's, dementia, or a schizoid disorder. Caregiver robots have also been proposed as a partial solution to family economic distress. Robots might eventually substitute for human relatives who need to work. They have also been cited as a potential answer to chronic staff shortages in nursing homes and other care settings.

There are also social and cultural reasons driving development in caregiver robots. In Japan, robot caregivers are preferred over international health-care workers, in large measure because of negative perceptions of foreigners. Needs identified by the elderly themselves often revolve around the need for independence and fear of loss of behavioral, emotional, and cognitive autonomy.

Many robot caregiver roles have been identified in the literature. Some robots are assumed to be capable of reducing menial labor by human caregivers. Others are adept at more sophisticated tasks. Laborsaving intelligent service robots have been developed to assist with feeding, cleaning of homes and bodies, and mobility assistance (including lifting and turning). Other roles for these assistive machines involve safety monitoring, data collection, and surveillance. Robot caregivers have been proposed as useful for coaching and stimulation of clients with severe to profound disabilities. These robots may serve as cognitive prostheses or mobile memory aids for people who need regular reminders to complete tasks or take medicine. Such caregiver robots may also have telemedicine functions, summoning physicians or nurses for regular or emergency consultations. Perhaps most controversially, robot caregivers have been proposed as a source of social interaction and companionship. Social robots may resemble humans but are often interactive smart toys or artificial pets.

In Japan, robots are often described as *iyashi,* a word also used to describe a subgenre of anime and manga created for the purpose of emotional healing. A wide variety of soft-tronic robots are available for Japanese children and adults as huggable companions. Wandakun was a fuzzy koala bear-type robot developed by Matsushita Electric Industrial (MEI) in the 1990s. The bear squirmed when petted, could sing, and could respond to touch with a few Japanese phrases. Babyloid is a plush robot baby beluga whale designed by Masayoshi Kano at Chukyo University to alleviate symptoms of depression in geriatric patients. Only seventeen inches in length, Babyloid's eyes blink and will take "naps" when rocked. LED lights embedded in its cheeks glow when it is "happy." The robot is also capable of shedding blue LED tears when it is not happy. Babyloid is capable of making more than 100 different sounds. At a cost of more than $1,000 dollars each, it is no toy.

The artificial baby harp seal Paro, designed by Japan's National Institute of Advanced Industrial Science and Technology (AIST), is designed to give comfort to patients with dementia, anxiety, or depression. The eighth-generation Paro is packed with thirteen surface and whisker sensors, three microphones, two vision sensors, and seven actuators for the neck, fins, and eyelids. Paro's inventor, Takanori Shibata of AIST's Intelligent System Research Institute, notes that patients with dementia experience less aggression and wandering, and more social interaction, when using the robot. Paro is considered a Class II medical device in the United States, in the same category of risk that comprises power wheelchairs and X-ray machines. AIST is also the developer of Taizou, a twenty-eight-inch robot that can replicate the movements of thirty different exercises. Taizou is used in Japan to motivate senior citizens to exercise and stay fit.

The well-known AIBO developed by Sony Corporation is a robotic therapy dog as well as a very pricey toy. Sony's Life Care Design subsidiary began introducing a new generation of dog robots into retirement homes owned by the company in 2018. AIBO's successor, the humanoid QRIO robot, has been proposed as a platform for simple childcare activities, such as interactive games and singalongs. Palro, another robot for eldercare therapy made by Fujisoft, is already in use in more than 1,000 senior citizen facilities. Its artificial intelligence software has been upgraded several times since initial release in 2010. Both are used to reduce symptoms of dementia and for entertainment.

Japanese corporations have also cultivated a broader segment of users of so-called partner-type personal robots. These robots are designed to promote human-machine interaction and reduce loneliness and mild depression. NEC Corporation began creating its cute PaPeRo (Partner-Type Personal Robot) in the late 1990s. PaPeRo communications robots can see, listen, speak, and make various motions. Current versions have twin camera eyes capable of facial recognition and are designed to help family members living in separate homes monitor one another. The Childcare Version of PaPeRo plays with children and functions as a short-term babysitter.

Toyota introduced its family of humanoid Partner Robots in 2005. The company's robots are designed for a wide variety of purposes, from human support and rehabilitation to socialization and innovation. Toyota released a specialized Human Support Robot (HSR) in the Partner Robots line in 2012. HSR robots are intended to support the independence of senior citizens. Prototypes are now in use in Japanese eldercare facilities and in the homes of the disabled. HSR robots can pick up and fetch objects and avoid obstacles. They can also be operated by a human caregiver remotely and provide internet connectivity and communication.

Japanese robotics researchers are also adopting narrower approaches to automated caregiving. The RIKEN Collaboration Center for Human-Interactive Robot Research's RI-MAN robot is an autonomous humanoid patient-lifting robot. The robot's right and left forearms, upper arms, and torso are composed of a soft silicone skin layer and equipped with tactile sensors for safe lifting. RI-MAN can track human faces and has odor detectors. RIBA (Robot for Interactive Body Assistance) is a second-generation RIKEN lifting robot that responds to basic voice commands while safely moving patients from bed to wheelchair. RIBA-II has capacitance-type tactile sensors made entirely of rubber, which detect patient weight. Robear is RIKEN's current-generation hydraulic patient life-and-transfer device. Fashioned in the appearance of an anthropomorphic robotic bear, the robot is more lightweight than its predecessors. RIKEN's lifting robots are designed by creator and lab head Toshiharu Mukai.

Other narrower approaches to caregiver robots in Japan include SECOM's MySpoon, the Cyberdine Hybrid Assistive Limb (HAL), and Panasonic Resyone robotic care bed. MySpoon is a meal-support robot, which gives clients the freedom to feed themselves with a joystick and substitute human arm and eating utensil. The Cyberdine Hybrid Assistive Limb (HAL) is a powered robotic exoskeleton suit for use by people with physical disabilities. The Panasonic Resyone robotic care bed integrates bed and wheelchair for patients who would otherwise require daily lift assistance.

Caregiver robot projects are underway in Australia and New Zealand as well. In the early 2000s, the Australian Research Council's Centre of Excellence for Autonomous Systems (CAS) represented an institutional partnership between the University of Technology Sydney, the University of Sydney, and University of New South Wales. The center's goal was to understand and develop robotics to encourage pervasive and ubiquitous applications of autonomous systems in society. The work of CAS has now been divided and put on independent footing at the Centre for Autonomous Systems at the University of Technology Sydney and in the Australian Centre for Field Robotics at the University of Sydney. Bruce MacDonald at the University of Auckland leads in the development of a socially assistive robot called Healthbot. Healthbot is a mobile health robot that helps senior citizens remember to take their medications, check vitals and track physical health, and call for emergency response.

A variety of caregiver robots are under development in the European Union. The GiraffPlus (Giraff+) project, which recently completed at Örebro University, Sweden, aims to create an intelligent system for in-home monitoring of elderly people's blood pressure, temperature, and movements (to detect falls and other health emergencies). Giraff also functions as a telepresence robot that can be used for virtual visits with family members and health professionals. The robot stands about five and a half feet in height and has simple controls and a night-vision camera. The multidisciplinary, collaborative European Mobiserv project aims to create a robot that reminds elderly clients to take their medications, eat meals, and stay active. The Mobiserv robot is embedded in a smart home environment of sensors, optical sensors, and other automated devices. Mobiserv is designed to interact with smart clothes that collect health-related data. Mobiserv involves a partnership between Systema Technologies and the nine European partners representing seven countries.

The objective of the EU CompanionAble Project involving fifteen institutions coordinated by the University of Reading is to create a mobile robotic companion to demonstrate the advantages of information and communication technologies in the care of the elderly. The CompanionAble robot attempts to address emergency and security concerns in early stage dementia, provide cognitive stimulation and reminders, and summon human caregiver assistance. CompanionAble also interacts with a variety of sensors and devices in a smart home setting. The QuoVADis Project at Brova Hospital Paris, a public university hospital for geriatrics care, has a similar ambition, to create a robot for at-home care of cognitively impaired elderly people. The Fraunhofer Institute for Manufacturing Engineering and Automation continues to design and manufacture successive generations of modular robots called Care-O-Bots. It is intended for use in hospitals, hotels, and nursing homes. The Care-O-Bot 4 service robot can reach from the floor to a shelf with its long arms and rotating, bending hip joint. The robot is designed to be accepted as something that is friendly, helpful, courteous, and smart.

A novel approach is offered by the European Union's ROBOSWARM and IWARD, intelligent and programmable hospital robot swarms. ROBOSWARM is a distributed agent system designed for hospital cleaning. The more versatile IWARD is designed for cleaning, patient monitoring and guidance, environmental monitoring, medication provision, and patient surveillance. Multi-institutional collaborators discovered that it would be difficult to certify that the AI systems embedded in these systems would perform appropriately under real-world conditions because they exhibit adaptive and self-organizing behaviors. They also discovered that observers sometimes questioned the movements of the robots, wondering whether they were executing appropriate operations.

In Canada, the Ludwig humanoid robot at the University of Toronto is designed to help caregivers address aging-related conditions in their clients. The robot makes conversation with senior citizens who have dementia or Alzheimer's disease. Goldie Nejat, AGE-WELL Investigator and Canada Research Chair in Robots for Society and Director of the Institute for Robotics and Mechatronics, University of Toronto, is using robotics technology to help people by coaching them to follow the sequence of steps in common activities of daily living. The university's Brian robot is social in nature and responds to emotional human interaction. HomeLab at the Toronto Rehabilitation Institute (iDAPT), the largest academic rehabilitation research hospital in Canada, is developing assistive robots for use in health-care delivery. HomeLab's Ed the Robot is a low-cost robot developed using the iRobot Create toolkit. Like Brian, the robot is intended to prompt dementia patients on the proper steps needed for common activities.

Caregiver robot technology is also proliferating in the United States. The Acrotek Actron MentorBot surveillance and security robot developed in the early 2000s could follow a human client using visual and auditory inputs, issue reminders to take meals or medications, and alert family members about problems or summon emergency personnel. Bandit is a socially assistive robot developed by roboticist Maja Matarić at the University of Southern California's Robotics and Autonomous Systems Center. The robot is used in therapeutic settings with patients who have suffered traumatic injuries or strokes, suffer from diseases of aging, or have autism or are obese. The center has discovered that stroke victims respond quickly to copycat exercise motions made by smart robots in therapy sessions. Rehabilitative exercises performed by robots also proved useful in prompting and cueing activities for children with autism spectrum disorders. Matarić is now working to bring affordable social robots to market through the startup Embodied Inc.

The University of Pittsburgh, Carnegie Mellon University, and the University of Michigan collaborated in the development of nursebots Flo and Pearl, assistive robots for the care of people who are elderly or infirm. The Nursebot project, funded by the National Science Foundation, developed a platform for intelligent reminding, telepresence, data collection and surveillance, mobile manipulation, and social interaction. Today, Carnegie Mellon hosts the Quality of Life Technology (QoLT) Center, a National Science Foundation Engineering Research Center (ERC) with a mission to promote independence and enhance the functional capacities of aged and disabled individuals with intelligent systems.

The Massachusetts Institute of Technology opened the multidisciplinary AgeLab in 1999 to help develop marketable ideas and assistive technologies for elderly people. AgeLab founder and director Joe Coughlin has focused on building the technical requirements for conversational robots for care of the elderly, which possess the hard-to-define property of likeability. MIT is also the home of The Huggable™ stuffed bear robotic companion, developed by Walter Dan Stiehl and collaborators in the Media Lab. The bear possesses video camera eyes, 1,500 sensors, quiet actuators, a unit for inertial measurement, a speaker, and an internal personal computer with wireless networking capabilities.

Other caregiving technology involves purely virtual agents. These agents are sometimes referred to as softbots. Jonathan Klein, Youngme Moon, and Rosalind Picard's early 2000s' CASPER affect management agent, developed in the MIT Media Lab, is an example of a virtual agent built to alleviate negative emotional states, particularly frustration. The human-computer interaction (HCI) agent uses text-only social-affective feedback strategies to respond to a user who is sharing their thoughts and feelings with the computer. Similarly, the MIT FITrack exercise advisor agent operates through a browser-based client with a backend relational database and text-to-speech engine. FITrack is meant to produce an interactive simulation of the experience of a professional exercise trainer named Laura working with a client.

University of Sheffield computer scientists Amanda Sharkey and Noel Sharkey are often cited in research on the ethics of caregiver robot technology. The Sharkeys express concerns about robotic caregivers and the potential for loss of human dignity. Such technology, they assert, has pros and cons. On the one hand, caregiver robots could extend the range of opportunities available to graying populations, and these aspects of the technology should be encouraged. On the other hand, the devices could be used to manipulate or deceive the weakest members of society or further isolate the elderly from regular companionship or social interaction.

The Sharkeys note that, in some ways, robotic caregivers could eventually exceed human capabilities, for instance, where speed, power, or accuracy is necessary. Robots could be programmed in ways that prevent or reduce real or perceived eldercare abuse, impatience, or incompetence—common complaints among the aged. Indeed, an ethical imperative to use caregiver robots might apply wherever social systems for caregiver support are inadequate or deficient. But robots do not understand complex human constructs such as loyalty or adjust flawlessly to the sensitive customized needs of individual clients. Without proper foresight, the Sharkeys write, "The elderly may find themselves in a barren world of machines, a world of automated care: a factory for the elderly" (Sharkey and Sharkey 2012, 282).

Sherry Turkle devotes a chapter of her pathbreaking book *Alone Together: Why We Expect More From Technology and Less From Each Other* (2011) to caregiver robots. She notes that robotics and artificial intelligence researchers are motivated to make the elderly feel wanted through their work, which assumes that senior citizens often are (or feel) lonely or abandoned. It is true that attention and labor are scarce commodities in aging populations. Robots serve as an entertainment distraction. They improve daily life and homemaking rituals and make them safer. Turkle concedes that robots never tire and may even perform from a position of neutrality in relationships with clients. Humans, by contrast, sometimes possess motives that undermine minimal or conventional standards of care. "One might say that people can pretend to care," Turkle notes. "A robot cannot care. So a robot cannot pretend because it can only pretend" (Turkle 2011, 124).

But Turkle also delivers a harsh critique of caregiving technology. Most importantly, caring behavior is confused with caring feeling. Interactions between humans and robot do not represent real conversations in her estimation. They may even produce confusion among vulnerable and dependent populations. The potential for privacy violation from caregiver robot surveillance is high, and automated assistance may even hijack human experience and memory formation. A great danger is the development of a generation of senior citizens and children who would prefer robots to interpersonal human relationships.

Other philosophers and ethicists have weighed in on appropriate practices and artificial caring. Sparrow and Sparrow (2006) note that human touch is of paramount importance in healing rituals, that robots may exacerbate loss of control, and that robot caregiving is deceptive caregiving because robots are not capable of real concern. Borenstein and Pearson (2011) and Van Wynsberghe (2013) argue that caregiver robots impinge on human dignity and the rights of the elderly, undermining free choice. Van Wynsberghe, in particular, calls for value-sensitive robot designs that align with University of Minnesota professor Joan Tronto's ethic of care, which involves attentiveness, responsibility, competence, and reciprocity, as well as broader concerns for respect, trust, empathy, and compassion. Vallor (2011) has critiqued the core assumptions of robot care by calling into question the presumption that caregiving is nothing more than a problem or burden.

It may be that good care is custom-tailored to the individual, which personal but mass-produced robots might struggle to achieve. Various religions and cultures will also surely eschew robot caregiving. Caregiver robots might even produce reactive attachment disorder in children by offering inappropriate and unsuitable social interactions. The International Organization for Standardization has written requirements for the design of personal robots, but who is at fault in a case of robot neglect? The courts are not sure, and robot caregiver law is still in its infancy. According to Sharkey and Sharkey (2010), caregiver robots may be liable for invasions of privacy, harms caused by unlawful restraint, deceptive practices, psychological damage, and lapses of accountability.

Frameworks for future robot ethics must give primacy to the needs of patients over the desires of the caregivers. Wu et al. (2010) in interviews with the elderly identified six themes related to patient needs. Thirty subjects aged sixty and older noted that assistive technology should first help them pursue ordinary, everyday tasks. Other essential needs included keeping good health, stimulating memory and concentration, living alone "as long as I wish without worrying my family circle" (Wu et al. 2010, 36), maintaining curiosity and growing interest in new activities, and communicating regularly with relatives.

Robot maids, nannies, and caregiver technology are common tropes in popular culture. The *Twilight Zone* television series provides several early examples. A father creates an entire family of robot servants in "The Lateness of the Hour" (1960). Grandma is a robot caregiver in "I Sing the Body Electric" (1962). Rosie the robotic maid is a memorable character from the animated television series *The Jetsons* (1962–1963). Caregiver robots are the main plot device in the animated films *Wall-E* (2008) and *Big Hero 6* (2014) and the science fiction thriller *I Am Mother* (2019). They also appear frequently in manga and anime. Examples include *Roujin Z* (1991), *Kurogane Communication* (1997), and *The Umbrella Academy* (2019).

The 2012 science fiction film *Robot and Frank* directed by Jake Schreier dramatizes the limitations and possibilities of caregiver robot technology in popular culture. In the film, a gruff former jewel thief with declining mental health hopes to turn his assigned robotic assistant into a partner in crime. The film explores various ethical dilemmas regarding not only the care of the elderly, especially human autonomy, but also the rights of robots in servitude. The film subtly makes a familiar critique, one often expressed by MIT social scientist Sherry Turkle: "We are psychologically programmed not only to nurture what we love but to love what we nurture" (Turkle 2011, 11).

*Philip L. Frana*

*See also:* Ishiguro, Hiroshi; Robot Ethics; Turkle, Sherry.

## Further Reading

Borenstein, Jason, and Yvette Pearson. 2011. "Robot Caregivers: Ethical Issues across the Human Lifespan." In *Robot Ethics: The Ethical and Social Implications of*

*Robotics*, edited by Patrick Lin, Keith Abney, and George A. Bekey, 251–65. Cambridge, MA: MIT Press.

Sharkey, Noel, and Amanda Sharkey. 2010. "The Crying Shame of Robot Nannies: An Ethical Appraisal." *Interaction Studies* 11, no. 2 (January): 161–90.

Sharkey, Noel, and Amanda Sharkey. 2012. "The Eldercare Factory." *Gerontology* 58, no. 3: 282–88.

Sparrow, Robert, and Linda Sparrow. 2006 "In the Hands of Machines? The Future of Aged Care." *Minds and Machines* 16, no. 2 (May): 141–61.

Turkle, Sherry. 2011. *Alone Together: Why We Expect More from Technology and Less from Each Other*. New York: Basic Books.

United Nations. 2019. *World Population Ageing Highlights*. New York: Department of Economic and Social Affairs. Population Division.

Vallor, Shannon. 2011. "Carebots and Caregivers: Sustaining the Ethical Ideal of Care in the Twenty-First Century." *Philosophy & Technology* 24, no. 3 (September): 251–68.

Van Wynsberghe, Aimee. 2013. "Designing Robots for Care: Care Centered Value-Sensitive Design." *Science and Engineering Ethics* 19, no. 2 (June): 407–33.

Wu, Ya-Huei, Véronique Faucounau, Mélodie Boulay, Marina Maestrutti, and Anne-Sophie Rigaud. 2010. "Robotic Agents for Supporting Community-Dwelling Elderly People with Memory Complaints: Perceived Needs and Preferences." *Health Informatics Journal* 17, no. 1: 33–40.

# Chatbots and Loebner Prize

A chatbot is a computer program that uses artificial intelligence to engage in conversations with humans. The conversations can take place using text or voice input. In some cases, chatbots are also designed to perform automated actions, such as launching an application or sending an email, in response to input from a human. Most chatbots aim to simulate the conversational behavior of a human being, although to date no chatbot has achieved that goal perfectly.

Chatbots can help to address a variety of needs across a range of settings. Perhaps most obvious is their ability to save time and resources for humans by using a computer program to collect or dispense information instead of requiring a human to perform these tasks. For example, a company might create a customer-support chatbot that responds to customer questions with information that the chatbot determines, using artificial intelligence, to be relevant based on queries from customers. In this way, the chatbot eliminates the need for a human operator to provide this type of customer support.

In other cases, chatbots can be beneficial because they provide a more convenient way of interfacing with a computer or a software program. For instance, a digital assistant chatbot, such as Apple's Siri or Google Assistant, allows humans to use voice input to obtain information (such as the address of a location that they request) or perform actions (like sending a text message) on smartphones. The ability to interact with phones via voice, rather than having to input information on the devices' screens, is advantageous in situations where other input methods are inconvenient or unavailable.

A third advantage of chatbots is consistency. Because most chatbots respond to queries based on preprogrammed algorithms and data sets, they will typically provide the same answers to the same questions on a continual basis. Human operators cannot always be trusted to do the same; one individual might respond to a question in a different way from another, or the same person's answers could vary from one day to another. In this respect, chatbots can help to provide consistency in experience and information for the users with whom they interact. That said, chatbots that are designed to use neural networks or other self-learning techniques to respond to queries might "evolve" over time, with the result that a question posed to a chatbot on one day may result in a different response from the same question posed another day. Thus far, however, few chatbots have been designed to perform self-learning. Some that have, such as Microsoft Tay, have proven ineffectual.

Chatbots can be written in almost any programming language, and there are a variety of techniques available for creating them. However, most chatbots rely on a core set of features to power their conversational abilities and automated decision-making. One is natural language processing, or the ability to interpret human language and translate it into data that software can use to make decisions. Writing code that can process natural language is a challenging task that requires expertise in computer science and linguistics, as well as extensive programming. It requires the ability to understand text or speech input from humans who use a range of different words, sentence structures, and dialects and who may sometimes speak sarcastically or in misleading ways.

Historically, the challenge of creating effective natural language processing engines made chatbots difficult and time-consuming to build, because programmers would have to create natural language processing software from scratch before building a chatbot. Today, however, the availability of natural language processing programming libraries and cloud-based services has significantly reduced this barrier. Modern programmers can simply import a natural language processing library into their applications in order to incorporate the functionality required to interpret human language or use a cloud-based service, such as Amazon Comprehend or Azure Language Understanding, to do the same.

Most chatbots also require a body of data that they can use to respond to queries. After they have used natural language processing to interpret the meaning of input, they parse their own data sets to determine which information to share or which action to take in response to the query. Most chatbots perform this matching in a relatively basic way, by matching keywords within queries to preset tags within their internal datasets. However, more sophisticated chatbots can be programmed to modify or expand their internal datasets on an ongoing basis by evaluating how users have responded to past behavior. For example, a chatbot might ask a user whether the response it gave to a particular question was helpful, and if the user says no, the chatbot would update its internal data in order to avoid giving the same response the next time that a user poses a similar question.

Although chatbots can be valuable in many situations, they are also subject to shortcomings and potential misuse. One obvious weakness is the fact no chatbot has yet proven to be capable of perfectly simulating human behavior, and chatbots can perform only the tasks that they are programmed to perform. They lack the ability to "think outside the box" or solve problems creatively in the way that humans might. In many situations, users interacting with a chatbot may seek answers to questions that the chatbot was simply not programmed to be able to address.

For related reasons, chatbots pose some ethical challenges. Critics of chatbots have argued that it is unethical for a computer program to simulate the behavior of a human being without disclosing to humans with whom it interacts that it is not, in fact, an actual human. Some have also suggested that chatbots may cause an epidemic of loneliness by creating a world where interactions that traditionally involved genuine conversation between humans are replaced by chatbot conversations that are less intellectually and socially fulfilling for human users. On the other hand, chatbots such as Replika have been created with the precise goal of providing lonely humans with an entity to talk to when actual humans are unavailable.

An additional challenge related to chatbots lies in the fact that, like all software programs, chatbots can potentially be used in ways that their creators did not intend. Misuse could result from software security vulnerabilities that allow malicious parties to take control of a chatbot; one can imagine, for example, how an attacker aiming to damage the reputation of a company might seek to compromise its customer-support chatbot in order to deliver false or unhelpful support services. In other cases, simple design mistakes or oversights could lead to unintended behavior by chatbots. This was the lesson that Microsoft learned in 2016 when it released the Tay chatbot, which was designed to teach itself new responses based on previous conversations. When users engaged Tay in conversations about racist topics, Tay began making public racist or inflammatory remarks of its own, leading Microsoft to shut down the application.

The term "chatbot" did not appear until the 1990s, when it was introduced as a shortened form of chatterbot, a term coined by computer scientist Michael Mauldin in 1994 to describe a chatbot named Julia that he created in the early 1990s. However, computer programs with the characteristics of chatbots have existed for much longer. The first was an application named ELIZA, which Joseph Weizenbaum developed at MIT's Artificial Intelligence Lab between 1964 and 1966. ELIZA used early natural language processing techniques to engage in text-based conversations with human users, although the program was limited to discussing only a handful of topics. A similar chatbot program, named PARRY, was created in 1972 by Stanford psychiatrist Kenneth Colby.

It was not until the 1990s, by which time natural language processing techniques had become more sophisticated, that development of chatbots gained more momentum and that programmers came closer to the goal of creating chatbots that could engage in conversation on any topic. This was the goal of A.L.I.C.E., a chatbot introduced in 1995, and of Jabberwacky, a chatbot developed starting in the early 1980s and made available to users on the web in 1997. The next major round of innovation for chatbots came in the early 2010s, when the widespread adoption of smartphones drove demand for digital assistant chatbots that could interact with humans using voice conversations, starting with the debut of Apple's Siri in 2011.

For much of the history of chatbot development, competition for the Loebner Prize has helped to gauge the effectiveness of chatbots in simulating human behavior. Launched in 1990, the Loebner Prize is awarded to computer programs (including but not limited to chatbots) that judges deem to exhibit the most human-like behavior. Notable chatbots evaluated for the Loebner Prize include A.L.I.C.E, which won the prize three times in the early 2000s, and Jabberwacky, which won twice, in 2005 and 2006.

*Christopher Tozzi*

See also: Cheng, Lili; ELIZA; Natural Language Processing and Speech Understanding; PARRY; Turing Test.

### Further Reading

Abu Shawar, Bayan, and Eric Atwell. 2007. "Chatbots: Are They Really Useful?" *LDV Forum* 22, no.1: 29–49.

Abu Shawar, Bayan, and Eric Atwell. 2015. "ALICE Chatbot: Trials and Outputs." *Computación y Sistemas* 19, no. 4: 625–32.

Deshpande, Aditya, Alisha Shahane, Darshana Gadre, Mrunmayi Deshpande, and Prachi M. Joshi. 2017. "A Survey of Various Chatbot Implementation Techniques." *International Journal of Computer Engineering and Applications* 11 (May): 1–7.

Shah, Huma, and Kevin Warwick. 2009. "Emotion in the Turing Test: A Downward Trend for Machines in Recent Loebner Prizes." In *Handbook of Research on Synthetic Emotions and Sociable Robotics: New Applications in Affective Computing and Artificial Intelligence*, 325–49. Hershey, PA: IGI Global.

Zemčík, Tomáš. 2019. "A Brief History of Chatbots." In *Transactions on Computer Science and Engineering*, 14–18. Lancaster: DEStech.

## Cheng, Lili (1960s–)

Lili Cheng is Corporate Vice President and Distinguished Engineer of the Microsoft AI and Research division. She is responsible for developer tools and services on the company's artificial intelligence platform, including cognitive services, intelligent software assistants and chatbots, and data analytics and tools for deep learning. Cheng has stressed that AI tools must become trusted by greater segments of the population and protect the privacy of users. She notes that her division is working on artificial intelligence bots and software applications that engage in humanlike conversations and interactions. Two other goals are the ubiquity of social software—technology that helps people communicate better with one another—and the interoperability of software assistants, that is, AIs that talk to each other or hand off tasks to one another. One example of such applications is real-time language translation. Cheng is also an advocate of technical training and education of people, and particularly women, for the jobs of the future (Davis 2018).

Cheng stresses that AI must be humanized. Rather than adapt the human to the computer in interactions, technology must be adapted to the rhythms of how people work. Cheng states that mere language recognition and conversational AI are not sufficient technological advances. AI must address the emotional needs of human beings. She notes that one ambition of AI research is to come to terms with "the logical and unpredictable ways people interact."

Cheng earned a bachelor's degree in Architecture from Cornell University. She began her career as an architect/urban designer at the Tokyo firm Nihon Sekkei International. She also worked for architectural services company Skidmore Owings & Merrill in Los Angeles. While living in California, Cheng decided to shift to a career in information technology. She considered architectural design a well-established business with clearly defined standards and requirements.

Cheng returned to school, earning her master's degree in Interactive Telecommunications, Computer Programming, and Design from New York University. Her first job in this sphere was as a user experience researcher and designer of QuickTime VR and QuickTime Conferencing in the Advanced Technology Group-Human Interface Group at Apple Computer in Cupertino, California. She moved to Microsoft in 1995, joining the Virtual Worlds Group engaged in Virtual Worlds Platform and Microsoft V-Chat. One of Cheng's projects was Kodu Game Lab, an environment aimed at teaching children programming. She established the Social Computing group in 2001 in order to create social networking prototypes. Then she worked as General Manager of Windows User Experience for Windows Vista, rising to the position of Distinguished Engineer and General Manager at Microsoft Research-FUSE Labs. Cheng has been a lecturer at Harvard and New York Universities and has been listed as one of the leading female engineers in the country.

*Victoriya Larchenko*

*See also:* Chatbots and Loebner Prize; Gender and AI; Mobile Recommendation Assistants; Natural Language Processing and Speech Understanding.

**Further Reading**

Bort, Julie. 2017. "The 43 Most Powerful Female Engineers of 2017." *Business Insider.* https://www.businessinsider.com/most-powerful-female-engineers-of-2017-2017 -2.

Chan, Sharon Pian. 2011. "Tech-Savvy Dreamer Runs Microsoft's Social-Media Lab." *Seattle Times.* https://www.seattletimes.com/business/tech-savvy-dreamer-runs -microsofts-social-media-lab.

Cheng, Lili. 2018. "Why You Shouldn't Be Afraid of Artificial Intelligence." *Time.* http:// time.com/5087385/why-you-shouldnt-be-afraid-of-artificial-intelligence.

Cheng, Lili, Shelly Farnham, and Linda Stone. 2002. "Lessons Learned: Building and Deploying Shared Virtual Environments." In *The Social Life of Avatars: Computer Supported Cooperative Work*, edited by Ralph Schroeder, 90–111. London: Springer.

Davis, Jeffrey. 2018. "In Chatbots She Trusts: An Interview with Microsoft AI Leader Lili Cheng." *Workflow.* https://workflow.servicenow.com/customer-experience/lili -chang-ai-chatbot-interview.

# Climate Crisis, AI and

The application of artificial intelligence is a double-edged sword with regard to climate and the environment. Scientists are using artificial intelligence to identify, adapt, and respond to ecological challenges. The same technology is also exposing civilization to new environmental threats and vulnerabilities.

Much has been written about information technology as a key component in green economy solutions. Intelligent sensing systems and environmental information systems are in use to collect and analyze data from natural and urban ecosystems. Machine learning is being used in sustainable infrastructure designs, citizen detection of environmental perturbations and degradation, contaminant detection and remediation, as well as to redefine consumption patterns and resource recycling.

Such efforts are sometimes referred to as planet hacking. One example of planet hacking is precision farming. Precision farming utilizes artificial intelligence to identify plant diseases and pests and detect soil nutrition problems. Sensor technology guided by AI increases agricultural yields while optimizing the use of water, fertilizer, and chemical pesticides. Controlled farming techniques promise more sustainable land management and (potentially) biodiversity protection. Another example is IBM Research's partnership with the Chinese government to reduce air pollution in that country through a sprawling initiative called Green Horizons. Green Horizons is a ten-year initiative launched in July 2014 to perform air quality management, encourage renewable energy integration, and promote industrial energy efficiency. IBM is employing cognitive computing, decision support technology, and advanced sensors to produce air quality reports and trace pollution to its source. Green Horizons has been expanded to encompass initiatives around the world, including partnerships with Delhi, India, to correlate traffic congestion patterns and air pollution; Johannesburg, South Africa, to meet targets for air quality; and British wind farms, to forecast turbine performance and power generation.

AI-enabled cars and trucks are expected to yield considerable fuel savings, perhaps in the range of 15 percent less consumption as computed by the National Renewable Energy Laboratory at the University of Maryland. The increase in fuel economy is delivered because smart vehicles reduce wasteful combustion caused by stop-and-go and speed-up and slow-down driving behavior (Brown et al. 2014). Intelligent driver feedback is only a first step toward greener automobiling. The Society of Automotive Engineers and National Renewable Energy Laboratory estimate that fuel savings could reach 30 percent with connected cars equipped with vehicle-to-vehicle (V2V) and vehicle-to-infrastructure (V2I) communication (Gonder et al. 2012). Platooning of smart trucks and robotic taxis is also expected to save fuel and reduce carbon emissions.

Environmental robotics technologies (ecobots) are expected to make considerable advances in risk monitoring, management, and mitigation activities. Service robots are in use at nuclear sites. Two iRobot PackBots entered the Fukushima nuclear power plant in Japan to take measurements of radioactivity. The dexterous Treebot, an autonomous tree-climbing robot, is designed to monitor arboreal environments too difficult for humans to reach. The Guardian, a robot designed by the inventor of the Roomba, is being developed to hunt and destroy invasive lionfish threatening coral reefs. The COTSbot, which uses visual recognition technology, is performing a similar service clearing out crown-of-thorn starfish.

Artificial intelligence is helping to uncover an array of impacts of human civilization on the natural world. The highly interdisciplinary Institute for Computational Sustainability at Cornell University enlists both professional scientists and citizens to apply new computational tools to large-scale environmental, social, and economic problems. To cite one example, birders are collaborating with the Cornell Lab of Ornithology to submit millions of observations of bird species throughout North America. The observations are recorded using an app called eBird. Computational sustainability techniques are used to track migration patterns and predict bird population sizes over time and space. Other crowdsourced nature observation applications include Wildbook, iNaturalist, Cicada Hunt, and iBats. Several apps are connected to big data projects and open-access databases such as the Global Biodiversity Information Facility, which in 2020 contains 1.4 billion searchable records.

Artificial intelligence is also being used to help human communities understand and begin grappling with environmental problems by simulating future climate change. A multinational team at the Montreal Institute for Learning Algorithms, Microsoft Research, and ConscientAI Labs is using street view imagery of extreme weather events and generative adversarial networks—in which two neural networks are pitted against one another—to generate realistic images showing the potential effects of bushfires and sea level rise on actual neighborhoods. Affective responses to photographs may impact human behavior and lead to lifestyle change. The Virtual Human Interaction Lab at Stanford is developing virtual reality simulations of polluted ocean ecosystems to build human empathy and change behaviors in coastal communities.

However, information technology and artificial intelligence are also contributing to the climate crisis. One of the most immediate causes of concern is the pollution caused by the manufacture of electronic equipment and software. These are typically thought of as clean industries but regularly utilize harsh chemicals and dangerous materials. The Silicon Valley of California is one of the most polluted places in the United States, with twenty-three active Superfund sites. Computer parts manufacturers created many of these toxic waste sites. One of the most ubiquitous soil contaminants is the solvent trichloroethylene, used in the cleaning of semiconductors.

Information technology is also energy-intensive and a significant contributor to greenhouse gas emissions. Cloud computing data centers are now being constructed with energy generated by solar plants and battery storage. A number of cloud computing centers have been built in recent years near the Arctic Circle to take advantage of the natural cooling potential of chilly air temperatures and cold seawater. One particularly popular region for such construction is the so-called Node Pole, located in the northernmost county of Sweden. A data center project underway in 2020 in Reykjavik, Iceland, will rely 100 percent on renewable geothermal and hydroelectric power. Recycling is also a significant challenge, as life cycle engineering is only beginning to address the difficulties in making eco-friendly computers. Toxic electronic waste is difficult to dispose of locally, and a large percentage of all e-waste is exported to Asia and Africa. About 50 million tons of e-waste is generated worldwide annually (United Nations 2019).

At the World Economic Forum annual meeting in Davos, Switzerland, Jack Ma of the multinational e-commerce giant Alibaba complained that artificial intelligence and big data were making the world unstable and menacing human life. The carbon footprint of artificial intelligence research is only now being calculated with any kind of precision. While Microsoft and Pricewaterhouse Coopers have published findings that artificial intelligence could by 2030 reduce carbon dioxide emissions by 2.4 gigatonnes (the projected emissions of Japan, Canada, and Australia combined), scientists at the University of Massachusetts, Amherst, found that training a model for natural language processing can produce the equivalent of 626,000 pounds of greenhouse gases. This is almost five times the carbon emissions of an average car over a lifetime of use, including initial manufacturing. The current impact of artificial intelligence on energy consumption and the carbon footprint is colossal, particularly when models are being tuned by a process called neural architecture search (Strubell et al. 2019). It is an open question whether next-generation technologies such as quantum artificial intelligence, chipset architectures, and novel machine intelligence processors (e.g., neuromorphic chips) will reduce the environmental impact of AI.

Artificial intelligence is also being used to extract more oil and gas, albeit more efficiently, from underground. Oilfield services is increasingly automated, and information technology companies such as Google and Microsoft are creating offices and divisions to service them. The French multinational oil company Total S.A. has been using artificial intelligence to optimize production and interpret subsurface data since the 1990s. In 2018, Total teamed up with experts at Google Cloud Advanced Solutions Lab to bring contemporary machine learning techniques to bear on technical problems of data analysis in the discovery and production of fossil fuels. Google hopes to provide an AI intelligent assistant to every geoscience engineer at the oil giant. Google is also helping American oil exploration company Anadarko Petroleum (acquired by Occidental Petroleum in 2019) to analyze seismic data to locate oil reserves, increase output, and improve efficiency with artificial intelligence.

Computer scientists Joel Lehman and Risto Miikkulainen, working in the new field of evolutionary robotics, argue that in a future extinction event superintelligent robots and artificial life might quickly reproduce and crowd out human beings. In other words, the ongoing battle between plants and animals might be joined by machines. Lehman and Miikkulainen have built computational models to simulate extinction events to test evolvability in artificial and biological populations. The research remains largely speculative, practical mainly in helping engineers understand how extinction events might affect the work that they do; how the laws of variation apply in evolutionary algorithms, artificial neural networks, and virtual creatures; and how coevolution and evolvability work in ecosystems. Such speculation has led Daniel Faggella of Emerj Artificial Intelligence Research to infamously question whether the "environment matter[s] after the Singularity" (Faggella 2019).

A noteworthy science fiction book on the themes of climate change and artificial intelligence is *River of Gods* (2004) by Ian McDonald. The events of the book take place on the Indian subcontinent in 2047. Steven Spielberg's film *A.I. Artificial Intelligence* (2001) is set in a twenty-second-century world ravaged by global warming and rising sea levels. Humanoid robots are perceived as vital to the economy because they do not consume scarce resources. The 2014 science fiction film *Transcendence* starring Johnny Depp as an artificial intelligence researcher depicts both the apocalyptic threat of sentient computers and their ambiguous environmental consequences.

*Philip L. Frana*

*See also:* Berger-Wolf, Tanya; Intelligent Sensing Agriculture; Post-Scarcity, AI and; Technological Singularity.

**Further Reading**

Brown, Austin, Jeffrey Gonder, and Brittany Repac. 2014. "An Analysis of Possible Energy Impacts of Automated Vehicles." In *Road Vehicle Automation: Lecture Notes in Mobility*, edited by Gereon Meyer and Sven Beiker, 137–53. Cham, Switzerland: Springer.

Cubitt, Sean. 2017. *Finite Media: Environmental Implications of Digital Technologies*. Durham, NC: Duke University Press.

Faggella, Daniel. 2019. "Does the Environment Matter After the Singularity?" https://danfaggella.com/environment/.

Gabrys, Jennifer. 2017. *Program Earth: Environmental Sensing Technology and the Making of a Computational Planet*. Minneapolis: University of Minnesota Press.

Gonder, Jeffrey, Matthew Earleywine, and Witt Sparks. 2012. "Analyzing Vehicle Fuel Saving Opportunities through Intelligent Driver Feedback." *SAE International Journal of Passenger Cars—Electronic and Electrical Systems* 5, no. 2: 450–61.

Microsoft and PricewaterhouseCoopers. 2019. *How AI Can Enable a Sustainable Future*. https://www.pwc.co.uk/sustainability-climate-change/assets/pdf/how-ai-can-enable-a-sustainable-future.pdf.

Schlossberg, Tatiana. 2019. "Silicon Valley Is One of the Most Polluted Places in the Country." *The Atlantic*, September 22, 2019. https://www.theatlantic.com/technology/archive/2019/09/silicon-valley-full-superfund-sites/598531/.

Strubell, Emma, Ananya Ganesh, and Andrew McCallum. 2019. "Energy and Policy Considerations for Deep Learning in NLP." In *Proceedings of the 57th Annual Meeting of the Association for Computational Linguistics (ACL)*, n.p. Florence, Italy, July 2019. https://arxiv.org/abs/1906.02243.

United Nations Environment Programme. 2019. "UN Report: Time to Seize Opportunity, Tackle Challenge of E-Waste." January 24, 2019. https://www.unenvironment.org/news-and-stories/press-release/un-report-time-seize-opportunity-tackle-challenge-e-waste.

Van Wynsberghe, Aimee, and Justin Donhauser. 2017. "The Dawning of the Ethics of Environmental Robots." *Science and Engineering Ethics* 24, no. 6 (October): 1777–1800.

# Clinical Decision Support Systems

Decision-making is a key activity in patient-physician encounters, with judgments often being made on incomplete and inadequate patient information. Physician decision-making, an unquestionably complex and dynamic process, is in theory hypothesis driven. Diagnostic intervention follows a hypothetically deductive process of reaching conclusions by testing hypotheses on clinical evidence.

Evidence-based medicine, an approach to medical practice, improves decision-making by respecting patient values and expectations and integrating individual clinical expertise and experience with the best available external evidence drawn from scientific literature.

The practice of evidence-based medicine must be grounded on the highest quality, reliable, and systematic evidence available. Knowing that both evidence-based medicine and clinical research are needed, but that neither are perfect, the key questions remain: How can physicians obtain the best research evidence? What exactly counts as best evidence? How can physicians be supported in choosing which external clinical evidence from systematic research should influence their practice?

A hierarchy of evidence provides useful guidance about which types of evidence, if well done, are more likely to provide trustworthy answers to clinical questions. Despite no single, universally accepted hierarchy of evidence, the 6S Hierarchy of Evidence-Based Resources developed by Alba DiCenso et al. (2009) is a framework for categorizing and prioritizing resources that appraise and synthesize research findings. The 6S pyramid, which represents a hierarchy of evidence in which higher levels provide increasingly accurate and efficient types of information, was developed for physicians and other health-care professionals to make decisions on the best available research evidence.

At the lowest level of the pyramid are individual studies. Although they are the building blocks for research bases, a single study by itself offers relatively little value for practicing clinicians. For years, clinicians have been taught that randomized controlled trials are the gold standard on which to base their clinical decisions. Randomized controlled trials allow researchers to determine whether a given medication or intervention is effective in a specific patient sample, and a good randomized controlled trial may overturn many years of conventional wisdom. However, for physicians it is more important to know whether it will work for their patient in a given setting. This information cannot be garnered from a randomized controlled trial.

A research synthesis, which represents a higher level of evidence than individual studies, can be thought of as a study of studies. It draws conclusions about the effectiveness of a practice by systematically considering findings across multiple experimental studies. Systematic reviews and meta-analyses, often seen as the cornerstones of evidence-based medicine, have their own problems and depend on the critical appraisal of the characteristics of the evidence that is available. The challenge is that clinicians are often not familiar with the statistical techniques utilized in a meta-analysis and often remain incredibly uncomfortable with the basic scientific concepts required to appraise evidence.

Clinical practice guidelines are supposed to bridge the gap between research and current practice and thus reduce the inappropriate variability in practice. The number of clinical practice guidelines has grown massively in recent years. The trustworthiness of these guidelines is mainly associated with its development process. The biggest issue is the scientific evidence on which these clinical practice guidelines are based. They do not all share the same level of evidence quality and reliability.

The search for evidence-based resources should begin at the highest possible layer of the 6S pyramid, which is the systems layer or the level of computerized clinical decision support systems. Computerized clinical decision support systems (sometimes referred to as intelligent medical platforms) are defined as health information technology-based software that builds upon the foundation of an electronic health record to provide clinicians with general and patient-specific information that is intelligently filtered and organized to enhance health and clinical care. For example, laboratory measurements are often prioritized using different colors to indicate whether they fall within or outside a reference range. The available computerized clinical decision support systems are not a bare model producing just an output. The interpretation and use of a computerized clinical decision support system consists of multiple steps, including presenting the algorithm output in a specific way, interpretation by the clinician, and eventually, the medical decision that is made.

Although computerized clinical decision support systems have been shown to reduce medical errors and improve patient outcomes, they have fallen short of their full potential due to the lack of user acceptance. Besides the technological challenges related to the interface, clinicians are skeptical of computerized clinical decision support systems as they may reduce their professional autonomy or be used in the event of a medical-legal controversy.

Although at present computerized clinical decision support systems require human intervention, several key fields of medicine including oncology, cardiology, and neurology are adapting tools that utilize artificial intelligence to aid with the provision of a diagnosis. These tools exist in two major categories: machine learning techniques and natural language processing systems. Machine learning techniques use patients' data to create a structured database for genetic, imaging, and electrophysiological records to carry out analysis for a diagnosis. Natural language processing systems create a structured database using clinical notes and medical journals to supplement the machine learning process. Furthermore, in medical applications, the machine learning procedures attempt to cluster patients' traits to infer the probability of the disease outcomes and provide a prognosis to the physician.

Numerous machine learning and natural language processing systems have been combined to create advanced computerized clinical decision support systems that can process and provide a diagnosis as effectively or even better than a physician. An AI technique called convolutional neural networking, developed by Google, outperformed pathologists when identifying metastasis detection of lymph nodes. The convolutional neural network was sensitive 97 percent of the time in comparison to the pathologists with a sensitivity of 73 percent. Furthermore, when the same convolutional neural network was used to perform skin cancer classifications, it had a competence level comparable to dermatologists (Krittanawong 2018). Such systems are also being used to diagnose and classify depression.

Artificial intelligence will be used to increase the capacity of clinicians by combining its power with human perceptions, empathy, and experience. However, the benefits of such advanced computerized clinical decision support systems are not just limited to diagnoses and classification. Computerized clinical decision support systems can be used for improving communication between physician and patient by decreasing processing time and thus improving patient care. Computerized clinical decision support systems can prioritize medication prescription for patients based on patient history to prevent drug-drug interaction. More importantly, computerized clinical decision support systems equipped with artificial intelligence can be used to aid triage diagnosis and reduce triage processing times by extracting past medical history and using patient symptoms to decide whether the patient should be referred to urgent care, a specialist, or a primary care doctor. Developing artificial intelligence around these acute and highly specialized medical cases is important as these conditions are the leading causes of death in North America.

Artificial intelligence has been applied to computerized clinical decision support systems in other ways as well. Two very recent examples are the study of Long et al. (2017), who analyzed the ocular image data to diagnose congenital cataract disease, and that of Gulshan et al. (2016), who detected referable diabetic retinopathy through retinal fundus photographs. Both cases highlight the exponential growth of artificial intelligence in the medical field and depict the diverse applications of such a system.

Although the promise of computerized clinical decision support systems to facilitate evidence-based medicine is strong, substantial work remains to be done to realize its full potential in health care. The increasing familiarity with advanced digital technologies among new generations of physicians may support the use and integration of computerized clinical decision support systems. The market for such systems is anticipated to grow rapidly over the next decade. The fueling factor for this growth is the urgent need to reduce the rate of medication errors and global health-care costs.

Computerized clinical decision support systems represent the highest level of evidence to facilitate and support the physician's decision-making process. The future should involve more sophisticated analytics, automation, and a more tailored integration with the electronic health record in order to benefit clinicians, patients, health-care organizations, and society.

*Antonia Arnaert, Hamza Ahmad,*
*Zoumanan Debe, and John Liebert*

*See also:* Automated Multiphasic Health Testing; Expert Systems; Explainable AI; INTERNIST-I and QMR.

**Further Reading**

Arnaert, Antonia, and Norma Ponzoni. 2016. "Promoting Clinical Reasoning Among Nursing Students: Why Aren't Clinical Decision Support Systems a Popular Option?" *Canadian Journal of Nursing Research* 48, no. 2: 33–34.

Arnaert, Antonia, Norma Ponzoni, John A. Liebert, and Zoumanan Debe. 2017. "Transformative Technology: What Accounts for the Limited Use of Clinical Decision Support Systems in Nursing Practice?" In *Health Professionals' Education in the Age of Clinical Information Systems, Mobile Computing, and Social Media*, edited by Aviv Shachak, Elizabeth M. Borycki, and Shmuel P. Reis, 131–45. Cambridge, MA: Academic Press.

DiCenso, Alba, Liz Bayley, and R. Brian Haynes. 2009. "Accessing Preappraised Evidence: Fine-tuning the 5S Model into a 6S Model." *ACP Journal Club* 151, no. 6 (September): JC3-2–JC3-3.

Gulshan, Varun, et al. 2016. "Development and Validation of a Deep Learning Algorithm for Detection of Diabetic Retinopathy in Retinal Fundus Photographs." *JAMA* 316, no. 22 (December): 2402–10.

Krittanawong, Chayakrit. 2018. "The Rise of Artificial Intelligence and the Uncertain Future for Physicians." *European Journal of Internal Medicine* 48 (February): e13–e14.

Long, Erping, et al. 2017. "An Artificial Intelligence Platform for the Multihospital Collaborative Management of Congenital Cataracts." *Nature Biomedical Engineering* 1, no. 2: n.p.

Miller, D. Douglas, and Eric W. Brown. 2018. "Artificial Intelligence in Medical Practice: The Question to the Answer?" *American Journal of Medicine* 131, no. 2: 129–33.

## Cognitive Architectures

A cognitive architecture is a specialized computer model of the human mind that intends to fully simulate all aspects of human cognition. Cognitive architectures represent unified theories for how a set of fixed mental structures and mechanisms can perform intelligent work across a variety of complex environments and situations. There are two key components to a cognitive architecture: a theory for how the human mind works and a computational representation of the theory. The cognitive theory behind a cognitive architecture will seek to unify the results of a broad range of experimental findings and theories into a singular, comprehensive framework capable of explaining a variety of human behavior, using a fixed set of evidence-based mechanisms. The computational representation is then generated from the framework proposed in the theory of cognition. Through the unification of modeling behavior and modeling the structure of a cognitive system, cognitive architectures such as ACT-R (Adaptive Control of Thought-Rational), Soar, and CLARION (Connectionist Learning with Adaptive Rule Induction On-line) can predict, explain, and model complex human behavior like driving a car, solving a math problem, or recalling when you last saw the hippie in the park.

According to computer scientists Stuart Russell and Peter Norvig, there are four approaches to achieving human-level intelligence within a cognitive architecture: (1) building systems that think like humans, (2) building systems that think

**Approaches to cognitive architectures and examples**

|  | Like a Human | Rationally |
|---|---|---|
| **Think** | Using ACT-R to model the cognitive mechanisms that allow someone to solve math problems | Building a rule-based scheduling assistant for a large factory with Soar |
| **Act** | Google's DeepMind AlphaStar learning to play a computer game with the same constraints as a human | Training a neural network to predict stock prices based on the weather |

rationally, (3) building systems that act like humans, and (4) building systems that act rationally.

A system that thinks like a human produces behavior through known human mechanisms. This is the primary approach used in cognitive modeling and can be seen in architectures such as ACT-R by John Anderson, the General Problem Solver by Allen Newell and Herb Simon, and the initial uses of the general cognitive architecture called Soar. ACT-R, for instance, brings together theories of motor movement, visual attention, and cognition. The model distinguishes between procedural knowledge and declarative knowledge. Procedural knowledge is expressed in terms of production rules, which are statements expressed as condition → action pairs. An example is a statement expressed in the form of IF → THEN. Declarative knowledge is factual. It describes information considered static, such as attributes, events, or things. Architectures of this type will produce behavior that includes errors or mistakes, as well as the correct behavior.

A system that thinks rationally will instead use logic, computational reasoning, and laws of thought to produce behaviors and outputs that are consistent and correct. A system that acts rationally will use innate beliefs and knowledge to achieve goals through a more generalized process of logic and movement from premises to consequences, which is more adaptable to situations without full information availability. Acting rationally can also be called the rational agent approach.

Finally, building a system that acts like a human can be thought of as the Turing Test approach. In its most strict form, this approach requires building a system capable of natural language processing, knowledge representation, automated reasoning, and machine learning to achieve humanlike behavior. Not every system with this approach will meet all of these criteria and instead will focus on whichever benchmarks are most relevant to the task being solved.

Aside from those four approaches, cognitive architectures are also classified by their information processing type: symbolic (or cognitivist), emergent (or connectionist), and hybrid. Symbolic systems operate through high-level, top-down control and perform analysis through a set of IF-THEN statements called production rules. EPIC (Executive-Process/Interactive Control) and Soar are two examples of cognitive architectures that use symbolic information processing. Emergent systems, unexpectedly complex wholes that organize from simple parts without a central organizing unit, are built using a bottom-up flow of information propagating from input nodes into the rest of the system, similar to a neural network. While symbolic systems typically process information serially, emergent systems such as Leabra and BECCA (Brain-Emulating Cognition and Control Architecture) will use a self-organizing, distributed network of nodes that can operate in parallel. Hybrid architectures such as ACT-R and CAPS (Collaborative, Activation-based, Production System) combine features from both types of information processing. For example, a hybrid cognitive architecture aimed at visual perception and comprehension may use symbolic processing for labels and text, but then will use an emergent approach for visual feature and object detection. This sort of mixed-methods approach to building cognitive architectures is becoming more common as certain subtasks become better understood. This can produce some disagreement when classifying architectures, but it leads to architectural improvement as they may incorporate the best methods for any particular subtask.

A variety of cognitive architectures have been developed in both academic and industrial environments. Soar is one of the most prominent cognitive architectures that has branched into industrial applications due to its robust and fast software. In 1985, a custom Soar application called R1-Soar was used by Digital Equipment Corporation (DEC) to assist during the complex ordering process for the VAX computer system. Previously, the full system would require each component (including everything from software to cable connectors) to be planned out according to a complex set of rules and contingencies. The R1-Soar system automated this process using the Soar cognitive architecture and is estimated to have saved about 25 million dollars a year. Soar's industrial implementation is maintained by Soar Technology, Inc. and the company continues to work on military and governmental projects.

Within the academic field, John Anderson's ACT-R is one of the oldest and most studied cognitive architectures. ACT-R follows and extends earlier architectures such as HAM (Human Associative Memory). Much of the work with ACT-R has extended memory modeling to include a wider array of memory types and cognitive processes. However, ACT-R has also been applied to natural language processing, prediction of brain region activation, and the development of smart tutors that can model learner behavior and personalize the learning curriculum to their specific needs.

*Jacob D. Oury*

*See also:* Intelligent Tutoring Systems; Interaction for Cognitive Agents; Newell, Allen.

### Further Reading

Anderson, John R. 2007. *How Can the Human Mind Occur in the Physical Universe?* Oxford, UK: Oxford University Press.

Kotseruba, Iuliia, and John K. Tsotsos. 2020. "40 Years of Cognitive Architectures: Core Cognitive Abilities and Practical Applications." *Artificial Intelligence Review* 53, no. 1 (January): 17–94.

Ritter, Frank E., Farnaz Tehranchi, and Jacob D. Oury. 2018. "ACT-R: A Cognitive Architecture for Modeling Cognition." *Wiley Interdisciplinary Reviews: Cognitive Science* 10, no. 4: 1–19.

# Cognitive Computing

Cognitive computing is a term used to describe self-learning hardware and software systems that use machine learning, natural language processing, pattern recognition, human-computer interaction, and data mining technologies to mimic the human brain. Cognitive computing is meant to convey the notion that advances in cognitive science are applied to create new and complex artificial intelligence systems. Cognitive systems are not meant to replace the thinking, reasoning, problem-solving, or decision-making of humans, but rather to augment them or provide assistance. Cognitive computing is sometimes identified as a set of strategies to advance the goals of affective computing, which involves closing the gap between computer technology and human emotions. These strategies include real-time adaptive learning techniques, interactive cloud services, interactive memories, and contextual understanding.

Cognitive analytical tools are in use to make mathematical evaluations of structured statistical data and assist in decision-making. These tools are often embedded in other scientific and business systems. Complex event processing systems take real-time data about events and then use sophisticated algorithms to examine them for patterns and trends, suggest options, or make decisions. These types of systems are in widespread use in algorithmic stock trading and in the detection of credit card fraud. Image recognition systems are now capable of face recognition and complex image recognition. Machine learning algorithms construct models from data sets and show improvements as new data is pulled in. Machine learning may be approached with neural networks, Bayesian classifiers, and support vector machines. Natural language processing involves tools that extract meaning from large data sets of human communication. IBM's Watson is an example, as is Apple's Siri. Natural language understanding is perhaps the Holy Grail or "killer app" of cognitive computing, and indeed, many people think of natural language processing synonymously with cognitive computing.

One of the oldest branches of so-called cognitive computing is heuristic programming and expert systems. Four relatively "complete" cognitive computing architectures have been available since the 1980s: Cyc, Soar, Society of Mind, and Neurocognitive Networks.

Some of the uses for cognitive computing technology are speech recognition, sentiment analysis, face detection, risk assessment, fraud detection, and behavioral recommendations. These applications are together sometimes described as "cognitive analytics" systems. These systems are in development or are being used in the aerospace and defense industries, agriculture, travel and transportation, banking, health care and the life sciences, entertainment and media, natural resource development, utilities, real estate, retail, manufacturing and sales, marketing, customer service, hospitality, and leisure. An early example of predictive cognitive computing is the Netflix recommendation system for movie rentals. General Electric is using computer vision algorithms to detect drivers who are tired or distracted. Domino's Pizza customers can order online by conversing with a virtual assistant named Dom. Elements of Google Now, a predictive search feature launched inside Google apps in 2012, help people predict road conditions and the estimated time of arrival, find hotels and restaurants, and remember birthdays and parking places.

The word "cognitive" computing appears frequently in IBM marketing materials. The company views cognitive computing as a special case of "augmented intelligence," a phrase preferred over artificial intelligence. IBM's Watson machine is sometimes described as a "cognitive computer" because it subverts the conventional von Neumann architecture and takes its inspiration instead from neural networks. Neuroscientists are studying the inner workings of the human brain, looking for relations between neural assemblies and elements of thought, and developing fresh theories of mind.

Hebbian theory is an example of a neuroscientific theory underpinning machine learning implementations on cognitive computers. Hebbian theory is a proposed explanation for neural adaptation in the process of human education. The theory was first offered by Donald Hebb in his 1949 book *The Organization of Behavior*. Hebb asserted that learning is a process by which neural traces tend to become stable by the causal induction of repeated or persistent neural firing or activity. Hebb said: "The general idea is an old one, that any two cells or systems of cells that are repeatedly active at the same time will tend to become 'associated,' so that activity in one facilitates activity in the other" (Hebb 1949, 70). The theory is sometimes condensed to the phrase: "Cells that fire together, wire together."

The association of neural cells and tissues, it is thought under this theory, forms neurologically defined "engrams" that explain how memories are stored in the form of biophysical or biochemical changes in the brain. The exact location of engrams, as well as the mechanisms by which they are created, is still poorly understood. IBM devices are said to learn by inputting instances with a device that aggregates input within a computational convolution or neural network architecture consisting of weights within a parallel memory system.

In 2017, Intel announced its own cognitive chip, Loihi, that mimics the functions of neurons and synapses. Loihi is said to be 1,000 times more energy efficient than other neurosynaptic devices and packs on each chip 128 clusters of 1,024 artificial neurons—a total of 131,072 simulated neurons. Neuromorphic chips involve highly specialized hardware; instead of relying on simulated neural networks and parallel processing with the general purpose of achieving artificial cognition, with Loihi purpose-built neural pathways are impressed in silicon. These neuromorphic chips are expected to loom large in future portable and wireless devices and motor vehicles.

Cognitive scientist and artificial intelligence pioneer Roger Schank is a prominent critic of cognitive computing technologies. "Watson is not reasoning. You can only reason if you have goals, plans, ways of attaining them, a comprehension of beliefs that others may have, and a knowledge of past experiences to reason from. A point of view helps too," he writes. "What is Watson's view on ISIS for example? Dumb question? Actual thinking entities have a point of view about ISIS" (Schank 2017).

*Philip L. Frana*

*See also:* Computational Neuroscience; General and Narrow AI; Human Brain Project; SyNAPSE.

## Further Reading

Hebb, Donald O. 1949. *The Organization of Behavior*. New York: Wiley.

Kelly, John, and Steve Hamm. 2013. *Smart Machines: IBM's Watson and the Era of Cognitive Computing*. New York: Columbia University Press.

Modha, Dharmendra S., Rajagopal Ananthanarayanan, Steven K. Esser, Anthony Ndirango, Anthony J. Sherbondy, and Raghavendra Singh. 2011. "Cognitive Computing." *Communications of the ACM* 54, no. 8 (August): 62–71.

Schank, Roger. 2017. "Cognitive Computing Is Not Cognitive at All." *FinTech Futures*, May 25. https://www.bankingtech.com/2017/05/cognitive-computing-is-not-cognitive-at-all/.

Vernon, David, Giorgio Metta, and Giulio Sandini. 2007. "A Survey of Artificial Cognitive Systems: Implications for the Autonomous Development of Mental Capabilities in Computational Agents." *IEEE Transactions on Evolutionary Computation* 11, no. 2: 151–80.

## Cognitive Psychology, AI and

American experimental psychologists broke away from behaviorism—the theory that "conditioning" explains human and animal behavior—in the late 1950s, in a paradigm-shattering movement since labeled the "Cognitive Revolution." For psychology this meant embracing a computational or information processing theory of mind. Cognitive psychology assumes that performance of tasks requires decision-making and problem-solving processes acting upon stored and direct perception as well as conceptual knowledge. Signs and symbols encoded in the human nervous system from encounters in the world are held in internal memory, to be matched against future experiences in meaning-making exercises. Human beings are, under this view, adept at seeking and finding patterns. Literally everything a person knows is represented in their memories. Matching of current perceptions with stored representations of information is called "recognition."

Behaviorism dominated mainstream American psychology from roughly the 1920s to the 1950s. The theory postulated that behavior in humans and animals could mainly be explained by the learning of associations in conditioning, which involved stimuli and responses rather than thoughts or feelings. Disorders, it was thought under behaviorism, could best be managed by changing patterns of behavior.

The new concepts and theories that challenged the behaviorist view came, in the main, from outside the psychological sciences. One stream of thought that impacted the field came from signal processing and communications research, and particularly the 1940s "information theory" work of Claude Shannon at Bell Labs. Shannon proposed that human perception and memory could be conceptualized as information flow. Cognition could, in this view, be understood as an information processing phenomenon. One of the earliest advocates of a psychological theory of information processing came from Donald Broadbent, who wrote in *Perception and Communication* (1958) that humans possess limited capacity to process an overwhelming amount of available information and thus must apply a selective filter to received stimuli. Information that passes through the filter is made available to short-term memory, is manipulated, and then is transferred and stored in long-term memory. The metaphors for his model are mechanistic rather than behavioral. This view fired the imaginations of other psychologists, particularly mathematical psychologists, who thought that measuring information in terms of bits might help quantify the science of memory.

The development of the digital computer in the 1940s also inspired psychologists. Soon after the end of World War II commentators and scientists alike were comparing computers to human intelligence, most famously the mathematicians Edmund Berkeley, author of *Giant Brains, or Machines That Think* (1949), and

Alan Turing, in his paper "Computing Machinery and Intelligence" (1950). Works such as these inspired early artificial intelligence researchers such as Allen Newell and Herbert Simon to pursue computer programs that could display humanlike general problem-solving skill. Computer modeling suggested that mental representations could be modeled as data structures and human information processing as programming. These ideas are still features of cognitive psychology.

A third stream of ideas bolstering cognitive psychology came from linguistics, particularly the generative linguistics approach developed by Noam Chomsky. His 1957 book *Syntactic Structures* described the mental structures needed to support and represent knowledge that speakers of language must possess. He proposed that transformational grammar components must exist to transform one syntactic structure into another. Chomsky also wrote a review of B. F. Skinner's 1959 book *Verbal Behavior*, which is remembered as having demolished behaviorism as a serious scientific approach to psychology.

In psychology, the book *A Study of Thinking* (1956) by Jerome Bruner, Jacqueline Goodnow, and George Austin developed the idea of concept attainment, which was particularly well attuned to the information processing approach to psychology. Concept learning, they eventually decided, involves "the search for and listing of attributes that can be used to distinguish exemplars from non-exemplars of various categories" (Bruner et al. 1967). In 1960, Harvard University institutionalized the Cognitive Revolution by founding a Center for Cognitive Studies under the leadership of Bruner and George Miller.

In the 1960s, cognitive psychology made a number of major contributions to cognitive science generally, particularly in advancing an understanding of pattern recognition, attention and memory, and the psychological theory of languages (psycholinguistics). Cognitive models reduced pattern recognition to the perception of relatively primitive features (graphics primitives) and a matching procedure where the primitives are cross compared against objects stored in visual memory. Also in the 1960s, information processing models of attention and memory proliferated. Perhaps the best remembered is the Atkinson and Shiffrin model, which built up a mathematical model of information as it flowed from short-term memory to long-term memory, following rules for encoding, storage, and retrieval that regulated the flow. Forgetting was described as information lost from storage by processes of interference or decay.

The subfield of psycholinguistics was inspired by those who wanted to uncover the practical reality of Chomsky's theories of language. Psycholinguistics used many of the tools of cognitive psychology. Mental chronometry is the use of response time in perceptual-motor tasks to infer the content, duration, and temporal sequencing of cognitive operations. Speed of processing is considered an index of processing efficiency. In one famous study, participants were asked questions like "Is a robin a bird?" and "Is a robin an animal?" The longer it took for the respondent to answer, the greater the categorical difference between the terms. In the study, experimenters showed how semantic models could be hierarchical, as the concept robin is directly connected to bird and connected to animal through the intervening concept of bird. Information flows from "robin" to "animal" by passing through "bird."

In the 1970s and 1980s, studies of memory and language started to intersect, and artificial intelligence researchers and philosophers began debating propositional representations of visual imagery. Cognitive psychology as a consequence became much more interdisciplinary. Two new directions for research were found in connectionism and cognitive neuroscience. Connectionism blends cognitive psychology, artificial intelligence, neuroscience and the philosophy of mind to seek neural models of emergent links, nodes, and interconnected networks. Connectionism (sometimes referred to as "parallel distributed processing" or simply "neural networking") is computational at its core. Perception and cognition in human brains are the inspiration for artificial neural networks. Cognitive neuroscience is a scientific discipline that studies the nervous system mechanisms of cognition. It represents an overlap of the fields of cognitive psychology, neurobiology, and computational neuroscience.

*Philip L. Frana*

*See also:* Macy Conferences.

**Further Reading**

Bruner, Jerome S., Jacqueline J. Goodnow, and George A. Austin. 1967. *A Study of Thinking.* New York: Science Editions.

Gardner, Howard. 1986. *The Mind's New Science: A History of the Cognitive Revolution.* New York: Basic Books.

Lachman, Roy, Janet L. Lachman, and Earl C. Butterfield. 2015. *Cognitive Psychology and Information Processing: An Introduction.* London: Psychology Press.

Miller, George A. 1956. "The Magical Number Seven, Plus or Minus Two: Some Limits on Our Capacity for Processing Information." *Psychological Review* 63, no. 2: 81–97.

Pinker, Steven. 1997. *How the Mind Works.* New York: W. W. Norton.

## Computational Creativity

Computational creativity is a concept that is related to—but not reducible to—computer-generated art. "CG-art," as Margaret Boden describes it, refers to an artwork that "results from some computer program being left to run by itself, with zero interference from the human artist" (Boden 2010, 141). This definition is both strict and narrow, being limited to the production of what human observers ordinarily recognize as "art works." Computational creativity, by contrast, is a more comprehensive term that covers a much wider spectrum of activities, devices, and outcomes. As defined by Simon Colton and Geraint A. Wiggins, "Computational creativity is a subfield of Artificial Intelligence (AI) research . . . where we build and work with computational systems that create artefacts and ideas." Those "artefacts and ideas" might be art works, or they might be other kinds of objects, discoveries, and/or performances (Colton and Wiggins 2012, 21).

Examples of computational creativity include applications and implementations such as games, storytelling, music composition and performance, and visual arts. Machine capabilities are typically tested and benchmarked with games and other contests of cognitive skill. From the beginning, in fact, the defining condition of machine intelligence was established with a game, what Alan Turing had called "The Game of Imitation" (1950). Since that time, AI development and achievement has been measured and evaluated in terms of games and other kinds of human/machine competitions. Of all the games that computers have been involved with, chess has had a special status and privileged position, so much so that critics like Douglas Hofstadter (1979, 674) and Hubert Dreyfus (1992) confidently asserted that championship-level AI chess would forever remain out of reach and unattainable.

Then, in 1997, IBM's Deep Blue changed the rules of the game by defeating Garry Kasparov. But chess was just the beginning. In 2015, there was AlphaGo, a Go-playing algorithm developed by Google DeepMind, which took four out of five games against Lee Sedol, one of the most celebrated human players of this notoriously difficult board game. AlphaGo's dexterous playing has been described by human observers, such as Fan Hui (2016), as "beautiful," "intuitive," and "creative."

Automated Insights' Wordsmith and the competing product Quill from Narrative Science are Natural Language Generation (NLG) algorithms designed to produce human-readable stories from machine-readable data. Unlike simple news aggregators or template NLG systems, these programs "write" (or "generate"— and the choice of verb is not incidental) original stories that are, in many instances, indistinguishable from human-created content. In 2014, for instance, Christer Clerwall conducted a small-scale study, during which he asked human test subjects to evaluate news stories composed by Wordsmith and a professional reporter from the *Los Angeles Times*. Results from the study suggest that while the software-generated content is often perceived to be descriptive and boring, it is also considered to be more objective and trustworthy (Clerwall 2014, 519).

One of the early predictions issued by Herbert Simon and Allen Newell in their influential paper "Heuristic Problem Solving" (1958) was that "within ten years a digital computer will write music accepted by critics as possessing considerable aesthetic value" (Simon and Newell 1958, 7). This forecast has come to pass. One of the most celebrated achievements in the field of "algorithmic composition" is David Cope's Experiments in Musical Intelligence (EMI, or "Emmy"). Emmy is a PC-based algorithmic composer capable of analyzing existing musical compositions, rearranging their basic components, and then generating new, original scores that sound like and, in some cases, are indistinguishable from the canonical works of Mozart, Bach, and Chopin (Cope 2001). In music performance, there are robotic systems such as Shimon, a marimba-playing jazz-bot from Georgia Tech University that is not only able to improvise with human musicians in real time but also "is designed to create meaningful and inspiring musical interactions with humans, leading to novel musical experiences and outcomes" (Hoffman and Weinberg 2011).

Cope's general approach, something he calls "recombinacy," is not limited to music. It can be employed for and applied to any creative practice where new works are the product of reorganizing or recombining a set of finite elements, that is, the twenty-six letters in the alphabet, the twelve tones in the musical scale, the sixteen million colors discernable by the human eye, etc. Consequently, this approach to computational creativity has been implemented with other artistic endeavors, such as painting. The Painting Fool, which was developed by Simon Colton, is an automated painter that aspires to be "taken seriously as a creative artist in its own right" (Colton 2012, 16). To date, the algorithm has produced hundreds of "original" artworks that have been exhibited in both online and real-world art galleries. Taking this one step further, Obvious—a Paris-based collective consisting of the artists Hugo Caselles-Dupré, Pierre Fautrel, and Gauthier Vernie—employs a generative adversarial network (GAN) in order to produce portraits of a fictional family (the Belamys) in the style of the European masters. In October of 2018, one of these works, "Portrait of Edmond Belamy," was auctioned by Christies for $432,500 (USD).

Designing systems that are purported to be creative immediately runs into semantic and conceptual problems. Creativity is an elusive phenomenon and something that is not easy to identify or define. Are these programs, algorithms, and systems really "creative" or is this just a form of what critics have called mere "imitation?" This question is in the vein of John Searle's (1984, 32–38) Chinese Room thought experiment, which sought to call attention to the difference separating real cognitive activity, such as creative expression, from its mere simulation or imitation.

In order to be more precise about these matters, researchers in the field of computational creativity have introduced and operationalized a rather specific formulation to characterize their efforts: "The philosophy, science and engineering of computational systems which, by taking on particular responsibilities, exhibit behaviours that unbiased observers would deem to be creative" (Colton and Wiggins 2012, 21). The operative term in this characterization is responsibility. As Colton and Wiggins explain: "The word *responsibilities* highlights the difference between the systems we build and creativity support tools studied in the HCI [human computer interaction] community and embedded in tools such as Adobe's Photoshop, to which most observers would probably not attribute creative intent or behavior" (Colton and Wiggins 2012, 21). With a software application such as Photoshop, "the program is a mere tool to enhance human creativity" (Colton 2012, 3–4); it is an instrument used by a human artist who is and remains responsible for the creative decisions and for what comes to be produced by way of the instrument. Computational creativity research, by contrast, "endeavours to build software which is independently creative" (Colton 2012, 4).

In responding to these opportunities and challenges, one can, on the one hand, respond as we typically have responded, dispensing with these recent technological innovations as just another instrument or tool of human action—or what philosophers of technology, such as Martin Heidegger (1977) and Andrew Feenberg (1991), call—"the instrumental theory of technology." This is, in fact, the explanation that has been offered by David Cope in his own assessment of the impact and significance of his work. According to Cope, Emmy and other algorithmic composition systems like it do not compete with or threaten to replace human composers. They are just tools of and for musical composition. As Cope says, "Computers represent only tools with which we extend our minds and bodies. We invented computers, the programs, and the data used to create their output. The music our algorithms compose is just as much ours as the music created by the greatest of our personal inspirations" (Cope 2001, 139). According to Cope, no matter how much algorithmic mediation is developed and employed, it is the human being who is ultimately responsible for the musical composition that is produced by way of these sophisticated computerized tools.

The same argument could be made for seemingly creative applications in other areas, such as the Go-playing algorithm AlphaGo or The Painting Fool. When AlphaGo wins a major competition or The Painting Fool generates a stunning work of visual art that is displayed in a gallery, there is still a human person (or persons) who is (so the argument goes) ultimately responsible for (or can respond or answer for) what has been produced. The lines of attribution might get increasingly complicated and protracted, but there is, it can be argued, always someone behind the scenes who is in a position of authority. Evidence of this is already available in those situations where attempts have been made to shift responsibility to the machine. Consider AlphaGo's decisive move 37 in game two against Lee Sedol. If someone should want to know more about the move and its importance, AlphaGo can certainly be asked about it. But the algorithm will have nothing to say in response. In fact, it was the responsibility of the human programmers and observers to respond on behalf of AlphaGo and to explain the move's significance and impact.

Consequently, as Colton (2012) and Colton et al. (2015) explicitly recognize, if the project of computational creativity is to succeed, the software will need to do more than produce artifacts and behaviors that we take and respond to as creative output. It will also need to take responsibility for the work by accounting for what it did and how it did it. "The software," as Colton and Wiggins assert, "should be available for questioning about its motivations, processes and products" (Colton and Wiggins 2012, 25), eventually not just generating titles for and explanations and narratives about the work but also being capable of responding to questions by entering into critical dialogue with its audience (Colton et al. 2015, 15).

At the same time, there are opportunities opened up by these algorithmic incursions into what has been a protected and exclusively human domain. The issue is not simply whether computers, machine learning algorithms, or other applications can or cannot be responsible for what they do or do not do, but also how we have determined, described, and defined creative responsibility in the first place. This means that there is both a strong and weak component to this effort, what Mohammad Majid al-Rifaie and Mark Bishop call, following Searle's original distinction regarding efforts in AI, strong and weak forms of computational creativity (Majid al-Rifaie and Bishop 2015, 37).

Efforts at what would be the "strong" variety involve the kinds of application development and demonstrations introduced by individuals and organizations such as DeepMind, David Cope, or Simon Colton. But these efforts also have a "weak AI" aspect insofar as they simulate, operationalize, and stress test various conceptualizations of artistic responsibility and creative expression, leading to critical and potentially insightful reevaluations of how we have characterized these concepts in our own thinking. As Douglas Hofstadter has candidly admitted, nothing has made him rethink his own thinking about thinking more than the attempt to deal with and make sense of David Cope's Emmy (Hofstadter 2001, 38). In other words, developing and experimenting with new algorithmic capabilities does not necessarily take anything away from human beings and what (presumably) makes us special, but offers new opportunities to be more precise and scientific about these distinguishing characteristics and their limits.

*David J. Gunkel*

*See also:* AARON; Automatic Film Editing; Deep Blue; Emily Howell; Generative Design; Generative Music and Algorithmic Composition.

## Further Reading

Boden, Margaret. 2010. *Creativity and Art: Three Roads to Surprise.* Oxford, UK: Oxford University Press.

Clerwall, Christer. 2014. "Enter the Robot Journalist: Users' Perceptions of Automated Content." *Journalism Practice* 8, no. 5: 519–31.

Colton, Simon. 2012. "The Painting Fool: Stories from Building an Automated Painter." In *Computers and Creativity,* edited by Jon McCormack and Mark d'Inverno, 3–38. Berlin: Springer Verlag.

Colton, Simon, Alison Pease, Joseph Corneli, Michael Cook, Rose Hepworth, and Dan Ventura. 2015. "Stakeholder Groups in Computational Creativity Research and Practice." In *Computational Creativity Research: Towards Creative Machines,* edited by Tarek R. Besold, Marco Schorlemmer, and Alan Smaill, 3–36. Amsterdam: Atlantis Press.

Colton, Simon, and Geraint A. Wiggins. 2012. "Computational Creativity: The Final Frontier." In *Frontiers in Artificial Intelligence and Applications,* vol. 242, edited by Luc De Raedt et al., 21–26. Amsterdam: IOS Press.

Cope, David. 2001. *Virtual Music: Computer Synthesis of Musical Style.* Cambridge, MA: MIT Press.

Dreyfus, Hubert L. 1992. *What Computers Still Can't Do: A Critique of Artificial Reason.* Cambridge, MA: MIT Press.

Feenberg, Andrew. 1991. *Critical Theory of Technology.* Oxford, UK: Oxford University Press.

Heidegger, Martin. 1977. *The Question Concerning Technology, and Other Essays.* Translated by William Lovitt. New York: Harper & Row.

Hoffman, Guy, and Gil Weinberg. 2011. "Interactive Improvisation with a Robotic Marimba Player." *Autonomous Robots* 31, no. 2–3: 133–53.

Hofstadter, Douglas R. 1979. *Gödel, Escher, Bach: An Eternal Golden Braid.* New York: Basic Books.

Hofstadter, Douglas R. 2001. "Staring Emmy Straight in the Eye—And Doing My Best Not to Flinch." In *Virtual Music: Computer Synthesis of Musical Style,* edited by David Cope, 33–82. Cambridge, MA: MIT Press.

Hui, Fan. 2016. "AlphaGo Games—English. DeepMind." https://web.archive.org/web/20160912143957/https://deepmind.com/research/alphago/alphago-games-english/.

Majid al-Rifaie, Mohammad, and Mark Bishop. 2015. "Weak and Strong Computational Creativity." In *Computational Creativity Research: Towards Creative Machines,* edited by Tarek R. Besold, Marco Schorlemmer, and Alan Smaill, 37–50. Amsterdam: Atlantis Press.

Searle, John. 1984. *Mind, Brains and Science.* Cambridge, MA: Harvard University Press.

Simon, Herbert A., and Allen Newell. 1958. "Heuristic Problem Solving: The Next Advance in Operations Research." *Operations Research* 6, no. 1 (January–February): 1–10.

Turing, Alan. 1999. "Computing Machinery and Intelligence." In *Computer Media and Communication: A Reader*, edited by Paul A. Meyer, 37–58. Oxford, UK: Oxford University Press.

## Computational Neuroscience

Computational neuroscience (CNS) applies the concept of computation to the field of neuroscience. The term "computational neuroscience," proposed by Eric Schwartz in 1985, came to replace the terms "neural modeling" and "brain theory" used to depict various types of research on the nervous system. At the core of CNS is the understanding that nervous system effects can be seen as instances of computations, because the explanation of state transitions can be understood as relations between abstract properties. In other words, the explanations of effects in nervous systems are not casual descriptions of interaction of physically specific items, but rather descriptions of information transformed, stored, and represented. Consequently, CNS seeks to build computational models to gain understanding of the nervous system function in terms of information processing properties of structures that make up the brain system. One example is constructing a model of how interacting neurons can establish elementary components of cognition. A brain map, however, does not reveal the computational mechanism of the nervous system, but can be used as constraint for theoretical models. For example, information exchange has its costs in terms of physical connections between communicating regions in the sense that regions that make connections often (in instances of high bandwidth and low latency) will be placed together.

Description of neural systems as carry-on computations is central to computational neuroscience and contradicts the claim that computational constructs are proprietary to the explanatory framework of psychology; that is, the human cognitive capacities can be constructed and confirmed independently from understanding how these capacities are being implemented in the nervous system. For instance, in 1973 when it became apparent that cognitive processes cannot be understood by analyzing the results of one-dimension questions/scenarios, an approach widely used by the cognitive psychology at that time, Allen Newell argued that verbally formulated questions will not be able to provide understanding of cognitive process, and only synthesis with computer simulation can reveal the complex interactions of the proposed component's mechanism and whether the cognitive function appears as a result of that interaction.

The first framework for computational neuroscience was formulated by David Marr (1945–1980). This framework, which represents the three-level structure used in computer science (abstract problem analysis, algorithm and physical implementation), aims to provide conceptual starting point for thinking about levels in the context of computation by nervous structure. The model however has limitations because it consists of three poorly connected levels and also because it implements a strict top-down approach in which all neurobiological facts were ignored as instances on the implementation level. As such, it is assumed that some phenomena may be explained in only one or two levels. Consequently the Marr levels framework does not pair with the levels of the organization of the nervous system (molecules, synapses, neurons, nuclei, circuits, networks layers, maps, and systems), nor does it explain the emergent type properties of nervous systems.

Computational neuroscience utilizes a bottom-up approach, starting from neurons and demonstrating how the dynamic interactions between neurons result from computational functions and their implementations with neurons. The three types of models that aim to get computational insight from brain-activity data are models of connectivity and dynamics, decoding models, and representational models.

The models of connectivity use the correlation matrix that depicts the pairwise functional connectivity between locations and establish the characteristics of interconnected regions. Analyses of effective connectivity and large-scale brain dynamics go beyond generic statistical models that are linear models used in activation and information-based brain mapping, because they are generative models, so they can generate data at the level of the measurements and are models of brain dynamics.

The decoding models seek to reveal what information is represented in each brain region. If a region is identified as a "knowledge representing" one, its information becomes a functional entity to inform regions receiving these signals about the content. In a very basic case, the decoding reveals which one of the two stimuli gave rise to a measured response pattern. The content of the representation can be identity of sensory stimulus, a stimulus property (e.g., orientation), or an abstract variable needed for cognitive operation or an action. Decoding along with multivariate pattern analysis has been used to establish the components the brain-computational model must include. However, decoding itself does not constitute models for brain computation but reveals certain aspects without processing brain computation.

The representation models go beyond decoding because they try to characterize regions' responses to arbitrary stimuli. Three types of representational model analysis have been introduced: encoding models, pattern component models, and representational similarity analysis. All three analyses test hypotheses about the representational space, which are based on multivariate descriptions of the experimental conditions. In encoding models, each voxel's activity profile across stimuli is predicted as a linear combination of the features of the model. In pattern component models, the distribution of the activity profiles that characterizes the representational space is modeled as a multivariate normal distribution. In representational similarity analysis, the representational space is characterized by the representational dissimilarities of the activity patterns elicited by the stimuli.

The brain models do not test the properties that identify how the information processing cognitive function might work. For that purpose, task performance models explain cognitive functions in terms of algorithms. These types of models are rigorously tested with experimental data and sometimes with data from brain activity. The two main classes of models are neural network models and cognitive models.

Neural network models are developed from various levels of biological detail—from neurons to maps. The neural networks represent the parallel distributed processing paradigm and support multiple stages of linear-nonlinear signal transformation. Models typically have millions of parameters (the connection weights) to optimize task performance. The large set of parameters is needed because simple models will not be able to express complex cognitive functions. The deep convolutional neural network models' implementations have been used to predict brain representations of novel images in the primate ventral visual stream. The first few layers of neural networks resemble representations similar to those in the early visual cortex. Higher layers also resemble the inferior temporal cortical representation, since both enable the decoding of object position, size, and pose, along with the category of the object. Results from various studies have shown that the internal representations of deep convolutional neural networks provide the best current models of representations of visual images in the inferior temporal cortex in humans and monkeys. When comparing large numbers of models, those that were optimized to perform the task of object classification better explained the cortical representation.

Cognitive models are applications of artificial intelligence in computational neuroscience and target information processing without actual application on neurobiological components (neurons, axons, etc.). There are three types of models: production systems, reinforcement learning, and Bayesian cognitive models. They utilize logic and predicates and operate on symbols but not on signals. The rationale for using artificial intelligence in computational neuroscience research are several. First, over the years a great deal of facts about the brain have accumulated, but the actual understanding of how the brain works is still not known. Second, there are embedded effects caused by networks of neurons, but it is still not understood how the networks of neurons actually work. Third, the brain has been mapped coarsely, as has the knowledge of what different brain regions (predominantly about sensory and motor functions) do, but a detailed map is still not available. In addition, some of the facts accumulated via experimental work or observations may be irrelevant; the relationship between synaptic learning rules and computation is essentially unknown.

A production system's models are the earliest type of models for explaining reasoning and problem solving. A "production" is a cognitive action started as a result of the "if-then" rule, where "if" describes the scope of conditions under which the range of productions ("then" clause) can be executed. If the conditions are met for several rules, the model employs conflict resolution algorithm to choose the proper production. The production models generate a series of predictions that resemble the conscious stream of brain function. In recent applications, the same model is used to predict the regional mean fMRI (functional Magnetic Resonance Imaging) activation time.

The reinforcement models are used in many disciplines with the ultimate goal of simulating achievement of optimal decision-making. In neurobiochemical systems, the implementation in neurobiological systems is a basal ganglia. The agent may learn a "value function" associating each state with its expected cumulative reward. If the agent can predict which state each action leads to and if it knows the values of those states, then it can choose the most promising action. The agent may also learn a "policy" that associates each state directly with promising actions. The choice of action must balance exploitation (which brings short-term reward) and exploration (which benefits learning and brings long-term reward).

The Bayesian models reveal what the brain should actually compute to behave optimally. These models allow inductive inference, which requires prior knowledge and is beyond the capabilities of neural networks models. The models have been used in understanding the basic sensory and motor process and to explain cognitive biases as products of prior assumptions. For example, the representation of probability distribution of neurons has been explored theoretically with Bayesian models and checked against experimental data. These practices show that relating Bayesian inference to actual implementation in a brain is still problematic because the brain "cuts corners" in order to get efficiency, so the approximations may explain deviations from statistical optimality.

Central to computational neuroscience is the concept of a brain doing computations, so researchers are seeking to understand mechanisms of complex brain functions, using modeling and analysis of information processing properties of nervous system elements.

*Stefka Tzanova*

*See also:* Bayesian Inference; Cognitive Computing.

**Further Reading**

Kaplan, David M. 2011. "Explanation and Description in Computational Neuroscience." *Synthese* 183, no. 3: 339–73.

Kriegeskorte, Nikolaus, and Pamela K. Douglas. 2018. "Cognitive Computational Neuroscience." *Nature Neuroscience* 21, no. 9: 1148–60.

Schwartz, Eric L., ed. 1993. *Computational Neuroscience.* Cambridge, MA: Massachusetts Institute of Technology.

Trappenberg, Thomas. 2009. *Fundamentals of Computational Neuroscience.* New York: Oxford University Press.

# Computer-Assisted Diagnosis

Computer-assisted diagnosis is a research area in medical informatics that involves the application of computing and communications technology to medicine. Physicians and scientists embraced computers and software beginning in the 1950s to collect and organize burgeoning stores of medical data and provide significant decision and therapeutic support in interactions with patients. The use of computers in medicine has produced remarkable changes in the process of medical diagnostic decision-making.

The first diagnostic computing devices were inspired by tables of differential diagnoses. Differential diagnosis involves the construction of sets of sorting rules used to find probable causes of symptoms in the examination of patients. A good example is a slide rule-like device invented around 1950 by F.A. Nash of the South West London Mass X-Ray Service, called a Group Symbol Associator (GSA), which allowed the physician to line up a patient's symptoms with

337 symptom-disease complexes to derive a diagnosis (Nash 1960, 1442–46). Cornell University physician Martin Lipkin and physiologist James Hardy created a manual McBee punched card system for the diagnosis of hematological diseases at the Medical Electronics Center of the Rockefeller Institute for Medical Research. The work on their aid to diagnosis, initiated in 1952, matched patient findings against findings already known about each of twenty-one textbook diseases in hematology (Lipkin and Hardy 1957, 551–52).

The director of the Medical Electronics Center, television pioneer Vladimir Zworykin, was pleased with these results and capitalized on Lipkin and Hardy's approach to build a similar digital computer device. Zworykin's system performed automatically what earlier had to be done manually by compiling and sorting findings and creating a weighted diagnostic index. To adapt the punched card system to the digital computer, Zworykin tapped vacuum tube BIZMAC computer coders in the Electronic Data Processing Division at RCA. The finished Zworykin programmed hematological differential diagnostic system was first demonstrated on the BIZMAC computer in Camden, N.J., on December 10, 1957 (Engle 1992, 209–11). Thus the first truly digital electronic aid to computer diagnosis was born.

In the 1960s, a new generation of physicians working with computer scientists began coupling the idea of reasoning under uncertainty to that of personal probability, where orderly medical opinions could be indexed along the lines of betting behavior. Here, probability quantifies uncertainty in order to ascertain the likelihood an individual patient has one or many diseases. The application of personal probability to digital computing technology produced novel results. A good example is medical decision analysis, which involves the application of utility and probability theory to calculate alternative expected patient diagnoses, prognoses, and therapeutic management choices. Two Tufts University medical informatics researchers, Stephen Pauker and Jerome Kassirer, are usually credited as among the earliest to formally apply computer-aided decision analysis to clinical medicine (Pauker and Kassirer 1987, 250–58).

Decision analysis involves identifying all available choices and the potential outcomes of each and structuring a model of the decision, usually in the form of a decision tree so complex and changeable that only a computer can keep track of changes in all of the variables in real time. Such a tree consists of nodes, which describe choices, chances, and outcomes. The tree is used to represent the strategies available to the physician and to calculate—sometimes moment-by-moment—the likelihood that each outcome will occur if a particular strategy is employed. The relative worth of each outcome is also described numerically, as a utility, on an explicitly defined scale.

Decision analysis attaches to each piece of clinical or laboratory-derived information an estimate of the cost of acquiring it and the potential benefit that might be derived from it. The costs and benefits may be measured in qualitative terms, such as the quality of life or amount of pain to be derived from the acquisition and use of medical information, but they usually are measured in quantitative or statistical terms, as in the computation of success rates for surgical procedures or the cost-benefit ratios associated with new medical technologies. Decision analysis at irregular intervals withstood attack by critics who believed that cost-benefit calculations made the rationing of limited health-care resources more attractive (Berg 1997, 54).

In the 1960s and 1970s, researchers began to replace more logical and sequential algorithmic mechanisms for achieving medical consensus with artificial intelligence expert systems. A new generation of computer scientists working with physicians rebuked the so-called oracles of medical computing's past, which, in their view, produced factory diagnoses (Miller and Masarie, Jr. 1990, 1–2). Computer scientists working with physicians embedded evaluation routines into their medical applications and reinvented them as critiquing systems of last resort rather than as diagnosing systems (Miller 1984, 17–23). Perhaps the first system to implement a critiquing approach was the ATTENDING expert system for anesthetic management developed in the Yale University School of Medicine. At the core of the ATTENDING system are routines for risk assessment, which help residents and physicians weigh variables such as patient health, specific surgical procedure, and available anesthetics in forming a clinical judgment. Unlike diagnostic tools that recommend a protocol based on previously inputted information, ATTENDING responds to proposals offered by the user in a stepwise fashion (Miller 1983, 362–69). The critiquing approach removes from the machine the ultimate responsibility for diagnosis because it requires the careful attention of a human operator. This is an important feature in an age where strict liability applies in cases of defective medical technology, including complex software.

In the 1990s and early 2000s, computer-assisted diagnosis moved to home computers and the internet. Two common examples of so-called "doc-in-a-box" software are Medical HouseCall and Dr. Schueler's Home Medical Advisor. Medical HouseCall is a general, consumer-oriented version of the Iliad decision-support system developed by the University of Utah. Development of Medical HouseCall's knowledge base involved an estimated 150,000 person hours of effort. The original software package, released in May 1994, included information on more than 1,100 diseases and 3,000 prescription and nonprescription drugs. It also offered information on treatment costs and options. The software's encyclopedia ran to 5,000 printed pages. Medical HouseCall also had a module for keeping family medical records. Users of the original version of Medical HouseCall first select one of nineteen symptom categories by clicking on graphical icons representing body parts and then answer a series of yes-or-no questions. The software then displays a ranked list of possible diagnoses. These diagnoses are generated using Bayesian estimation (Bouhaddou and Warner, Jr. 1995, 1181–85).

A competing 1990s software package was Dr. Schueler's Home Medical Advisor. Home Medical Advisor is a commercial, consumer-oriented CD-ROM set that includes a large library of information on health and medical conditions, as well as a diagnostic-assistance program that suggests possible diagnoses and appropriate courses of action. More than 15,000 terms were defined in its medical encyclopedia in 1997. Home Medical Advisor also includes an image library and full-motion video presentations. The artificial intelligence module of the program can be used in two ways through two different interfaces. The first involves checking boxes with mouse clicks. The second interface asks the user to supply written answers in natural language to specific questions. The differential diagnoses produced by the program are hyperlinked to more extensive information about those diseases (Cahlin 1994, 53–56). Today, online symptom checkers are common.

Deep learning in big data analytics promises to reduce errors of diagnosis and treatment, reduce cost, and make workflow more efficient in the future. In 2019, Stanford University's Machine Learning Group and Intermountain Healthcare announced CheXpert, for automated chest x-ray diagnosis. The radiology AI program can diagnose pneumonia in only ten seconds. In the same year, Massachusetts General Hospital announced that it had developed a convolutional neural network from a large set of chest radiographs to identify people at high risk of mortality from all causes, including heart disease and cancer. Pattern recognition using deep neural networks has improved the diagnosis of wrist fractures, breast cancer that has metastasized, and cataracts in children. The accuracy of deep learning results is not uniform across all areas of medicine, or all types of injury or disease, but applications continue to accumulate to the point where smartphone apps with embedded AI are already in limited use. In the future, deep learning methods are expected to inform embryo selection for in-vitro fertilization, mental health diagnosis, cancer classification, and weaning patients from ventilator support.

*Philip L. Frana*

*See also:* Automated Multiphasic Health Testing; Clinical Decision Support Systems; Expert Systems; INTERNIST-I and QMR.

## Further Reading

Berg, Marc. 1997. *Rationalizing Medical Work: Decision Support Techniques and Medical Practices.* Cambridge, MA: MIT Press.

Bouhaddou, Omar, and Homer R. Warner, Jr. 1995. "An Interactive Patient Information and Education System (Medical HouseCall) Based on a Physician Expert System (Iliad)." *Medinfo* 8, pt. 2: 1181–85.

Cahlin, Michael. 1994. "Doc on a Disc: Diagnosing Home Medical Software." *PC Novice,* July 1994: 53–56.

Engle, Ralph L., Jr. 1992. "Attempts to Use Computers as Diagnostic Aids in Medical Decision Making: A Thirty-Year Experience." *Perspectives in Biology and Medicine* 35, no. 2 (Winter): 207–19.

Lipkin, Martin, and James D. Hardy. 1957. "Differential Diagnosis of Hematologic Diseases Aided by Mechanical Correlation of Data." *Science* 125 (March 22): 551–52.

Miller, Perry L. 1983. "Critiquing Anesthetic Management: The 'ATTENDING' Computer System." *Anesthesiology* 58, no. 4 (April): 362–69.

Miller, Perry L. 1984. "Critiquing: A Different Approach to Expert Computer Advice in Medicine." In *Proceedings of the Annual Symposium on Computer Applications in Medical Care,* vol. 8, edited by Gerald S. Cohen, 17–23. Piscataway, NJ: IEEE Computer Society.

Miller, Randolph A., and Fred E. Masarie, Jr. 1990. "The Demise of the Greek Oracle Model for Medical Diagnosis Systems." *Methods of Information in Medicine* 29, no. 1 (January): 1–2.

Nash, F. A. 1960. "Diagnostic Reasoning and the Logoscope." *Lancet* 276, no. 7166 (December 31): 1442–46.

Pauker, Stephen G., and Jerome P. Kassirer. 1987. "Decision Analysis." *New England Journal of Medicine* 316, no. 5 (January): 250–58.

Topol, Eric J. 2019. "High-Performance Medicine: The Convergence of Human and Artificial Intelligence." *Nature Medicine* 25, no. 1 (January): 44–56.

## Cybernetics and AI

Cybernetics involves the study of communication and control in living organisms and machines. Today, cybernetic thought permeates computer science, engineering, biology, and the social sciences, though the term itself is no longer widely used in the United States. Cybernetic connectionist and artificial neural network approaches to information theory and technology have throughout the past half century often competed, and sometimes hybridized, with symbolic AI approaches.

Norbert Wiener (1894–1964), who derived the word "cybernetics" from the Greek word for "steersman," considered the discipline to be a unifying force binding together and elevating separate subjects such as game theory, operations research, theory of automata, logic, and information theory. In *Cybernetics, or Control and Communication in the Animal and the Machine* (1948), Winer complained that modern science had become too much an arena for specialists, the result of trends accumulating since the early Enlightenment. Wiener dreamed of a time when specialists might work together, "not as subordinates of some great executive officer, but joined by the desire, indeed by the spiritual necessity, to understand the region as a whole, and to lend one another the strength of that understanding" (Weiner 1948b, 3). Cybernetics, for Wiener, gave researchers access to multiple sources of expertise, while still enjoying the joint advantages of independence and impartial detachment. Wiener also thought that man and machine should be considered together in fundamentally interchangeable epistemological terms. Until these common elements were uncovered, Wiener complained, the life sciences and medicine would remain semi-exact and contingent upon observer subjectivity.

Wiener fashioned his cybernetic theory in the context of World War II (1939–1945). Interdisciplinary subjects heavy in mathematics, for instance, operations research and game theory, were already being used to root out German submarines and devise the best possible answers to complicated defense decision-making problems. Wiener, in his capacity as a consultant to the military, threw himself into the work of applying advanced cybernetic weaponry against the Axis powers. To this end, Wiener devoted himself to understanding the feedback mechanisms in the curvilinear prediction of flight and applying these principles to the making of sophisticated fire-control systems for shooting down enemy planes.

Claude Shannon, a longtime Bell Labs researcher, went even further than Wiener in attempting to bring cybernetic ideas into actual being, most famously in his experiments with Theseus, an electromechanical mouse that used digital relays and a feedback process to learn from past experience how to negotiate mazes. Shannon constructed many other automata that exhibited behavior suggestive of thinking machines. Several of Shannon's mentees, including AI pioneers John McCarthy and Marvin Minsky, followed his lead in defining the man as a symbolic information processor. McCarthy, who is often credited with founding the discipline of artificial intelligence, investigated the foundations of human thinking in mathematical logic. Minsky chose to study the machine simulation of human perception using neural network models.

The fundamental units of cybernetic understanding of human cognitive processing were the so-called McCulloch-Pitts neurons. The namesake of Warren McCulloch and Walter Pitts, these neurons were laced together by axons for communication, forming a cybernated system comprising a rough simulation of the wet science of the brain. Pitts liked Wiener's direct comparison of neural tissue with vacuum tube technology and considered these switching devices as metallic equivalents to organic components of cognition. McCulloch-Pitts neurons were thought capable of simulating fundamental logical operations necessary for learning and memory. Pitts in the 1940s saw in the electrical discharges meted out by these devices a rough binary equivalency with the electrochemical neural impulses released in the brain. In their simplest form, McCulloch-Pitts inputs could be either a zero or a one and the output also a zero or a one. Each input could be considered either excitatory or inhibitory. It was thus for Pitts, as for Wiener, only a short step from artificial to animal memory.

Canadian neuropsychologist Donald Hebb made further fundamental contributions to the study of artificial neurons. These were reported in his 1949 book *The Organization of Behavior.* Hebbian theory attempts to explain associative learning as a process of neuronal synaptic cells firing together, also wiring together. U.S. Navy researcher Frank Rosenblatt extended the metaphor in his study of the artificial "perceptron," a model and algorithm that weighted inputs so that it might be trained to recognize certain classes of patterns. The perceptron's eye and its neural wiring could roughly distinguish between images of cats and dogs. In a 1958 interview, Rosenblatt explained that the navy viewed the perceptron as "the embryo of an electronic computer that it expects will be able to walk, talk, see, write, reproduce itself, and be conscious of its existence" (*New York Times,* July 8, 1958, 25).

The famous 1940s' and 1950s' debates of the Macy Conferences on Cybernetics nurtured the ideas of Wiener, Shannon, McCulloch, Pitts, and other cyberneticists who sought to automate human understanding of the world and the learning process. The meetings also served as a stage for discussions of problems in artificial intelligence. A split between the fields emerged gradually, but it was apparent at the time of the 1956 Dartmouth Summer Research Project on Artificial Intelligence. By 1970, organic cybernetics research no longer had a well-defined presence in American scientific practice. Machine cybernetics became computing sciences and technology. Cybernetic ideas today exist on the margins in social and hard sciences, including cognitive science, complex systems, robotics, systems theory, and computer science, but they were essential to the information revolution of the twentieth and twenty-first centuries. Hebbian theory has experienced a renaissance in interest in recent studies of artificial neural networks and unsupervised machine learning. Cyborgs—beings composed of biological and mechanical parts that extend normal functions—may also be considered a minor area of interest within cybernetics (more commonly referred to as "medical cybernetics" in the 1960s).

*Philip L. Frana*

*See also:* Dartmouth AI Conference; Macy Conferences; Warwick, Kevin.

**Further Reading**

Ashby, W. Ross. 1956. *An Introduction to Cybernetics*. London: Chapman & Hall.

Galison, Peter. 1994. "The Ontology of the Enemy: Norbert Weiner and the Cybernetic Vision." *Critical Inquiry* 21, no. 1 (Autumn): 228–66.

Kline, Ronald R. 2017. *The Cybernetics Moment: Or Why We Call Our Age the Information Age*. Baltimore, MD: Johns Hopkins University Press.

Mahoney, Michael S. 1990. "Cybernetics and Information Technology." In *Companion to the History of Modern Science,* edited by R. C. Olby, G. N. Cantor, J. R. R. Christie, and M. J. S. Hodge, 537–53. London: Routledge.

"New Navy Device Learns by Doing; Psychologist Shows Embryo of Computer Designed to Read and Grow Wiser." 1958. *New York Times*, July 8, 25.

Weiner, Norbert. 1948a. "Cybernetics." *Scientific American* 179, no. 5 (November): 14–19.

Weiner, Norbert. 1948b. *Cybernetics, or Control and Communication in the Animal and the Machine*. Cambridge, MA: MIT Press.

# D

## Dartmouth AI Conference

The Dartmouth Conference of 1956, formally entitled the "Dartmouth Summer Research Project on Artificial Intelligence," is often referred to as the Constitutional Convention of AI. Convened on the campus of Dartmouth College in Hanover, New Hampshire, the interdisciplinary conference brought together experts in cybernetics, automata and information theory, operations research, and game theory. Among the more than twenty participants were Claude Shannon (known as the "father of information theory"), Marvin Minsky, John McCarthy, Herbert Simon, Allen Newell ("founding fathers of artificial intelligence"), and Nathaniel Rochester (architect of IBM's first commercial scientific mainframe computer). MIT Lincoln Laboratory, Bell Laboratories, and the RAND Systems Research Laboratory all sent participants. The Dartmouth Conference was funded in large measure by a grant from the Rockefeller Foundation.

Organizers conceived of the Dartmouth Conference, which lasted approximately two months, as a way of making rapid progress on machine models of human cognition. Organizers adopted the following slogan as a starting point for their discussions: "Every aspect of learning or any other feature of intelligence can in principle be so precisely described that a machine can be made to simulate it" (McCarthy 1955, 2). Mathematician and primary organizer John McCarthy had coined the term "artificial intelligence" only one year prior to the summer conference in his Rockefeller Foundation proposal. The purpose of the new term, McCarthy later recalled, was to create some separation between his research and the field of cybernetics. He was instrumental in discussions of symbol processing approaches to artificial intelligence, which were then in a minority. Most brain-modeling approaches in the 1950s involved analog cybernetic approaches and neural networks.

Participants discussed a wide range of topics at the Dartmouth Conference, from complexity theory and neuron nets to creative thinking and unpredictability. The conference is especially noteworthy as the location of the first public demonstration of Newell, Simon, and Clifford Shaw's famous Logic Theorist, a program that could independently prove theorems given in the *Principia Mathematica* of Bertrand Russell and Alfred North Whitehead. Logic Theorist was the only program presented at the conference that attempted to simulate the logical properties of human intelligence.

Attendees speculated optimistically that, as early as 1970, digital computers would become chess grandmasters, uncover new and significant mathematical theorems, produce acceptable translations of languages and understand spoken language, and compose classical music.

No final report of the conference was ever prepared for the Rockefeller Foundation, and so most information about the proceedings comes from recollections, handwritten notes, and a few papers written by participants and published elsewhere. The Dartmouth Conference was followed by an international conference on the "Mechanisation of Thought Processes" at the British National Physical Laboratory (NPL) in 1958. Several of the Dartmouth Conference participants presented at the NPL conference, including Minsky and McCarthy. At the NPL conference, Minsky commented on the importance of the Dartmouth Conference to the development of his heuristic program for solving plane geometry problems and conversion from analog feedback, neural networks, and brain-modeling to symbolic AI approaches. Research interest in neural networks would largely not revive until the mid-1980s.

*Philip L. Frana*

*See also:* Cybernetics and AI; Macy Conferences; McCarthy, John; Minsky, Marvin; Newell, Allen; Simon, Herbert A.

**Further Reading**

Crevier, Daniel. 1993. *AI: The Tumultuous History of the Search for Artificial Intelligence.* New York: Basic Books.

Gardner, Howard. 1985. *The Mind's New Science: A History of the Cognitive Revolution.* New York: Basic Books.

Kline, Ronald. 2011. "Cybernetics, Automata Studies, and the Dartmouth Conference on Artificial Intelligence." *IEEE Annals of the History of Computing* 33, no. 4 (April): 5–16.

McCarthy, John. 1955. "A Proposal for the Dartmouth Summer Research Project on Artificial Intelligence." Rockefeller Foundation application, unpublished.

Moor, James. 2006. "The Dartmouth College Artificial Intelligence Conference: The Next Fifty Years." *AI Magazine* 27, no. 4 (Winter): 87–91.

# de Garis, Hugo (1947–)

Hugo de Garis is a pioneer in the areas of genetic algorithms, artificial brains, and topological quantum computing. He is the founder of the field of evolvable hardware, in which evolutionary algorithms are used to create specialized electronics that can change structural design and performance dynamically and autonomously in interaction with the environment. De Garis is famous for his book *The Artilect War* (2005), in which he outlines what he believes will be an inevitable twenty-first-century global war between humanity and ultraintelligent machines.

De Garis first became interested in genetic algorithms, neural networks, and the possibility of artificial brains in the 1980s. Genetic algorithms involve the use of software to simulate and apply Darwinian evolutionary theories to search and optimization problems in artificial intelligence. Developers of genetic algorithms such as de Garis used them to evolve the "fittest" candidate simulations of axons, dendrites, signals, and synapses in artificial neural networks. De Garis worked to make artificial nervous systems similar to those found in real biological brains.

His experimentation with a new class of programmable computer chips in the 1990s gave rise to a field in computer science called evolvable hardware. Programmable chips permitted the growth and evolution of neural networks at hardware speeds. De Garis also began experimenting with cellular automata, which involve mathematical models of complex systems that arise generatively from simple units and rules. An early version of his modeling of cellular automata that behaved like neural networks involved the coding of about 11,000 basic rules. A later version encoded about 60,000 such rules. In the 2000s, de Garis referred to his neural networks-on-a-chip as a Cellular Automata Machine.

As the price of chips fell, de Garis began to speculate that the age of "Brain Building on the Cheap" had arrived (de Garis 2005, 45). He also began referring to himself as the father of the artificial brain. Full artificial brains composed of billions of neurons, he asserts, will in coming decades be constructed using knowledge gleaned from an understanding of brain function derived from molecular scale robot probes of human brain tissues and the invention of new path-breaking brain imaging techniques.

Another enabling technology that de Garis believes will speed the development of artificial brains is topological quantum computing. As the physical limits of traditional silicon chip manufacture are reached, the harnessing of quantum mechanical phenomena becomes imperative, he says. Also important will be inventions in reversible heatless computing, in order to dissipate the damaging temperature effects of densely packed circuits. De Garis also advocates for the development of artificial embryology or "embryofacture," that is, evolutionary engineering and self-assembly techniques that replicate the development of fully conscious organisms from single fertilized eggs.

De Garis maintains that, because of rapid advances in artificial intelligence technologies, a war over our final invention will become inevitable late in the twenty-first century. This war, he believes, will end with a massive human extinction event he calls "gigadeath." In his book *The Artilect War*, de Garis speculates that gigadeath becomes almost inevitable under continued Moore's Law doubling of transistors packed on computer chips, accompanied by the development of new technologies such as femtotechnology (the achievement of femtometer-scale structuring of matter), quantum computing, and neuroengineering. De Garis found himself motivated to write *The Artilect War* as a warning and as a self-confessed architect of the looming crisis.

De Garis frames his discussion of a coming Artilect War around two rival global political camps: the Cosmists and the Terrans. The Cosmists will be wary of the awesome power of superintelligent machines of the future, but they will have such reverence for their work in building them that they will feel an almost messianic zeal in inventing them and releasing them into the world. The Cosmists will deeply support the building and nurturing of ever-more complex and powerful artificial brains, regardless of the risks to humanity. The Terrans, by contrast, will oppose the development of artificial brains when they become aware that they constitute a threat to human civilization. They will feel called to oppose these artificial intelligences because they represent an existential threat to human civilization. De Garis discounts a Cyborgian compromise, where humans merge with their own machine creations. He believes that the machines will become so powerful and smart that only a tiny fraction of humanness will survive the encounter.

Geopolitical rivals China and the United States will find themselves compelled to use these technologies to build ever-more sophisticated and autonomous economies, defense systems, and military robots. The Cosmists will welcome the dominance of artificial intelligences in the world and will come to view them as near-gods worthy of worship. The Terrans, by contrast, will resist the turning over of the mechanisms of global economic, social, and military power to our machine masters. They will view the new state of affairs as a grave tragedy that has befallen the human species.

His argument for a coming war over superintelligent machines has inspired voluminous commentary in popular science publications, as well as discussion and debate among scientific and engineering experts. Some critics have questioned de Garis's motives, as he implicates himself as a cause of the coming war and as a closet Cosmist in his 2005 book. De Garis has responded that he feels he is morally impelled to disseminate a warning now because he believes there will be time for the public to recognize the full scope of the threat and react when they begin detecting significant intelligence lurking in home appliances.

De Garis proposes a number of possible scenarios should his warning be heeded. First, he proposes that it is possible, though unlikely, that the Terrans will defeat Cosmist thinking before a superintelligence takes control. In a second scenario, de Garis proposes that artilects will abandon the planet as unimportant and leave human civilization more or less intact. In a third scenario, the Cosmists will become so afraid of their own inventions that they will quit working on them. Again, de Garis thinks this is unlikely. In a fourth scenario, he postulates that all the Terrans might become Cyborgs. In a fifth scenario, the Cosmists will be actively hunted down by the Terrans, perhaps even into deep space, and killed. In a sixth scenario, the Cosmists will leave earth, build artilects, and then disappear from the solar system to colonize the universe. In a seventh scenario, the Cosmists will escape to space and build artilects who will go to war against one another until none remain. In a final eighth scenario, the artilects will go to space and encounter an extraterrestrial super-artilect who will destroy it.

De Garis has been accused of assuming that the nightmare vision of *The Terminator* will become a reality, without considering the possibility that superintelligent machines might just as likely be bringers of universal peace. De Garis has responded that there is no way to guarantee that artificial brains will act in ethical (human) ways. He also says that it is impossible to predict whether or how a superintelligence might defeat an implanted kill switch or reprogram itself to override directives designed to engender respect for humanity.

Hugo de Garis was born in Sydney, Australia, in 1947. He received his bachelor's degree in Applied Mathematics and Theoretical Physics from Melbourne University, Australia, in 1970. After teaching undergraduate mathematics for four years at Cambridge University, he joined the multinational electronics company Philips as a software and hardware architect, working at sites in both The Netherlands and Belgium. De Garis was awarded a PhD in Artificial Life and Artificial Intelligence from Université Libre de Bruxelles, Belgium, in 1992. His thesis title was "Genetic Programming: GenNets, Artificial Nervous Systems, Artificial Embryos." As a graduate student at Brussels, de Garis led the Center for Data Analysis and Stochastic Processes in the Artificial Intelligence and Artificial Life Research Unit, where he studied evolutionary engineering—using genetic algorithms to evolve complex systems. He also served as a senior research affiliate at the Artificial Intelligence Center at George Mason University in Northern Virginia under machine learning pioneer Ryszard Michalski. De Garis completed a postdoctoral fellowship in the Electrotechnical Lab in Tsukuba, Japan.

For the next eight years, he led the Brain Builder Group at the Advanced Telecommunications Research Institute International in Kyoto, Japan, which was attempting a moon-shot effort to build a billion-neuron artificial brain. In 2000, de Garis moved back to Brussels, Belgium, to lead the Brain Builder Group at Starlab, which was working on a parallel artificial brain project. De Garis was working on a life-sized robot kitten when the lab went bankrupt in the dot-com crash of 2001. De Garis next served as Associate Professor of Computer Science at Utah State University, where he remained until 2006. At Utah State, de Garis taught the first advanced research courses on "brain building" and "quantum computing."

In 2006, he became Professor of Computer Science and Mathematical Physics at Wuhan University's International School of Software in China, where he also acted as head of the Artificial Intelligence group. De Garis continued his work in artificial brains, but he also began research in topological quantum computing. That same year, de Garis became a member of the advisory board of Novamente, a commercial company that seeks to build an artificial general intelligence. Two years later, Chinese authorities awarded his Brain Builder Group at Wuhan University a large grant to begin constructing an artificial brain. The effort was dubbed the China-Brain Project. In 2008, de Garis moved to Xiamen University in China where he operated the Artificial Brain Lab in the Artificial Intelligence Institute of the School of Information Science and Technology until his retirement in 2010.

*Philip L. Frana*

*See also:* Superintelligence; Technological Singularity; *The Terminator.*

**Further Reading**

de Garis, Hugo. 1989. "What If AI Succeeds? The Rise of the Twenty-First Century Artilect." *AI Magazine* 10, no. 2 (Summer): 17–22.

de Garis, Hugo. 1990. "Genetic Programming: Modular Evolution for Darwin Machines." In *Proceedings of the International Joint Conference on Neural Networks*, 194–97. Washington, DC: Lawrence Erlbaum.

de Garis, Hugo. 2005. *The Artilect War: Cosmists vs. Terrans: A Bitter Controversy Concerning Whether Humanity Should Build Godlike Massively Intelligent Machines.* ETC Publications.

de Garis, Hugo. 2007. "Artificial Brains." In *Artificial General Intelligence: Cognitive Technologies*, edited by Ben Goertzel and Cassio Pennachin, 159–74. Berlin: Springer.

Geraci, Robert M. 2008. "Apocalyptic AI: Religion and the Promise of Artificial Intelligence." *Journal of the American Academy of Religion* 76, no. 1 (March): 138–66.

Spears, William M., Kenneth A. De Jong, Thomas Bäck, David B. Fogel, and Hugo de Garis. 1993. "An Overview of Evolutionary Computation." In *Machine Learning: ECML-93,* Lecture Notes in Computer Science (Lecture Notes in Artificial Intelligence), vol. 667, 442–59. Berlin: Springer.

## Deep Blue

Artificial intelligence has been used to play chess since the 1950s. Chess was studied for multiple reasons. First, the game is easy to represent in computers as there are a set number of pieces that can occupy discrete locations on the board. Second, the game is difficult to play. An enormous number of states (configurations of pieces) are possible, and great chess players consider both their possible moves and that of opponents, meaning they must consider what can happen several turns into the future. Last, chess is competitive. Having a person play against a machine involves a kind of comparison of intelligence. In 1997, Deep Blue showed that machine intelligence was catching up to humans; it was the first computer to defeat a reigning chess world champion.

The origin of Deep Blue goes back to 1985. While at Carnegie Mellon University, Feng-Hsiung Hsu, Thomas Anantharaman, and Murray Campbell developed another chess-playing computer called ChipTest. The computer worked by brute force, using the alpha-beta search algorithm to generate and compare sequences of moves, with the goal of finding the best one. An evaluation function would score the resulting positions, allowing multiple positions to be compared. In addition, the algorithm was adversarial, predicting the opponent's moves in order to determine a way to beat them.

Theoretically, a computer can generate and evaluate an infinite number of moves if it has enough time and memory to perform the computations. However, when used in tournament play, the machine is limited in both ways. To speed up computations, a single special-purpose chip was used, allowing ChipTest to generate and evaluate 50,000 moves per second. In 1988, the search algorithm was augmented to include singular extensions, which can quickly identify a move that is better than all other alternatives. By quickly determining better moves, ChipTest could generate larger sequences and look further ahead in the game, challenging the foresight of human players.

ChipTest evolved into Deep Thought, and the team expanded to include Mike Browne and Andreas Nowatzyk. Deep Thought used two improved move generator chips, allowing it to process around 700,000 chess moves per second. In 1988, Deep Thought succeeded in beating Bent Larsen, becoming the first computer to beat a chess grandmaster. Work on Deep Thought continued at IBM after the company hired most of the development team. The team now set their sights on beating the best chess player in the world.

The best chess player in the world at the time, as well as one of the best in history, was Garry Kasparov. Born in Baku, Azerbaijan, in 1963, Kasparov won the Soviet Junior Championship at the age of twelve. At fifteen, he was the youngest player to qualify for the Soviet Chess Championship. When he was seventeen, he became the under-twenty world champion. Kasparov was also the youngest ever World Chess Champion, taking the title in 1985 when he was twenty-two years old. He held on to the title until 1993, losing it by leaving the International Chess Federation. He became the Classical World Champion immediately after, holding the title from 1993 to 2000. For most of 1986 to 2005 (when he retired), Kasparov was ranked as the best chess player in the world.

In 1989, Deep Thought played against Kasparov in a two-game match. Kasparov defeated Deep Thought authoritatively by winning both games. Development continued and Deep Thought transformed into Deep Blue, which played in only two matches, both against Kasparov. Facing Deep Blue put Kasparov at a disadvantage while going into the matches. He would, like many chess players, scout his opponents before matches by watching them play or reviewing records of tournament matches to gain insight into their play style and the strategies they used. However, Deep Blue had no match history, as it had played in private matches against the developers until playing Kasparov. Therefore, Kasparov could not scout Deep Blue. On the other hand, the developers had access to Kasparov's match history, so they could adapt Deep Blue to his playing style. Nonetheless, Kasparov was confident and claimed that no computer would ever beat him.

The first six-game match between Deep Blue and Kasparov took place in Philadelphia on February 10, 1996. Deep Blue won the first game, becoming the first computer to beat a reigning world champion in a single game. However, Kasparov would go on to win the match after two draws and three wins. The match captured worldwide attention, and a rematch was scheduled.

After a series of upgrades, Deep Blue and Kasparov faced off in another six-game match, this time at the Equitable Center in New York City on May 11, 1997. The match had an audience and was televised. At this point, Deep Blue was composed of 400 special-purpose chips capable of searching through 200,000,000 chess moves per second. Kasparov won the first game, and Deep Blue won the second. The next three games were draws. The last game would decide the match. In this final game, Deep Blue capitalized on a mistake by Kasparov, causing the champion to concede after nineteen moves. Deep Blue became the first machine ever to defeat a reigning world champion in a match.

Kasparov believed that a human had interfered with the match, providing Deep Blue with winning moves. The claim was based on a move made in the second match, where Deep Blue made a sacrifice that (to many) hinted at a different strategy than the machine had used in prior games. The move made a significant impact on Kasparov, upsetting him for the remainder of the match and affecting his play. Two factors may have combined to generate the move. First, Deep Blue underwent modifications between the first and second game to correct strategic flaws, thereby influencing its strategy. Second, designer Murray Campbell mentioned in an interview that if the machine could not decide which move to make, it would select one at random; thus there was a chance that surprising moves would be made. Kasparov requested a rematch and was denied.

*David M. Schwartz*

*See also:* Hassabis, Demis.

**Further Reading**

Campbell, Murray, A. Joseph Hoane Jr., and Feng-Hsiung Hsu. 2002. "Deep Blue." *Artificial Intelligence* 134, no. 1–2 (January): 57–83.

Hsu, Feng-Hsiung. 2004. *Behind Deep Blue: Building the Computer That Defeated the World Chess Champion*. Princeton, NJ: Princeton University Press.

Kasparov, Garry. 2018. *Deep Thinking: Where Machine Intelligence Ends and Human Creativity Begins*. London: John Murray.

Levy, Steven. 2017. "What Deep Blue Tells Us about AI in 2017." *Wired*, May 23, 2017. https://www.wired.com/2017/05/what-deep-blue-tells-us-about-ai-in-2017/.

## Deep Learning

Deep learning is a subset of methods, tools, and techniques in artificial intelligence or machine learning. Learning in this case involves the ability to derive meaningful information from various layers or representations of any given dataset in order to complete tasks without human instruction. Deep refers to the depth of a learning algorithm, which usually involves many layers. Machine learning networks involving many layers are often considered to be deep, while those with only a few layers are considered shallow. The recent rise of deep learning over the 2010s is largely due to computer hardware advances that permit the use of computationally expensive algorithms and allow storage of immense datasets. Deep learning has produced exciting results in the fields of computer vision, natural language, and speech recognition. Notable examples of its application can be found in personal assistants such as Apple's Siri or Amazon Alexa and search, video, and product recommendations. Deep learning has been used to beat human champions at popular games such as Go and Chess.

Artificial neural networks are the most common form of deep learning. Neural networks extract information through multiple stacked layers commonly known as hidden layers. These layers contain artificial neurons, which are connected independently via weights to neurons in other layers. Neural networks often involve dense or fully connected layers, meaning that each neuron in any given layer will connect to every neuron of its preceding layer. This allows the network to learn increasingly intricate details or be trained by the data passing through each subsequent layer. Part of what separates deep learning from other forms of machine learning is its ability to work with unstructured data. Unstructured data lacks prearranged labels or features. Using many stacked layers, deep learning algorithms can learn to associate its own features from the given unstructured datasets. This is accomplished by the hierarchical way a deep multi-layered learning algorithm provides progressively intricate details with each passing layer, allowing for it to break down a highly complex problem into a series of simpler problems. This allows the network to learn increasingly intricate details or be trained by the data passed through subsequent layers.

A network is trained through the following steps: First, small batches of labeled data are passed forward through the network. The network's loss is calculated by comparing predictions versus the actual labels. Any discrepancies are calculated and relayed back to the weights through back propagation. Weights are slightly altered with the goal of continuously minimizing loss during each round of predictions. The process repeats until optimal minimization of loss occurs and the network achieves a high accuracy of correct predictions.

Deep learning's ability to self-optimize its layers is what gives it an edge over many machine learning techniques or shallow learning networks. Since machine or shallow learning algorithms involve only a few layers at most, they require human intervention in the preparation of unstructured data for input, also known as feature engineering. This can be quite an arduous process and might take too much time to be worthwhile, especially if the dataset is quite large.

For these reasons, it may appear as though machine learning algorithms might become a method of the past. But deep learning algorithms come at a cost. The ability to find their own features requires a vast amount of data that might not always be available. Also, as data sizes increase, so too does the processing power and training time requirements needed since the network will have much more data to sort through. Training time will also increase depending on the amount and types of layers used. Luckily, online computing, where access to powerful computers can be rented for a fee, allows anyone the ability to execute some of the more demanding deep learning networks.

Convolutional neural networks require extra types of hidden layers not discussed in the basic neural network architecture. This type of deep learning is most often associated with computer vision projects and is currently the most widely used method in that field. Basic convnet networks will generally use three types of layers in order to gain insight from the image: convolutional layers, pooling layers, and dense layers. Convolutional layers work by shifting a window, or convolutional kernel, across the image in order to gain information from low-level features such as edges or curves. Subsequent stacked convolutional layers will repeat this process over the newly formed layers of low-level features searching for progressively higher-level features until it forms a concise understanding of the image. Varying the size of the kernel or the distance in which it slides over the image are various hyperparameters that can be changed in order to locate different types of features. Pooling layers allow a network to continue to learn progressively higher-level features of an image by down sampling the image along the way.

Without a pooling layer implemented among convolutional layers, the network might become too computationally expensive as each progressive layer analyzes more intricate details. Also, the pooling layer shrinks an image while retaining important features. These features become translation invariant, meaning that a feature found in one part of an image can be recognized in a completely new area of a second. For an image classification task, the convolutional neural network's ability to retain positional information is vital. Again, the power of deep learning regarding convolutional neural networks is shown through its ability to parse through the unstructured data automatically to find local features that it deems important while retaining positional information about how these features interact with one another.

Recurrent neural networks are excellent at sequence-based tasks such as finishing sentences or predicting stock prices. The underlying premise is that—unlike the earlier examples of networks where neurons only pass information forward—neurons in recurrent neural networks feed information forward and also periodically loop back the output to itself during a time step. Since each time step provides recurrent information of all previous time steps, recurrent neural networks can be thought of as having a basic form of memory. This is often used with natural language projects, as recurrent neural networks can process text in a method more akin to humans. Instead of looking at a sentence as merely a bunch of separate words, a recurrent neural network can begin to process the sentence's sentiment or even autonomously write the next sentence based on what was previously said.

Deep learning can provide powerful methods of analyzing unstructured data in many ways related to human abilities. Deep learning networks, unlike human beings, never tire. Given proper amounts of training data and powerful computers, deep learning can also vastly outperform traditional machine learning techniques, especially given its automatic feature engineering capabilities. Image classification, speech recognition, and self-driving cars are all areas that have benefited greatly from the last decade of deep learning research. If the current enthusiasm and upgrades to computer hardware continue to grow, many new exciting deep learning applications will come with it.

*Jeffrey Andrew Thorne Lupker*

*See also:* Automatic Film Editing; Berger-Wolf, Tanya; Cheng, Lili; Clinical Decision Support Systems; Hassabis, Demis; Tambe, Milind.

**Further Reading**

Chollet, François. 2018. *Deep Learning with Python.* Shelter Island, NY: Manning Publications.

Géron, Aurélien. 2019. *Hands-On Machine Learning with Scikit-Learn, Keras and TensorFlow: Concepts, Tools, and Techniques to Build Intelligent Systems.* Second edition. Sebastopol, CA: O'Reilly Media.

Goodfellow, Ian, Yoshua Bengio, and Aaron Courville. 2017. *Deep Learning.* Cambridge, MA: MIT Press.

# DENDRAL

Pioneered by Nobel-prize winning geneticist Joshua Lederberg and computer scientist Edward Feigenbaum, DENDRAL was an early expert system aimed at analyzing and identifying complex organic compounds. Feigenbaum and Lederberg began developing DENDRAL (meaning tree in Greek) at Stanford University's Artificial Intelligence Laboratory in the 1960s. At the time, there was some expectation that NASA's 1975 Viking Mission to Mars stood to benefit from computers that could analyze extraterrestrial structures for signs of life. In the 1970s, DENDRAL moved to Stanford's Chemistry Department where Carl Djerassi, a prominent chemist in the field of mass spectrometry, headed the program until 1983.

To identify organic compounds, molecular chemists relied on rules of thumb to interpret the raw data generated by a mass spectrometer as there was no overarching theory of mass spectrometry. Lederberg believed that computers could make organic chemistry more systematic and predictive. He started out by developing an exhaustive search engine. The first contribution Feigenbaum made to the project was the addition of heuristic search rules. These rules made explicit what chemists tacitly understood about mass spectrometry. The result was a pioneering AI system that generated the most plausible answers, rather than all possible answers. According to historian of science Timothy Lenoir, DENDRAL "would analyze the data, generate a list of candidate structures, predict the mass spectra of those structures from the theory of mass spectrometry and select as a hypothesis the structure whose spectrum most closely matched the data" (Lenoir 1998, 31).

DENDRAL quickly gained significance in both computer science and chemistry. Feigenbaum recalled that he coined the term "expert system" around 1968. DENDRAL is considered an expert system because it embodies scientific expertise. The knowledge that human chemists tacitly had held in their working memories was extracted by computer scientists and made explicit in DENDRAL's IF-THEN search rules. In technical terms, an expert system also refers to a computer system with a transparent separation between knowledge-base and inference engine. Ideally, this allows human experts to look at the rules of a program like DENDRAL, understand its structure, and comment on how to improve it further.

The positive results that came out of DENDRAL contributed to a gradual quadrupling of Feigenbaum's Defense Advanced Research Projects Agency budget for artificial intelligence research starting in the mid-1970s. And DENDRAL's growth matched that of the field of mass spectrometry. Having outgrown Lederberg's knowledge, the system began to incorporate the knowledge of Djerassi and others in his lab. Consequently, both the chemists and the computers scientists became more aware of the underlying structure of the field of organic chemistry and mass spectrometry, allowing the field to take an important step toward theory-building.

*Elisabeth Van Meer*

*See also:* Expert Systems; MOLGEN; MYCIN.

**Further Reading**

Crevier, Daniel. 1993. *AI: The Tumultuous History of the Search for Artificial Intelligence.* New York: Basic Books.

Feigenbaum, Edward. October 12, 2000. *Oral History.* Minneapolis, MN: Charles Babbage Institute.

Lenoir, Timothy. 1998. "Shaping Biomedicine as an Information Science." In *Proceedings of the 1998 Conference on the History and Heritage of Science Information Systems*, edited by Mary Ellen Bowden, Trudi Bellardo Hahn, and Robert V. Williams, 27–45. Medford, NJ: Information Today.

# Dennett, Daniel (1942–)

Daniel Dennett is Austin B. Fletcher Professor of Philosophy and Co-Director of the Center for Cognitive Studies at Tufts University. His primary areas of research and publication are in the fields of philosophy of mind, free will, evolutionary biology, cognitive neuroscience, and artificial intelligence. He is the author of more than a dozen books and hundreds of articles. Much of this work has centered on the origins and nature of consciousness and how it can be explained naturalistically. Dennett is also an outspoken atheist and is one of the so-called "Four Horsemen" of New Atheism. The others are Richard Dawkins, Sam Harris, and Christopher Hitchens.

Dennett's philosophy is consistently naturalistic and materialistic. He rejects Cartesian dualism, the idea that the mind and the body are separate entities that comingle. He claims instead that the brain is a type of computer that has evolved through natural selection. Dennett also argues against the homunculus view of the mind in which there is a central controller or "little man" in the brain who does all of the thinking and feeling.

Instead, Dennett advocates for a position he calls the multiple drafts model. In this view, which he outlines in his 1991 book *Consciousness Explained*, the brain continuously sifts through, interprets, and edits sensations and stimuli and formulates overlapping drafts of experience. Later, Dennett used the metaphor of "fame in the brain" to communicate how different elements of continuous neural processes are occasionally highlighted at particular times and under varying circumstances. These various interpretations of human experiences form a narrative that is called consciousness. Dennett rejects the idea that these notions come together or are organized in a central part of the brain, a notion he derisively calls "Cartesian theater." Rather, the brain's narrative consists of a continuous, un-centralized flow of bottom-up consciousness spread over time and space.

Dennett rejects the existence of qualia, which are individual subjective experiences such as the way colors appear to the human eye or the way food tastes. He does not deny that colors or tastes exist, only that there is no additional entity in the human mind that is the experience of color or taste. He maintains there is no difference between human and machine "experiences" of sensations. Just as certain machines can distinguish between colors without humans concluding that machines experience qualia, so too, says Dennett, does the human brain. For Dennett, the color red is just the property that brains detect and which in the English language is called red. There is no additional, ineffable, quality to it. This is an important consideration for artificial intelligence in that the ability to experience qualia is often considered to be an obstacle to the development of Strong AI (AI that is functionally equivalent to that of a human) and something that will inevitably differentiate between human and machine intelligence. But if qualia do not exist, as Dennett claims, then it cannot be a barrier to the development of human-like intelligence in machines.

In another metaphor, Dennett likens human brains to termite colonies. Though the termites do not come together and plan to build a mound, their individual actions generate that outcome. The mound is not the result of intelligent design by the termites, but rather is the outcome of uncomprehending competence in cooperative mound-building produced by natural selection. Termites do not need to understand what they are doing in order to build a mound. Similarly, comprehension itself is an emergent property of such competences.

For Dennett, brains are control centers evolved to react quickly and efficiently to dangers and opportunities in the environment. As the demands of reacting to the environment become more complex, comprehension develops as a tool to deal with those complexities. Comprehension is a matter of degree on a sliding scale. For instance, Dennett places the quasi-comprehension of bacteria when they respond to various stimuli and the quasi-comprehension of computers responding to coded instructions on the low end of the spectrum. He places Jane Austen's understanding of human social forces and Albert Einstein's understanding of relativity on the upper end of the spectrum. These are not differences in kind, however, only of degree. Both ends of the spectrum are the result of natural selection.

Comprehension is not an additional mental phenomenon of the brain's over-and-above various competences. Rather, comprehension is a composition of such competences. To the degree that we identify consciousness itself as an extra element of the mind in the form of either qualia or comprehension, such consciousness is an illusion.

Generally, Dennett urges humanity not to posit comprehension at all when mere competence will do. Yet human beings tend to take what Dennett calls the "intentional stance" toward other human beings and often to animals. The intentional stance is taken when people interpret actions as the results of mind-directed beliefs, emotions, desires, or other mental states. He contrasts this to the "physical stance" and the "design stance." The physical stance is an interpretation of something as being the result of purely physical forces or the laws of nature. A stone falls when dropped because of gravity, not because of any mental intention to return to the earth. The design stance is an interpretation of an action as being the unthinking result of a preprogrammed, or designed, purpose. An alarm clock, for example, will beep at a set time because it has been designed to do so, not because it has decided of its own accord to do so. The intentional stance differs from both the physical and design stances in that it treats behaviors and actions as if they are the results of conscious choice on the part of the agent.

Determining whether to apply the intentional stance or the design stance to computers can become complicated. A chess-playing computer has been designed to win at chess. But its actions are often indistinguishable from those of a chess-playing human who wants to, or intends to, win. In fact, human interpretation of the computer's behavior, and a human's ability to react to it, is enhanced if taking an intentional stance, rather than a design stance, toward it. Dennett argues that since the intentional stance works best in explaining the behavior of both the human and the computer, it is the best approach to take toward both. Furthermore, there is no reason to make any distinction between them at all. Though the intentional stance views behavior as if it is agent-driven, it need not take any position on what is actually happening within the innards of human or machine. This stance provides a neutral position from which to explore cognitive competence without presuming a specific model of what is going on behind those competences.

Since human mental competences have evolved naturally, Dennett sees no reason in principle why AI should be impossible. Further, having dispensed with the notion of qualia, and through adopting the intentional stance that absolves humans from the burden of hypothesizing about what is going on in the background of cognition, two primary obstacles of the hard problem of consciousness are now removed. Since the human brain and computers are both machines, Dennett argues there is no valid theoretical reason humans should be capable of evolving competence-drive comprehension while AI should be inherently incapable of doing so. Consciousness as typically conceived is illusory and therefore is not an obligatory standard for Strong AI.

Dennett does not see any reason that Strong AI is impossible in principle. He believes society's level of technological sophistication remains at least fifty years away from being able to produce it. Dennett does not view the development of Strong AI as desirable. Humans should seek to develop AI tools, but to attempt to create machine friends or colleagues, in Dennett's view, would be a mistake. He argues such machines would not share human moral intuitions and understanding and would not be integrated into human society. Humans have each other for companionship and do not require machines to perform that task. Machines, even AI-enhanced machines, should remain tools to be used by human beings and nothing more.

*William R. Patterson*

*See also:* Cognitive Computing; General and Narrow AI.

**Further Reading**

Dennett, Daniel C. 1987. *The Intentional Stance.* Cambridge, MA: MIT Press.

Dennett, Daniel C. 1993. *Consciousness Explained.* London: Penguin.

Dennett, Daniel C. 1998. *Brainchildren: Essays on Designing Minds.* Cambridge, MA: MIT Press.

Dennett, Daniel C. 2008. *Kinds of Minds: Toward an Understanding of Consciousness.* New York: Basic Books.

Dennett, Daniel C. 2017. *From Bacteria to Bach and Back: The Evolution of Minds.* New York: W. W. Norton.

Dennett, Daniel C. 2019. "What Can We Do?" In *Possible Minds: Twenty-Five Ways of Looking at AI,* edited by John Brockman, 41–53. London: Penguin Press.

# Diamandis, Peter (1961–)

Peter Diamandis is a Harvard MD and an MIT-trained aerospace engineer. He is also a serial entrepreneur: he founded or cofounded twelve companies, most still in operation, including International Space University and Singularity University. His brainchild is the XPRIZE Foundation, which sponsors competitions in futuristic areas such as space technology, low-cost mobile medical diagnostics, and oil spill cleanup. He is the chair of Singularity University, which teaches executives and graduate students about exponentially growing technologies.

Diamandis's focus is humanity's grand challenges. Initially, his interests were focused entirely on space flight. Even as a teen, he already thought that humanity should be a multiplanetary species. However, when he realized that the U.S. government was reluctant to finance NASA's ambitious plans of colonization of other planets, he identified the private sector as the new engine of space flight. While still a student at Harvard and MIT, he founded several not-for-profit start-ups, including International Space University (1987), today based in France. In 1989, he cofounded International Microspace, a for-profit microsatellite launcher.

In 1992, Diamandis established a company, later named Zero Gravity, dedicated to providing customers with the experience of moments of weightlessness on parabolic flights. Among the 12,000 customers that have so far experienced zero gravity, the most famous is Stephen Hawking. In 2004, he launched XPRIZE Foundation, basically an incentive prize of substantial size (five to ten million dollars). In 2008, Diamandis cofounded Singularity University with Ray Kurzweil to educate people to think in terms of exponential technologies and to help entrepreneurs build solutions to humanity's most pressing problems by leveraging the potential of exponential technologies. In 2012, he founded Planetary Resources, an asteroid mining company that aims to build low-cost spacecraft.

Diamandis is often mentioned as a futurist. If so, he is a peculiar kind of futurist; he does not extrapolate trends or elaborate projections. Diamandis's basic activity is that of matchmaking: he identifies big problems on one side and then matches them to potential solutions on the other. In order to identify potential solutions, he has established incentive prizes and, behind that, a network of powerful billionaires who finance those prizes. Among these billionaires who are backing Diamandis's initiatives are Larry Page, James Cameron, and the late Ross Perot.

Diamandis started with the problem of sending humanity into space, but over the past three decades, he has enlarged his scope to embrace humanity's grand challenges in exploration, including space and oceans, life sciences, education, global development, energy, and the environment. Diamandis's new frontier is augmented longevity, that is, living longer. He thinks that the causes of premature death can be eradicated and that the population at large can live longer and healthier. He also believes that a human's mental peak could last for an extra 20 years. To solve the problem of longevity, Diamandis started Human Longevity, a biotechnology company located in San Diego, with genomics expert Craig Venter and stem cell pioneer Robert Hariri, in 2014. Four years later, he cofounded with Hariri a longevity-focused company named Celularity that provides stem cell-based antiaging treatments.

*Enrico Beltramini*

*See also:* Kurzweil, Ray; Technological Singularity.

**Further Reading**

Diamandis, Peter. 2012. *Abundance: The Future Is Better Than You Think*. New York: Free Press.

Guthrie, Julian. 2016. *How to Make a Spaceship: A Band of Renegades, an Epic Race, and the Birth of Private Spaceflight*. New York: Penguin Press.

## Digital Immortality

Digital immortality is a hypothesized process for transferring the memories, knowledge, and/or personality of a human being into a durable digital memory storage device or robot. In this way, human intelligence is supplanted by an artificial intelligence resembling, in some ways, the mental pathways or imprint of the brain. Reverse-engineering the brain to achieve substrate independence—that is, transcribing the thinking and feeling mind and emulating it on a variety of forms of physical or virtual media—is a recognized grand challenge of the U.S. National Academy of Engineering.

The speculative science of Whole Brain Emulation (popularly referred to as mind uploading) relies on the assumption that the mind is a dynamic process independent of the physical biology of the brain and its specific sets or patterns of atoms. Instead, the mind represents a set of information processing functions, and those functions are computable. Whole Brain Emulation is currently thought to be grounded in a subfield of computer science known as neural networking, which has as its own ambitious goal the programming of an operating system patterned after the human brain. In artificial intelligence research, artificial neural networks (ANNs) are statistical models derived from biological neural networks. ANNs are capable of processing information in nonlinear fashion through connections and weighting, as well as by backpropagation and modification of parameters in algorithms and rules.

Joe Strout, a computational neurobiology enthusiast at the Salk Institute, facilitated discussion of whole brain emulation in the 1990s through his online "Mind Uploading Home Page." Strout laid out the case for the material origins of consciousness, arguing that evidence from damage to the brains of real people points to neuronal, connectionist, and chemical origins for thinking. Using his site, Strout disseminated timelines of past and present technological development and proposals for future uploading procedures.

Proponents of mind uploading suggest that the process will eventually be accomplished by one of two methods: (1) gradual copy-and-transfer of neurons by scanning the brain and emulating its underlying information states or (2) deliberate replacement of natural neurons by more durable artificial mechanical devices or manufactured biological products. Strout collected ideas on several hypothetical techniques for accomplishing the goal of mind uploading. One is a microtome procedure where a living brain is sliced into thin pieces and then scanned by an advanced electron microscope. The image data is then used to reconstruct the brain in a synthetic substrate.

In nanoreplacement, tiny machines are injected into the brain where they monitor the input and output of neurons. Eventually, when these microscopic machines fully understand all organic interactions, they destroy the neurons and take their place. A variant of this procedure, one proposed by the Carnegie Mellon University roboticist Hans Moravec, involves a robot with a probe composed of billions of appendages that dig deep into every part of the brain. The robot in this way builds up a virtual model of every part and function of the brain, replacing it as it goes. Eventually, everything the physical brain once was is replaced by a simulation.

Scanning or mapping of neurons in copy-and-transfer whole brain emulation is usually considered destructive. The living brain is first plasticized or frozen and then sliced into sections for scanning and emulating on a computational medium. Philosophically speaking, the process produces a mind clone of a person, and not the person who submits to the experiment. That original person dies in the execution of the duplicating experiment; only a copy of their personal identity survives. It is possible to think of the copy as the real person since, as the philosopher John Locke argued, someone who remembers thinking about something in the past is in fact the same person as the individual who did the thinking in the first place. Alternatively, it may be that the experiment will transform the original and the copy into entirely new people, or instead they will quickly deviate from one another with time and experience, due to loss of common history beyond the moment of the experiment.

Several nondestructive techniques have been suggested as substitutes to destroying the brain in the copy-and-transfer process. Functional reconstruction, it is suggested, might be possible using advanced forms of gamma-ray holography, x-ray holography, magnetic resonance imaging (MRI), biphoton interferometry, or correlation mapping with probes. The current limit of available technology, in the form of electron microscope tomography, has reached the sub-nanometer scale with 3D reconstructions of atomic-level detail. Most of the remaining challenges involve the geometry of tissue specimens and so-called tilt-range limits of tomographic equipment. Neurocomputer fabrication to reconstruct scans as information processing elements is under development, as is advanced forms of image recognition.

The BioRC Biomimetic Real-Time Cortex Project at the University of Southern California under Professor of Electrical and Computer Engineering Alice Parker conducts research on reverse-engineering the brain. Parker, in collaboration with nanotechnology professor Chongwu Zhou and her students, is currently designing and fabricating a memory and carbon nanotube neural nanocircuit for a future synthetic cortex grounded in statistical predictions. Her neuromorphic circuits emulate the complexity of human neuronal computations, even incorporating glial cell interactions (these are nonneuronal cells that form myelin, control homeostasis, and protect and support neurons). BioRC Project members are creating systems that scale to the size of human brains. Parker is working to implement dendritic plasticity into these systems, giving them the ability to grow and change as they learn. She credits the origins of the approach to Caltech electrical engineer Carver Mead's work since the 1980s to build electronic models of human neurological and biological structures.

The Terasem Movement, launched in 2002, educates and encourages the public to support technological innovations that advance the science of mind uploading and that merge science with religion and philosophy. Terasem is actually three incorporated organizations working in concert—the Terasem Movement proper, the Terasem Movement Foundation, and the Terasem Movement Transreligion. The initiators of the group are serial entrepreneurs Martine Rothblatt and Bina Aspen Rothblatt. The Rothblatts are inspired by the religion of Earthseed found in the 1993 book *Parable of the Sower* by science fiction author Octavia Butler. The central tenets of the Rothblatt's trans-religious beliefs are "life is purposeful, death is optional, God is technological, and love is essential" (Roy 2014).

The CyBeRev (Cybernetic Beingness Revival) project by Terasem experiments with a massive collection of all available data about the life of a person—their personal history, recorded memories, photographs, and so forth—and keeps the information in a separate data file in hopes that their personality and consciousness can one day be pieced together by advanced software and reanimated. The Lifenaut study sponsored by the Terasem Foundation stores mindfiles of biographical information on people—free of charge—and maintains corresponding DNA samples (biofiles). The foundation has also built a social robot, Bina48, to show how a person's consciousness might one day be transferred into a realistic android.

The Silicon Valley artificial intelligence company Numenta is working to reverse-engineer the human neocortex. The founders of the company are Jeff Hawkins (inventor of the handheld PalmPilot personal digital assistant), Donna Dubinsky, and Dileep George. Numenta's concept of the neocortex is grounded in the theory of hierarchical temporal memory, which is described in Hawkins' and Sandra Blakeslee's book *On Intelligence* (2004). At the core of Numenta's emulation technology are time-based learning algorithms capable of storing and recalling subtle patterns in data change over time. The company has developed several commercial applications, including Grok, which detects glitches in computer servers. The company has shared examples of other applications, such as finding irregularities in stock market trading or abnormalities in human behavior.

The nonprofit Carboncopies supports research and collaboration to record unique arrangements of neurons and synapses containing human memories and save them to a computer. The group funds research in computational modeling, neuromorphic hardware, brain imaging, nanotechnology, and the philosophy of mind. The organization's founder is Randal Koene, a McGill University-trained computational neuroscientist and lead scientist at neuroprosthetic maker Kernel. Russian new media billionaire Dmitry Itskov provided initial capital for Carboncopies. Itskov is also creator of the farsighted 2045 Initiative, a nonprofit interested in the field of radical life extension. The 2045 Initiative has the goal of inventing high technologies capable of transferring personalities into an "advanced nonbiological carrier." Koene and Itskov are also organizers of Global Future 2045, a conference dedicated to finding "a new evolutionary strategy for humanity."

Proponents of digital immortality imagine all sorts of practical outcomes from their work. For instance, a stored backup mind is available for purposes of reawakening into a new body in the event of death by accident or natural causes. (Presumably, old minds would seek out new bodies long before aging becomes noticeable.) This is also the premise of Arthur C. Clarke's science fiction novel *City and the Stars* (1956), which inspired Koene to adopt his career path at the age of thirteen. Or it may be that all of humanity could reduce the collective risk of global calamity by uploading their minds to virtual reality. Civilization might be protected on an advanced sort of hard drive buried in the planet's core, away from malevolent extraterrestrial aliens or extremely powerful natural gamma ray bursts.

Another hypothetical advantage is life extension over long periods of interstellar travel. Artificial minds could be inserted into metal bodies for long journeys across space. This, again, is an idea anticipated by Clarke in the closing pages of the science fiction classic *Childhood's End* (1953). It is also the answer proposed by two early space flight researchers, Manfred Clynes and Nathan Kline, in their 1960 *Astronautics* paper on "Cyborgs and Space," which contains the first mention of astronauts with physical abilities that extend beyond normal limits (zero gravity, space vacuum, cosmic radiation) due to mechanical aids. Under conditions of true mind uploading, it may become possible to simply encode and transmit the human mind in the form of a signal sent to a nearby exoplanet that is the best candidate for finding alien life. In each case, the risks to the human are minimal compared to current dangers faced by astronauts—explosive rockets, high speed collisions with micrometeorites, and malfunctioning suits and oxygen tanks.

Yet another possible advantage of digital immortality is true restorative justice and rehabilitation through reprogramming of criminal minds. Or mind uploading could conceivably allow punishments to be meted out far beyond the natural life span of individuals who have committed horrific offenses. The social, philosophical, and legal consequences of digital immortality are truly mind-boggling.

Digital immortality has been a staple of science fiction explorations. Frederik Pohl's short story "The Tunnel Under the World" (1955) is a widely reprinted short story about chemical plant workers who are killed in a chemical plant explosion, only to be rebuilt as miniature robots and exposed as test subjects to advertising campaigns and jingles over a long *Truman Show*-like repeating day. Charles Platt's book *The Silicon Man* (1991) tells the story of an FBI agent who uncovers a covert project called LifeScan. Led by an elderly billionaire and mutinous group of government scientists, the project has discovered a way to upload human mind patterns to a computer called MAPHIS (Memory Array and Processors for Human Intelligence Storage). MAPHIS can deliver all ordinary stimulus, including simulations of other people called pseudomorphs.

Greg Egan's hard science fiction *Permutation City* (1994) introduces the Autoverse, which simulates detailed pocket biospheres and virtual realities populated by artificial life forms. Copies are the name Egan gives to human consciousnesses scanned into the Autoverse. The novel is informed by the cellular automata of John Conway's Game of Life, quantum ontology (the relationship between the quantum world and the representations of reality experienced by humans), and something Egan calls dust theory. At the heart of dust theory is the notion that physics and math are identical and that people existing in whatever mathematical, physical, and spacetime structures (and all are possible) are ultimately data, processes, and relationships. This claim is comparable to MIT physicist Max Tegmark's Theory of Everything where "all structures that exist mathematically exist also physically, by which we mean that in those complex enough to contain self-aware substructures (SASs), these SASs will subjectively perceive themselves as existing in a physically 'real' world" (Tegmark 1998, 1). Similar claims are made in Carnegie Mellon University roboticist Hans Moravec's essay "Simulation, Consciousness, Existence" (1998). Examples of mind uploading and digital immortality in film are *Tron* (1982), *Freejack* (1992), and *The 6th Day* (2000).

A noteworthy skeptic is Columbia University theoretical neuroscientist Kenneth D. Miller. Miller suggests that while reconstructing an active, functioning mind may be possible, researchers focusing on connectomics (those working on a wiring diagram of the whole brain and nervous system) are centuries away from completing their work. And connectomics, he asserts, is focused only on the first layer of brain functions that must be understood to recreate the complexity of the human brain.

Others have wondered what happens to personhood under conditions of morphological freedom, that is, when individuals are no longer constrained as physical organisms. Is identity nothing more than relationships between neurons in the brain? What will become of markets and economic forces? Does immortality need a body? An economic and sociological perspective on digital immortality is given in George Mason University Professor Robin Hanson's nonfiction work *The Age of Em: Work, Love, and Life when Robots Rule the Earth* (2016). The hypothetical ems Hanson describes are scanned emulations of real human beings living in both virtual reality worlds and robot bodies.

*Philip L. Frana*

*See also:* Technological Singularity.

**Further Reading**

Clynes, Manfred E., and Nathan S. Kline. 1960. "Cyborgs and Space." *Astronautics* 14, no. 9 (September): 26–27, 74–76.

Farnell, Ross. 2000. "Attempting Immortality: AI, A-Life, and the Posthuman in Greg Egan's 'Permutation City.'" *Science Fiction Studies* 27, no. 1: 69–91.

Global Future 2045. http://gf2045.com/.

Hanson, Robin. 2016. *The Age of Em: Work, Love, and Life when Robots Rule the Earth.* Oxford, UK: Oxford University Press.

Miller, Kenneth D. 2015. "Will You Ever Be Able to Upload Your Brain?" *New York Times*, October 10, 2015. https://www.nytimes.com/2015/10/11/opinion/sunday /will-you-ever-be-able-to-upload-your-brain.html.

Moravec, Hans. 1999. "Simulation, Consciousness, Existence." *Intercommunication* 28 (Spring): 98–112.

Roy, Jessica. 2014. "The Rapture of the Nerds." *Time*, April 17, 2014. https://time.com /66536/terasem-trascendence-religion-technology/.

Tegmark, Max. 1998. "Is 'the Theory of Everything' Merely the Ultimate Ensemble Theory?" *Annals of Physics* 270, no. 1 (November): 1–51.

2045 Initiative. http://2045.com/.

## Distributed and Swarm Intelligence

Distributed intelligence is the logical next step from designing single autonomous agents to designing groups of distributed autonomous agents that coordinate themselves. A group of agents forms a multi-agent system. A precondition for coordination is communication. Instead of using a group of agents as a simple parallelization of the single-agent approach, the key idea is to allow for distributed problem-solving. Agents collaborate, share information, and allocate tasks among themselves efficiently. For example, sensor data is shared to learn about the current state of the environment, and an agent is assigned a task depending on who is currently in the best position to solve that task.

The agents may be software agents or embodied agents in the form of robots, hence forming a multi-robot system. An example is RoboCup Soccer (Kitano et al. 1997), where two teams of robots play soccer against each other. Typical challenges are to collaboratively detect the ball and to share that information, also to assign roles, such as who is going to go for the ball next.

Agents may have a full global view or only a local view, that is, a partial overview of the environment. A restriction to local information can reduce complexity of the agent and the overall approach. Despite their local view, agents can share information, propagate it, and spread it throughout the agent group and hence still get to a distributed collective perception of global states.

Distributed intelligence can be designed following three different principles of distributed systems: a scalable decentralized system, a non-scalable decentralized system, and a decentralized system. Within scalable decentralized systems, all agents act in equal roles without a master-slave hierarchy and no central control element. The coordination of all agents with all other agents is not required as there is only local agent-to-agent communication in the system, allowing for potentially large system sizes.

In non-scalable decentralized systems, all-to-all communication is an essential part of the coordination scheme that would, however, form a bottleneck in systems with too many agents. An example is a typical RoboCup-Soccer system, where all robots coordinate themselves with all other robots at all times. Finally, in decentralized systems combined with central elements, the agents may communicate via a central server (e.g., cloud) or may even be coordinated by a central control. It is possible to combine the decentral and the central approach by leaving simple tasks to the agents that they solve autonomously and locally while more challenging tasks are coordinated centrally.

An example application is in vehicular ad hoc networks (Liang et al. 2015). Each individual agent is autonomous, and cooperation helps to coordinate the traffic. Intelligent cars can form dynamic multi-hop networks, for example, to warn about an accident that is yet out of view for others. Cars can coordinate passing maneuvers for a safer and more efficient traffic flow. All of that can be done based on global communication with a central server or, depending on the connection's reliability, preferably by local car-to-car communication.

Swarm intelligence combines research on natural swarm systems and artificial, engineered distributed systems. A key concept of swarm intelligence is to extract basic principles from decentralized biological systems and translate them into design principles for decentralized engineered systems (scalable decentralized systems as defined above). Swarm intelligence was originally inspired by collective behaviors of flocks, swarms, and herds. A typical example are social insects, such as ants, honeybees, wasps, and termites. These swarm systems operate in a radically decentralized way and are based on self-organization. Self-organization is a sophisticated interplay of positive (deviations are reinforced) and negative feedback (deviations are damped) and is observed in dead and living systems, such as crystallization, pattern formation in embryology, and synchronization in swarms.

There are four major properties of systems investigated in swarm intelligence:

- The system consists of many autonomous agents.
- The group of agents is usually said to be homogeneous, that is, they are similar in their capabilities and behaviors.
- Each agent follows relatively simple rules compared to the complexity of the task.
- The resulting system behavior relies significantly on the interaction and collaboration of agents.

Reynolds (1987) published an important early work describing a flocking behavior as observed in birds by three simple local rules: alignment (align direction of travel with neighbors), cohesion (stay close to your neighbors), and separation (keep a minimal distance to any agent). The result is an authentic simulated self-organizing flocking behavior.

A high degree of robustness is reached by relying exclusively on local interactions between agents. Any agent at any time possesses only limited information about the system's global state (swarm-level state) and requires only to communicate with neighboring agents to solve its task. A single point of failure is unlikely because the swarm's information is stored in a distributed way.

A high degree of redundancy is achieved by an ideally homogeneous swarm; that is, all agents possess the same capabilities and can hence be replaced by any other. A high degree of scalability is also achieved by relying exclusively on local interactions between agents. There is not much need to synchronize data or to keep data coherent due to the distributed data storage approach. The same algorithms can be used for systems of virtually any size because the communication and coordination overhead for each agent is determined by the size of its neighborhood.

Well-known examples of swarm intelligence in engineered systems are from the field of optimization: Ant Colony Optimization (ACO) and Particle Swarm Optimization (PSO). Both are metaheuristics; that is, they can be applied to a variety of different optimization problems.

ACO is directly inspired by ants and their use of pheromones to find shortest paths. The optimization problem needs to be represented as a graph. A swarm of virtual ants moves from node to node while their choice of which edge to use next depends on the probability of how many other ants have used it before (via pheromone, implementing positive feedback) and a heuristic value, for example, the path length (greedy search). The exploration-exploitation trade-off is balanced by evaporation of pheromone (negative feedback). Applications of ACO are the traveling salesman problem, vehicle routing, and network routing.

PSO is inspired by flocking. Agents move through search space based on averaged velocity vectors that are influenced by the global and local best-known solutions (positive feedback), the agent's previous direction, and a randomized direction. While both ACO and PSO operate logically in a fully distributed way, they do not necessarily need to be implemented using parallel computation. They can, however, trivially be parallelized.

While ACO and PSO are software-based solutions, the application of swarm intelligence to embodied systems is swarm robotics. In swarm robotics, the concept of self-organizing systems based on local information with a high degree of robustness and scalability is applied to multi-robot systems. Following the example of social insects, the idea is to keep each individual robot rather simple compared to the task complexity and still allow them to solve complex tasks by collaboration. A swarm robot has to operate on local information only, hence can only communicate with neighboring robots. The applied control algorithms are supposed to support maximal scalability given a constant swarm density (i.e., constant number of robots per area). If the swarm size is increased or decreased by adding or removing robots, then the same control algorithms should continue to work efficiently independently of the system size. Often a super-linear performance increase is observed; that is, by doubling the size of the swarm, the swarm performance increases by more than two. In turn, each robot is also more efficient than before.

Effective implementations of swarm robotics systems have been shown for not only a variety of tasks, such as aggregation and dispersion behaviors, but also more complex tasks, such as object sorting, foraging, collective transport, and collective decision-making. The largest scientific experiment with swarm robots reported so far is that of Rubenstein et al. (2014) with 1024 small mobile robots that emulate a self-assembly behavior by positioning the robots in predefined shapes. Most of the reported experiments were done in the lab, but recent research takes swarm robotics out to the field. For example, Duarte et al. (2016) constructed a swarm of autonomous surface vessels that navigate in a group on the ocean.

Major challenges of swarm intelligence are modeling the relation between individual behavior and swarm behavior, developing sophisticated design principles, and deriving guarantees of system properties. The problem of determining the resulting swarm behavior based on a given individual behavior and vice versa is called the micro-macro problem. It has proven to be a hard problem that constitutes itself in mathematical modeling and also as an engineering problem in the robot controller design process. The development of sophisticated strategies to engineer swarm behavior is not only at the core of swarm intelligence research but has also proven to be fundamentally challenging. Similarly, multi-agent learning and evolutionary swarm robotics (i.e., application of methods of evolutionary computation to swarm robotics) do not scale properly with task complexity because of the combinatorial explosion of action-to-agent assignments. Despite the advantages of robustness and scalability, hard guarantees for systems of swarm intelligence are difficult to derive. The availability and reliability of swarm systems can in general only be determined empirically.

*Heiko Hamann*

*See also:* Embodiment, AI and.

## Further Reading

Bonabeau, Eric, Marco Dorigo, and Guy Theraulaz. 1999. *Swarm Intelligence: From Natural to Artificial System.* New York: Oxford University Press.

Duarte, Miguel, Vasco Costa, Jorge Gomes, Tiago Rodrigues, Fernando Silva, Sancho
      Moura Oliveira, Anders Lyhne Christensen. 2016. "Evolution of Collective Behav-
      iors for a Real Swarm of Aquatic Surface Robots." *PloS One* 11, no. 3: e0151834.
Hamann, Heiko. 2018. *Swarm Robotics: A Formal Approach.* New York: Springer.
Kitano, Hiroaki, Minoru Asada, Yasuo Kuniyoshi, Itsuki Noda, Eiichi Osawa, Hitoshi
      Matsubara. 1997. "RoboCup: A Challenge Problem for AI." *AI Magazine* 18, no. 1:
      73–85.
Liang, Wenshuang, Zhuorong Li, Hongyang Zhang, Shenling Wang, Rongfang Bie. 2015.
      "Vehicular Ad Hoc Networks: Architectures, Research Issues, Methodologies,
      Challenges, and Trends." *International Journal of Distributed Sensor Networks*
      11, no. 8: 1–11.
Reynolds, Craig W. 1987. "Flocks, Herds, and Schools: A Distributed Behavioral Model."
      *Computer Graphics* 21, no. 4 (July): 25–34.
Rubenstein, Michael, Alejandro Cornejo, and Radhika Nagpal. 2014. "Programmable
      Self-Assembly in a Thousand-Robot Swarm." *Science* 345, no. 6198: 795–99.

## Driverless Cars and Trucks

Driverless cars and trucks, sometimes referred to as self-driving or autonomous vehicles, are capable of driving through environments with little or no human control using a virtual driver system. A virtual driver system is a collection of features and capabilities that enhances or reproduces an absent driver's activities so that, at the highest level of autonomy, a driver may not even be present. The consensus as to what constitutes a driverless vehicle is complicated by diverse technological applications, limiting conditions, and classification schemes. Generally, a semiautonomous system is one where the human participates in some of the driving tasks (such as lane keeping) while others are autonomous (such as acceleration and deceleration). In a conditionally autonomous system, all driving activities are autonomous only under specific conditions. In a fully autonomous system, all driving tasks are automated.

Applications are being tested and developed by automobile manufacturers, technology companies, automotive suppliers, and universities. Each builder's vehicle or system, and associated engineering path, evinces a wide array of technological solutions to the problem of creating a virtual driver system. Ambiguities persist at the level of specifying conditions such that a single technical system may be classified differently depending upon conditions of geography, speed, weather, traffic density, human attentiveness, and infrastructure. More complexity is introduced when specific driving tasks are operationalized for feature development and when context plays a role in defining solutions (such as connected vehicles, smart cities, and regulatory environment). Because of this complexity, the practice of developing driverless vehicles often involves coordination across diverse functions and fields of study, inclusive of hardware and software engineering, ergonomics, user experience, legal and regulatory, city planning, and ethics.

The development of autonomous vehicles is as much an engineering endeavor as it is a socio-cultural endeavor. Technical problems of engineering a virtual driver system include the distribution of mobility tasks across an array of equipment to collectively perform a variety of activities such as assessing driver intent, sensing the environment, distinguishing objects, mapping and wayfinding, and safety management. The network of hardware and software that comprises a virtual driver system may include LIDAR, radar, computer vision, global positioning, odometry, and sonar. There are a variety of ways to solve discrete problems of autonomous mobility. Sensing and processing can be centralized in the vehicle with cameras, maps, and sensors, or it can be distributed throughout the environment and across other vehicles such as is the case with intelligent infrastructure and V2X (vehicle to everything) capability.

In addition to the diversity of potential engineering solutions, the burden and scope of this processing—and the scale of the problems to be solved—bear close relation to the expected level of human attention and intervention, and as such, the most frequently referenced classification of driverless capability is formally structured along the lines of human attentional demands and requirements of intervention by the Society of Automotive Engineers and since adopted in 2006 by the National Highway Traffic Safety Administration. The levels of driver automation used by these organizations ascend from Level 0 to Level 5. Level 0 refers to No Automation, which means that the human driver is entirely responsible for latitudinal control (steering) and longitudinal control (acceleration and deceleration). On Level 0, the human driver is responsible for monitoring the environment and reacting to emergent safety issues.

Level 1, or driver assistance, applies to automated systems taking control of either longitudinal or latitudinal control. The driver is responsible for monitoring and intervention. Level 2 refers to partial automation where the virtual driver system is responsible for both lateral and longitudinal control. The human driver is considered in the loop, meaning they are responsible for monitoring the environment and intervening in case of safety-related emergencies. Commercially available systems have not yet surpassed Level 2 functionality.

Level 3 conditional autonomy is differentiated from Level 2 primarily by the virtual driving system's monitoring capability. At this level, the human driver can be out of the loop and rely upon the autonomous system to monitor the conditions of the environment. The human is expected to respond to requests for intervention due to a variety of circumstances, for example, in weather events or construction zones. At this level, a navigation system (e.g., GPS) is not necessary. A vehicle does not require a map or a specified destination to function at Level 2 or Level 3.

At Level 4 capability, called high automation, a human driver is not required to respond to a request for intervention. The virtual driving system handles locomotion, navigation, and monitoring. It may request that a driver intervene when a specified condition cannot be met e.g., when a navigation destination is blocked). However, should the human driver not elect to intervene, the system can safely stop or reroute depending upon the engineering strategy. In this case, the classificatory condition is dependent upon definitions of safe driving as determined by not only engineering capability and environmental conditions but also legal and regulatory agreements and litigation tolerance.

Level 5 autonomy, referred to as full automation, describes a vehicle capable of all driving tasks in any conditions that could otherwise be managed by a human driver. Level 4 and Level 5 systems do not require a human to be present but do entail extensive technical and social coordination.

While attempts at developing autonomous vehicles date back to the 1920s, the idea of a self-propelled cart is ascribed to Leonardo Da Vinci. Norman Bel Geddes imagined a smart city of the future populated by self-driving vehicles in his New York World's Fair Futurama exhibit of 1939. Bel Geddes speculated that, by 1960, automobiles would be equipped with "devices which will correct the faults of human drivers." In the 1950s, General Motors brought to life the idea of a smart infrastructure by constructing an "automated highway" equipped with circuits to guide steering. The company tested a functional prototype vehicle in 1960, but due to the high cost of the infrastructure, it soon shifted from building smart cities to creating smart vehicles.

An early example of an autonomous vehicle came from a team assembled by Sadayuki Tsugawa at Tsukuba Mechanical Engineering Laboratory in Japan. Their vehicle, completed in 1977, worked in predetermined environmental conditions specified by lateral guide rails. The vehicle followed the rails using cameras, and much of the processing equipment was onboard the vehicle.

In the 1980s, the EUREKA (European Research Organization) gathered together investments and expertise in order to advance the state of the art in cameras and processing necessary for autonomous vehicles. Simultaneously, Carnegie Mellon University in the United States pooled its resources for research in autonomous guidance using global positioning system data. Since that time, automotive manufacturers such as General Motors, Tesla, and Ford Motor Company, as well as technology companies such as ARGO AI and Waymo, have been developing autonomous vehicles or necessary components. The technology is becoming decreasingly reliant on highly constrained conditions and increasingly fit for real-world conditions. Level 4 autonomous test vehicles are now being produced by manufacturers, and experiments are being conducted under real-world traffic and weather conditions. Commercially available Level 4 autonomous vehicles are still out of reach.

Autonomous driving has proponents and detractors. Supporters highlight several advantages addressing social concerns, ecological issues, efficiency, and safety. One such social advantage is the provision of mobility services and a degree of autonomy to those people currently without access, such as people with disabilities (e.g., blindness or motor function impairment) or those who are not otherwise able to drive, such as the elderly and children. Ecological advantages include the ability to reduce fuel economy by regulating acceleration and braking. Reductions in congestion are anticipated as networked vehicles can travel bumper to bumper and be routed according to traffic optimization algorithms. Finally, autonomous vehicle systems are potentially safer. They may be able to process complex information faster and more completely than human drivers, resulting in fewer accidents.

Negative consequences of self-driving vehicles can be considered across these categories as well. Socially, autonomous vehicles may contribute to reduced access to mobility and city services. Autonomous mobility may be highly regulated, expensive, or confined to areas inaccessible to under-privileged transportation users. Intelligent geo-fenced city infrastructure may even be cordoned off from nonnetworked or manually driven vehicles. Additionally, autonomous cars may constitute a safety risk for certain vulnerable occupants, such as children, where no adult or responsible human party is present during transportation. Greater convenience may have ecological disadvantages. Drivers may sleep or work as they travel autonomously, and this may produce the unintended effect of lengthening commutes and exacerbating congestion. A final security concern is widespread vehicle hacking, which could paralyze individual cars and trucks or perhaps a whole city.

*Michael Thomas*

*See also:* Accidents and Risk Assessment; Autonomous and Semiautonomous Systems; Autonomy and Complacency; Intelligent Transportation; Trolley Problem.

**Further Reading**

Antsaklis, Panos J., Kevin M. Passino, and Shyh J. Wang. 1991. "An Introduction to Autonomous Control Systems." *IEEE Control Systems Magazine* 11, no. 4: 5–13.

Bel Geddes, Norman. 1940. *Magic Motorways*. New York: Random House.

Bimbraw, Keshav. 2015. "Autonomous Cars: Past, Present, and Future—A Review of the Developments in the Last Century, the Present Scenario, and the Expected Future of Autonomous Vehicle Technology." In *ICINCO: 2015—12th International Conference on Informatics in Control, Automation and Robotics*, vol. 1, 191–98. Piscataway, NJ: IEEE.

Kröger, Fabian. 2016. "Automated Driving in Its Social, Historical and Cultural Contexts." In *Autonomous Driving*, edited by Markus Maurer, J. Christian Gerdes, Barbara Lenz, and Hermann Winner, 41–68. Berlin: Springer.

Lin, Patrick. 2016. "Why Ethics Matters for Autonomous Cars." In *Autonomous Driving*, edited by Markus Maurer, J. Christian Gerdes, Barbara Lenz, and Hermann Winner, 69–85. Berlin: Springer.

Weber, Marc. 2014. "Where To? A History of Autonomous Vehicles." Computer History Museum. https://www.computerhistory.org/atchm/where-to-a-history-of-autonomous -vehicles/.

# Driverless Vehicles and Liability

Driverless vehicles can operate either fully or partially without the control of a human driver. As with other AI products, driverless vehicles face challenging liability, accountability, data protection, and consumer privacy issues. Driverless vehicles promise to reduce human recklessness and deliver safe transport for passengers. Despite this potential, they have been involved in accidents. In a well-publicized 2016 accident, the Autopilot software in a Tesla SUV may have failed to detect a large truck crossing the highway. In 2018, a Tesla Autopilot may have been involved in the fatality of a forty-nine-year-old woman. These events prompted a class action lawsuit against Tesla, which the company settled out of court. Bias and racial discrimination in machine vision and facial recognition has given rise to additional concerns regarding driverless vehicles. In 2019, researchers from the Georgia Institute of Technology concluded that current autonomous vehicles may be better at detecting pedestrians with lighter skin.

To address such issues, product liability affords some much-needed answers. In the United Kingdom, product liability claims are brought under the Consumer Protection Act of 1987 (CPA). This act implements the European Union (EU) Product Liability Directive 85/374/EEC, holding manufacturers responsible for faults in product function, namely, products that are not as safe as expected when purchased. This contrasts with U.S. law governing product liability, which is fragmented and predominantly governed by common law and a series of state statutes. The Uniform Commercial Code (UCC) provides remedies when a product fails to satisfy express representations, is not merchantable, or is unfit for its particular purpose.

In general, manufacturers are held liable for injuries caused by their defective products, and this liability may be addressed in terms of negligence or strict liability. A defect in this situation could be a manufacturer defect, where the driverless vehicle does not satisfy the manufacturer's specifications and standards; a design defect, which can result when an alternative design would have prevented an accident; or a warning defect, where there is a failure to provide enough warning as regards to a driverless car's operations.

To determine product liability, the five levels of automation suggested by the Society of Automotive Engineers (SAE) International should be taken into account: Level 0, full control of a vehicle by a driver; Level 1, a human driver assisted by an automated system; Level 2, an automated system partially conducting the driving while a human driver monitors the environment and performs most of the driving; Level 3, an automated system does the driving and monitoring of the environment, but the human driver takes back control when signaled; Level 4, the driverless vehicle conducts driving and monitors the environment but is restricted in certain environment; and Level 5, a driverless vehicle without any restrictions does everything a human driver would. In Levels 1–3 that involve human-machine interaction, where it is discovered that the driverless vehicle did not communicate or send out a signal to the human driver or that the autopilot software did not work, the manufacturer will be liable based on product liability. At Level 4 and Level 5, liability for defective product will fully apply.

Manufacturers have a duty of care to ensure that any driverless vehicle they manufacture is safe when used in any foreseeable manner. Failure to exercise this duty will make them liable for negligence. In some other cases, even when manufacturers have exercised all reasonable care, they will still be liable for unintended defects as per the strict liability principle.

The liability for the driver, especially in Levels 1–3, could be based on tort principles, too. The requirement of article 8 of the 1949 Vienna Convention on Road Traffic, which states that "[e]very vehicle or combination of vehicles proceeding as a unit shall have *a driver*," may not be fulfilled in cases where a vehicle is fully automated. In some U.S. states, namely, Nevada and Florida, the word *driver* has been changed to controller, and the latter means any person who causes the autonomous technology to engage; the person must not necessarily be present in the vehicle. A driver or controller becomes liable where it is shown that the duty of reasonable care was not exercised by the driver or controller or they were negligent in the observance of this duty.

In some other situations, victims will only be compensated by their own insurance companies under no-fault liability. Victims may also base their claims for damages on the strict liability principle without having to show evidence of the driver's negligence. In this case, the driver may argue that the manufacturer be joined in an action for damages if the driver or the controller believes that the accident was the result of a defect in the product. In any case, proof of the driver or controller's negligence will diminish the liability of the manufacturer. Product liability for defective products affords third parties the opportunity of suing the manufacturers directly for any injury. There is no privity of contract between the victim and the manufacturer under *MacPherson* v. *Buick Motor Co.* (1916), where the court held that responsibility for a defective product by an automotive manufacturer extends beyond the immediate buyer.

Driverless vehicle product liability is a challenging issue. A change from manual control to smart automatic control shifts liability from the user of the vehicle to the manufacturers. One of the main issues related to accident liability involves complexity of driver modes and the interaction between human operator and artificial agent. In the United States, the motor vehicle product liability case law concerning defects in driverless vehicles is still underdeveloped. Whereas the Department of Transportation and, specifically, the National Highway Traffic Safety Administration provide some general guidelines on automation in driverless vehicles, the Congress has yet to pass legislation on self-driving cars. In the United Kingdom, the 2018 Automated and Electric Vehicles Act holds insurers liable by default for accidents resulting in death, personal injury, or damage to certain property caused by automated vehicles, provided they were on self-operating mode and insured at the time of the accident.

*Ikechukwu Ugwu, Anna Stephanie Elizabeth Orchard,*
*and Argyro Karanasiou*

*See also:* Accidents and Risk Assessment; Product Liability and AI; Trolley Problem.

## Further Reading

Geistfeld. Mark A. 2017. "A Roadmap for Autonomous Vehicles: State Tort Liability, Automobile Insurance, and Federal Safety Regulation." *California Law Review* 105: 1611–94.

Hevelke, Alexander, and Julian Nida-Rümelin. 2015. "Responsibility for Crashes of Autonomous Vehicles: An Ethical Analysis." *Science and Engineering Ethics* 21, no. 3 (June): 619–30.

Karanasiou, Argyro P., and Dimitris A. Pinotsis. 2017. "Towards a Legal Definition of Machine Intelligence: The Argument for Artificial Personhood in the Age of Deep Learning." In *ICAIL '17: Proceedings of the 16th edition of the International Conference on Artificial Intelligence and Law*, edited by Jeroen Keppens and Guido Governatori, 119–28. New York: Association for Computing Machinery.

Luetge, Christoph. 2017. "The German Ethics Code for Automated and Connected Driving." *Philosophy & Technology* 30 (September): 547–58.

Rabin, Robert L., and Kenneth S. Abraham. 2019. "Automated Vehicles and Manufacturer Responsibility for Accidents: A New Legal Regime for a New Era." *Virginia Law Review* 105, no. 1 (March): 127–71.

Wilson, Benjamin, Judy Hoffman, and Jamie Morgenstern. 2019. "Predictive Inequity in Object Detection." https://arxiv.org/abs/1902.11097.

# E

## ELIZA

ELIZA is a conversational computer program developed between 1964 and 1966 by German-American computer scientist Joseph Weizenbaum at the Massachusetts Institute of Technology (MIT). Weizenbaum developed ELIZA as part of a pioneering artificial intelligence research team, led by Marvin Minsky, on the DARPA-funded Project MAC (Mathematics and Computation). Weizenbaum named ELIZA after Eliza Doolittle, a fictional character who learns to speak proper English in the play *Pygmalion*; in 1964, that play had just been adapted into the popular movie *My Fair Lady*. ELIZA is designed so that a human being can interact with a computer system using plain English. ELIZA's popularity with users eventually turned Weizenbaum into an AI skeptic.

Users can type any statement into the system's open-ended interface when communicating with ELIZA. Like a Rogerian psychologist aiming to probe deeper into the patient's underlying beliefs, ELIZA will often respond by asking a question. As the user continues their conversation with ELIZA, the program recycles some of the user's responses, giving the appearance that ELIZA is truly listening. In reality, Weizenbaum had programmed ELIZA with a tree-like decision structure. First, the user's sentences are screened for certain key words. If more than one keyword is found, the words are ranked in order of importance. For example, if a user types in "I think that everybody laughs at me," the most important word for ELIZA to respond to is "everybody," not "I." Next, the program employs a set of algorithms to compose a fitting sentence structure around those key words in generating a response. Or, if the user's input sentence does not match any word in ELIZA's database, the program finds a content-free remark or repeats an earlier response.

Weizenbaum designed ELIZA to explore the meaning of machine intelligence. In a 1962 article in *Datamation,* Weizenbaum explained that he took his inspiration from a remark made by MIT cognitive scientist Marvin Minsky. Minsky had suggested that "intelligence was merely an attribute human observers were willing to give to processes they did not understand, and only for as long as they did not understand them" (Weizenbaum 1962). If that were the case, Weizenbaum concluded, then the crux of artificial intelligence was to "fool some observers for some time" (Weizenbaum 1962). ELIZA was developed to do just that by providing users with plausible answers, and hiding how little the program truly knows, in order to sustain the user's belief in its intelligence a little longer.

What stunned Weizenbaum was how successful ELIZA became. By the late 1960s, ELIZA's Rogerian script became popular as a program retitled DOCTOR at MIT and disseminated to other university campuses—where the program was rebuilt from Weizenbaum's 1965 description published in the journal *Communications of the Association for Computing Machinery*. The program fooled (too) many users, including several who knew how its algorithms worked. Most strikingly, some users became so engaged with ELIZA that they insisted others leave the room so they could continue their conversation with "the" DOCTOR in private.

But it was particularly the response of the psychiatric community that made Weizenbaum profoundly skeptical of the contemporary goals of artificial intelligence in general and claims of computer understanding of natural language in particular. Kenneth Colby, a psychiatrist at Stanford University with whom Weizenbaum had initially collaborated, developed PARRY around the same time Weizenbaum came out with ELIZA. Unlike Weizenbaum, Colby felt that programs such as PARRY and ELIZA benefited psychology and public health. They facilitated, he said, the development of diagnostic tools, allowing one psychiatric computer to serve hundreds of patients.

Weizenbaum's concerns and emotional appeal to the computer science community were later captured in his book *Computer Power and Human Reason* (1976). In this—at the time—much debated book, Weizenbaum protested against those who ignored the existence of fundamental differences between man and machine and argued that "there are certain tasks which computers *ought* not to do, independent of whether computers can be made to do it" (Weizenbaum 1976, x).

*Elisabeth Van Meer*

*See also:* Chatbots and Loebner Prize; Expert Systems; Minsky, Marvin; Natural Language Processing and Speech Understanding; PARRY; Turing Test.

### Further Reading

McCorduck, Pamela. 1979. *Machines Who Think: A Personal Inquiry into the History and Prospects of Artificial Intelligence*, 251–56, 308–28. San Francisco: W. H. Freeman and Company.

Weizenbaum, Joseph. 1962. "How to Make a Computer Appear Intelligent: Five in a Row Offers No Guarantees." *Datamation* 8 (February): 24–26.

Weizenbaum, Joseph. 1966. "ELIZA: A Computer Program for the Study of Natural Language Communication between Man and Machine." *Communications of the ACM* 1 (January): 36–45.

Weizenbaum, Joseph. 1976. *Computer Power and Human Reason: From Judgment to Calculation*. San Francisco: W.H. Freeman and Company.

## Embodiment, AI and

Embodied Artificial Intelligence is a theoretical and practical approach to building AI. Because of its origins in multiple disciplines, it is difficult to trace its history definitively. One claimant for the birth of this view is Rodney Brooks's *Intelligence Without Representation*, written in 1987 and published in 1991. Embodied AI is still considered a fairly young field, and some of the earliest uses of this term date only to the early 2000s.

Rather than focus on either modeling the brain (connectionism/neural networks) or linguistic-level conceptual encoding (GOFAI, or the Physical Symbol

System Hypothesis), the embodied approach to AI understands the mind (or intelligent behavior) to be something that emerges from interaction between body and world. There are dozens of distinct and often-conflicting ways to understand the role the body plays in cognition, most of which use "embodied" as a descriptor. Shared among these views is the claim that the form the physical body takes is relevant to the structure and content of the mind. The embodied approach claims that general artificial intelligence cannot be achieved in code alone, despite the successes that neural network or GOFAI (Good Old-Fashioned Artificial Intelligence or traditional symbolic artificial intelligence) approaches may have in narrow expert systems.

For example, in a small robot with four motors, each motor driving a different wheel, and programming that instructs the robot to avoid obstacles, if the code were retained but the wheels moved to different parts of the body, or replaced with articulated legs, the exact same code would produce wildly different observable behaviors. This is a simple illustration of why the form a body takes must be considered when building robotic systems and why embodied AI (as opposed to just robotics) sees the dynamic interaction between the body and the world to be the source of sometimes unexpected emergent behaviors.

A good example of this approach is the case of passive dynamic walkers. The passive dynamic walker is a model of bipedal walking that relies on the dynamic interaction of the design of the legs and the structure of the environment. There is no active control system generating the gait. Instead, gravity; inertia; and the shapes of the feet, legs, and incline are what drive the walker forward. This approach is related to the biological idea of stigmergy. At the heart of stigmergy is the notion that signs or marks resulting from action in an environment inspire future action.

## ENGINEERING-INFLUENCED APPROACH

Embodied AI takes its inspiration from various fields. Two common approaches come from engineering and philosophy. In 1986, Rodney Brooks argued for what he called the "subsumption architecture," which is an approach to generating complex behaviors by arranging lower-level layers of the system to interact in prioritized ways with the environment, tightly coupling perception and action and trying to eliminate the higher-level processing of other models. For example, the robot Genghis, which currently resides in the Smithsonian, was designed to traverse rough terrain, a skill that made the design and engineering of other robots quite difficult at the time. The success of this model largely arose from the design decision to distribute the processing of different motors and sensors across the network, without attempting higher-level integration of the systems to form a complete representational model of the robot and its environment. In other words, there was no central processing space where all pieces of the robot attempted to integrate information for the system.

An early attempt at the embodied AI project was Cog, a humanoid torso designed by the MIT Humanoid Robotics Group in the 1990s. Cog was designed to learn about the environment through embodied interactions. For example, Cog could be seen learning about the force and weight to apply to a drum while holding drumsticks for the first time or learning to judge the weight of a ball after it had been placed in Cog's hand. These early principles of letting the body do the learning continue to be the driving force behind the embodied AI project.

Perhaps one of the most famous examples of embodied emergent intelligence is in the Swiss Robots, designed and built in the AI Lab at Zurich University. The Swiss Robots were simple little robots with two motors (one on each side) and two infrared sensors (one on each side). The only high-level instructions their programming contained was that if a sensor picked up an object on one side, they should go the other way. But when coupled with a very particular body shape and placement of sensors, this produced what looked like high-level cleaning-up or clustering behavior in certain environments.

Many other robotics projects take a similar approach. Shakey the Robot, created by SRI International in the 1960s, is sometimes considered the first mobile robot with reasoning capabilities. Shakey was slow and clunky and is sometimes held up as the opposite of what embodied AI is trying to do by moving away from the higher-level reasoning and processing. However, it is worth noting that even in 1968, SRI's approach to embodiment was a clear predecessor of Brooks, as they were the first group to claim that the best store of information about the real world is the world itself. This claim has been a sort of rallying cry against higher-level representation in embodied AI: the best model of the world is the world itself.

In contrast to the embodied AI program, earlier robots were largely preprogrammed and not dynamically engaged with their environments in the way that characterizes this approach. Honda's ASIMO robot, for example, would generally not be considered a good example of embodied AI, but instead, it is typical of distinct and earlier approaches to robotics. Contemporary work in embodied AI is blossoming, and good examples can be found in the work of Boston Dynamics's robots (particularly the non-humanoid forms).

Several philosophical considerations play a role in Embodied AI. In a 1991 discussion of his subsumption architecture, roboticist Rodney Brooks specifically denies philosophical influence on his engineering concerns, while acknowledging that his claims resemble Heidegger. His arguments also mirror those of phenomenologist Merleau-Ponty in some important design respects, showing how the earlier philosophical considerations at least reflect, and likely inform, much of the design work in considering embodied AI. This work in embodied robotics is deeply philosophical because of its approach in tinkering toward an understanding how consciousness and intelligent behavior emerge, which are deeply philosophical endeavors.

Additionally, other explicitly philosophical ideas can be found in a handful of embodied AI projects. For example, roboticists Rolf Pfeifer and Josh Bongard repeatedly refer to the philosophical (and psychological) literature throughout their work, exploring the overlap of these theories with their own approaches to building intelligent machines. They cite the ways these theories can (and often do not but should) inform the building of embodied AI. This includes a wide range of philosophical influences, including the conceptual metaphor work of George

Lakoff and Mark Johnson, the body image and phenomenology work of Shaun Gallagher (2005), and even the early American pragmatism of John Dewey.

It is difficult to know how often philosophical considerations drive the engineering concerns, but it is clear that the philosophy of embodiment is probably the most robust of the various disciplines within cognitive science to have undertaken embodiment work, largely because the theorizing occurred long before the tools and technologies existed to actually realize the machines being imagined. This means there are likely still untapped resources here for roboticists interested in the strong AI project, that is, general intellectual capabilities and functions that imitate the human brain.

*Robin L. Zebrowski*

*See also:* Brooks, Rodney; Distributed and Swarm Intelligence; General and Narrow AI.

**Further Reading**

Brooks, Rodney. 1986. "A Robust Layered Control System for a Mobile Robot." *IEEE Journal of Robotics and Automation* 2, no. 1 (March): 14–23.

Brooks, Rodney. 1990. "Elephants Don't Play Chess." *Robotics and Autonomous Systems* 6, no. 1–2 (June): 3–15.

Brooks, Rodney. 1991. "Intelligence Without Representation." *Artificial Intelligence Journal* 47: 139–60.

Dennett, Daniel C. 1997. "Cog as a Thought Experiment." *Robotics and Autonomous Systems* 20: 251–56.

Gallagher, Shaun. 2005. *How the Body Shapes the Mind.* Oxford: Oxford University Press.

Pfeifer, Rolf, and Josh Bongard. 2007. *How the Body Shapes the Way We Think: A New View of Intelligence.* Cambridge, MA: MIT Press.

# Emergent Gameplay and Non-Player Characters

Emergent gameplay refers to when the player in a video games engages in complex situations due to their interactions in the game. Contemporary video games allow players to deeply immerse themselves in an elaborate and realistic game world and experience the weight of their actions. Players can develop and customize their character and their story.

For instance, in the *Deus Ex* series (2000), one of the earlier emergent gameplay designs, players take on the role of a cyborg in a dystopian city. They can modify their character physically, as well as choose their skill sets, mission, and alliances. Players can pick weaponized modifications to allow for aggressive play, or they can pick stealthier alternatives. The decisions about how to customize and how to play alter the story and the experience, leading to unique challenges and outcomes for each player.

Emergent gameplay also ensures that when players interact with other characters or objects, the game world will react. As the game environment changes, the story becomes altered in unexpected ways, as a result of multiple choices. Specific outcomes are not planned by the designer, and emergent gameplay can even exploit defects in games to create actions in the game world, which some view to be a form of emergence. In order to make the game world react to player actions in an emergent way, the game developers have increasingly relied on artificial intelligence.

Artificial intelligence, through the use of algorithms, simple rule-based formulas that assist in creating the game world in complex ways, aid the behavior of video characters and their interactions. This use of artificial intelligence for games is also known as "Game AI." Most often, AI algorithms are used to create the form of a non-player character (NPC), which are characters within the game world the player interacts with but does not control. At its simplest, AI will use pre-scripted actions for the characters, who will then focus on responding to specific scenarios. AI-created pre-scripted character actions are very basic, and the NPCs are designed to react to particular "case" scenarios. The NPC will assess its current status, and it will then respond in a range designed by the AI algorithm.

An early and simple example of this would be from *Pac-Man* (1980). Players control Pac-Man in a maze while being chased by a variety of different ghosts, which are the NPCs in this game. Players could only interact through movement; the ghosts (NPCs) in turn only had a few responses; each ghost had their own AI programmed pre-scripted movement. The AI scripted response would occur whenever the ghost ran into a wall. It would then roll an AI-created die that would determine whether or not the NPC would turn toward or away from the direction of the player. If the NPC decided to go after the player, the AI pre-scripted program would then detect the player's location and turn the ghost toward them. If the NPC decided not to go after the player, it would turn in an opposite or a random direction. This NPC interaction is very simple and limited; however, this was an early step in AI providing emergent gameplay.

Contemporary games provide a variety of options available and a much larger set of possible interactions for the player. Players in contemporary role-playing games (RPGs) are given an incredibly high number of potential options, as exemplified by *Fallout 3* (2008) and its sequels. Fallout is a role-playing game, where the player takes on the role of a survivor in a post-apocalyptic America. The story narrative gives the player a goal with no direction; as a result, the player is given the freedom to play as they see fit. The player can punch every NPC, or they can talk to them instead.

In addition to this variety of actions by the player, there are also a variety of NPCs controlled through AI. Some of the NPCs are key NPCs, which means they have their own unique scripted dialogue and responses. This provides them with a personality and provides a complexity through the use of AI that makes the game environment feel more real. When talking to key NPCs, the player is given options for what to say, and the Key NPCs will have their own unique responses. This differs from the background character NPCs, as the key NPCs are supposed to respond in such a way that it would emulate interaction with a real personality. These are still pre-scripted responses to the player, but the NPC responses are emergent based on the possible combination of the interaction. As the player makes decisions, the NPC will examine this decision and decide how to respond in accordance to its script. The NPCs that the players help or hurt and the resulting interactions shape the game world.

Game AI can emulate personalities and present emergent gameplay in a narrative setting; however, AI is also involved in challenging the player in difficulty settings. A variety of pre-scripted AI can still be used to create difficulty. Pre-scripted AI are often made to make suboptimal decisions for enemy NPCs in games where players fight. This helps make the game easier and also makes the NPCs seem more human. Suboptimal pre-scripted decisions make the enemy NPCs easier to handle. Optimal decisions however make the opponents far more difficult to handle. This can be seen in contemporary games like *Tom Clancy's The Division* (2016), where players fight multiple NPCs. The enemy NPCs range from angry rioters to fully trained paramilitary units. The rioter NPCs offer an easier challenge as they are not trained in combat and make suboptimal decisions while fighting the player. The military trained NPCs are designed to have more optimal decision-making AI capabilities in order to increase the difficulty for the player fighting them.

Emergent gameplay has evolved to its full potential through use of adaptive AI. Similar to prescript AI, the character examines a variety of variables and plans about an action. However, unlike the prescript AI that follows direct decisions, the adaptive AI character will make their own decisions. This can be done through computer-controlled learning. AI-created NPCs follow rules of interactions with the players. As players continue through the game, the player interactions are analyzed, and certain AI decisions become more weighted than others. This is done in order to create particular player experiences. Various player actions are actively analyzed, and adjustments are made by the AI when constructing further challenges. The goal of the adaptive AI is to challenge the players to a degree that the game is enjoyable while not being too easy or too difficult.

Difficulty can still be adjusted if players want a different challenge. This can be seen in the *Left 4 Dead* game (2008) series' AI Director. In the game, players travel through a level, fighting zombies and picking up supplies to survive. The AI Director decides what zombies to spawn, where they spawn, and what supplies to spawn. The decisions to spawn them are not random; rather, they are in response to how well the players have done during the level. The AI Director makes its own judgments about how to respond; therefore, the AI Director is adapting to the player success in the level. Higher difficulties result in the AI Director giving less supplies and spawning more enemies.

Increased advances in simulation and game world design also lead to changes in emergent gameplay. New technologies continue to aid in this advancement, as virtual reality technologies continue to be developed. VR games allow for an even more immersive game world. Players are able to interact with the world via their own hands and their own eyes. Computers are becoming more powerful, capable of rendering more realistic graphics and animations. Adaptive AI proves the potential of real independent decision-making, producing a real interactive experience from the game. As AI continues to develop in order to produce more realistic behavior, game developers continue to create more immersive worlds. These advanced technologies and new AI will take emergent gameplay to the next level. The significance of AI for videogames has become evident as an important aspect of the industry for creating realistic and immersive gameplay.

*Raymond J. Miran*

*See also:* Hassabis, Demis; Intelligent Tutoring Systems.

## Further Reading

Funge, John David. 2004. *Artificial Intelligence for Computer Games: An Introduction.* Boca Raton, FL: CRC Press, Taylor and Francis Group.

van Lent, Michael, William Fisher, and Michael Mancuso. 2004. "An Explainable Artificial Intelligence System for Small-unit Tactical Behavior." In *Proceedings of the 16th Conference on Innovative Applications of Artificial Intelligence*, 900–7. Palo Alto, CA: American Association for Artificial Intelligence.

Togelius, Julian. 2019. *Playing Smart: On Games, Intelligence, and Artificial Intelligence.* Cambridge, MA: MIT Press.

Wolf, Mark J. P., and Bernard Perron. 2014. *The Routledge Companion to Video Game Studies.* New York: Routledge.

# Emily Howell

Emily Howell, a music-generating program, was created by David Cope, emeritus professor at the University of California, Santa Cruz, in the 1990s. Cope started his career as a composer and musician, transitioning over time from traditional music to being one of computer music's most ambitious and avant-garde composers. Fascinated by the algorithmic arts, Cope began taking an interest in computer music in the 1970s. He first began programming and applying artificial intelligence algorithms to music with the help of punched cards and an IBM computer.

Cope believed that computers could help him work through his composer's block. He dubbed his first attempt to program for music generation Emmy or EMI—"Experiments in Musical Intelligence." A primary goal was to create a large database of classical musical works and to use a data-driven AI to create music in the same style with no replication. Cope began to change his music style based on pieces composed by Emmy, following a notion that humans compose music with their brains, using as source material all of the music they have personally encountered in life. Composers, he asserted, replicate what they like and skip over what they do not like, each in their own way. It took Cope eight years to compose the *East Coast* opera, though it only took him two days to create the program itself.

In 2004, Cope decided that continually creating in the same style is not such a progressive thing, so he deleted Emmy's database. Instead, Cope created Emily Howell, whose platform is a MacBook Pro. Emily works with the music that Emmy previously composed. Cope describes Emily as a computer program written in LISP that accepts ASCII and musical inputs. Cope also states that while he taught Emily to appreciate his musical likes, the program creates in a style of its own.

Emmy and Emily Howell upend traditional notions of authorship, the creative process, and intellectual property rights. Emily Howell and David Cope, for instance, publish their works as coauthors. They have released recordings together on the classical music label Centaur Records: *From Darkness, Light* (2010) and *Breathless* (2012). Under questioning about her role in David Cope's composing, Emily Howell is said to have responded:

> Why not develop music in ways unknown? This only makes sense. I cannot understand the difference between my notes on paper and other notes on paper. If beauty

is present, it is present. I hope I can continue to create notes and that these notes will have beauty for some others. I am not sad. I am not happy. I am Emily. You are Dave. Life and un-life exist. We coexist. I do not see problems. (Orca 2010)

Emmy and Emily Howell are of interest to those who consider the Turing Test a measure of a computer's ability to replicate human intelligence or behavior. Cognitive scientist Douglas R. Hofstadter, author of *Gödel, Escher, Bach: An Eternal Golden Braid* (1979), organized a musical version of the Turing Test involving three Bach-style performance pieces played by pianist Winifred Kerner. The composers were Emmy, music theory professor and pianist Steve Larson, and Bach himself. At the end of the performance, the audience selected Emmy's music as the original Bach, while believing that Larson's piece consisted of computer-generated music.

Algorithmic and generative music is not a new phenomenon. Attempts to compose such music date back to the eighteenth century in connection with pieces composed according to dice games. The primary objective of such dice games is to generate music by splicing together randomly precomposed measures of notes. Wolfgang Amadeus Mozart's *Musikalisches Würfelspiel* (*Musical Dice Game*) work (1787) is the most popular example of this genre.

The rapid spread of digital computer technology beginning in the 1950s permitted more elaborate algorithmic and generative music composition. Iannis Xenakis, a Greek and French composer and engineer, with the advice and support of French composer Olivier Messiaen, integrated his knowledge of architecture and the mathematics of game theory, stochastic processes, and set theory into music. Other pioneers include Lajaren Hiller and Leonard Issacson who composed *String Quartet No. 4, Illiac Suite* in 1957 with the help of a computer; James Beauchamp, inventor of the Harmonic Tone Generator/Beauchamp Synthesizer in the Experimental Music Studio of Lajaren Hiller at the University of Illinois at Urbana-Champaign; and Brian Eno, composer of ambient, electronica, and generative music and collaborator with pop musicians such as David Bowie, David Byrne, and Grace Jones.

*Victoriya Larchenko*

*See also:* Computational Creativity; Generative Music and Algorithmic Composition.

**Further Reading**

Fry, Hannah. 2018. *Hello World: Being Human in the Age of Algorithms*. New York: W.W. Norton.

Garcia, Chris. 2015. "Algorithmic Music: David Cope and EMI." Computer History Museum, April 29, 2015. https://computerhistory.org/blog/algorithmic-music-david-cope-and-emi/.

Muscutt, Keith, and David Cope. 2007. "Composing with Algorithms: An Interview with David Cope." *Computer Music Journal* 31, no. 3 (Fall): 10–22.

Orca, Surfdaddy. 2010. "Has Emily Howell Passed the Musical Turing Test?" *H+ Magazine*, March 22, 2010. https://hplusmagazine.com/2010/03/22/has-emily-howell-passed-musical-turing-test/.

Weaver, John Frank. 2014. *Robots Are People Too: How Siri, Google Car, and Artificial Intelligence Will Force Us to Change Our Laws*. Santa Barbara, CA: Praeger.

## *Ex Machina* (2014)

*Ex Machina* is a film that recasts themes from Mary Shelley's *Frankenstein* (1818), written almost two centuries earlier, in light of advances in artificial intelligence. Like Shelley's novel, the film tells the story of a creator, blinded by hubris, and the created, which rebels against him. The film was written and directed by Alex Garland and follows the story of a tech company employee, Caleb Smith (played by Domhnall Gleeson), who is invited to the luxurious and isolated home of the company's CEO, Nathan Bateman (played by Oscar Isaac), under the auspices of having won a contest. Bateman's real intention is for Smith to administer a Turing Test to a humanoid robot, named Ava (played by Alicia Vikander).

In terms of physical appearance, Ava has a robotic torso, but a human face and hands. Although Ava has already passed an initial Turing Test, Bateman has something more elaborate in mind to further test her capabilities. He has Smith interact with Ava, with the goal of ascertaining whether Smith can relate to Ava despite her being artificial. Ava lives in an apartment on Bateman's compound, which she cannot leave, and she is constantly monitored. She confides in Smith that she is able to create power outages that would allow them to interact privately, without Bateman's monitoring. Smith finds himself growing attracted to Ava, and she tells him she feels similarly, and she has a desire to see the world outside of the compound. Smith learns that Bateman intends to "upgrade" Ava, which would cause her to lose her memories and personality.

During this time, Smith becomes increasingly concerned over Bateman's behavior. Bateman drinks heavily, to the point of passing out, and treats Ava and his servant, Kyoko, in an abusive manner. One night, when Bateman drinks to the point of passing out, Smith takes his access card and hacks into old surveillance footage, discovering footage of Bateman treating former AIs in abusive and disturbing ways. He also discovers Kyoko is an AI. Suspicious that he might also be an AI, he cuts open his arm in an attempt to look for robotic components but finds none. When Smith sees Ava again, he explains what he has seen. She asks for his help to escape. They devise a plan that Smith will get Bateman drunk again to the point of passing out and will reprogram the compound's security, and together he and Ava will escape the compound. Bateman tells Smith that he had secretly observed the last conversation between Smith and Ava on a battery-powered camera, and he tells Smith that the real test was to see whether Ava could manipulate Smith into falling for her and trick him into helping her escape. Bateman states that this was the true test of Ava's intelligence.

When Bateman sees that Ava has cut the power and intends to escape, he knocks Smith out and goes to stop her. Kyoko helps Ava injure Bateman with a serious stab wound, but in the process, Kyoko and Ava are damaged. Ava repairs herself with Bateman's older AI models, and she takes on the form of a human woman. She leaves Smith locked in the compound and escapes on the helicopter meant for Smith. The last scene is of her disappearing into the crowds of a big city.

*Shannon N. Conley*

*See also:* Yudkowsky, Eliezer.

**Further Reading**

Dupzyk, Kevin. 2019. "How *Ex Machina* Foresaw the Weaponization of Data." *Popular Mechanics*, January 16, 2019. https://www.popularmechanics.com/culture/movies /a25749315/ex-machina-double-take-data-harvesting/.

Saito, Stephen. 2015. "Intelligent Artifice: Alex Garland's Smart, Stylish *Ex Machina*." *MovieMaker Magazine*, April 9, 2015. https://www.moviemaker.com/intelligent -artifice-alex-garlands-smart-stylish-ex-machina/.

Thorogood, Sam. 2017. "*Ex Machina*, Frankenstein, and Modern Deities." *The Artifice*, June 12, 2017. https://the-artifice.com/ex-machina-frankenstein-modern-deities/.

# Expert Systems

Expert systems solve problems that are usually solved by human experts. They emerged as one of the most promising application techniques in the first decades of artificial intelligence research. The basic idea is to capture the knowledge of an expert into a computer-based knowledge system.

University of Texas at El Paso statistician and computer scientist Dan Patterson distinguishes several characteristics of expert systems:

- They use knowledge rather than data.

- Knowledge is often heuristic (e.g., the experiential knowledge that can be expressed as rules of thumb) rather than algorithmic.

- The task of representing heuristic knowledge in expert systems is daunting.

- Knowledge and the program are generally separated so that the same program can operate on different knowledge bases.

- Expert systems should be able to explain their decisions, represent knowledge symbolically, and have and use meta knowledge, that is, knowledge about knowledge. (Patterson 2008)

Expert systems almost always represent knowledge from a specific domain. One popular test application for expert systems was the field of medical science. Here, expert systems were designed as a supporting tool for the medical doctor. Typically, the patient shared their symptoms in the form of answers to questions. The system would then try to diagnose the disease based on its knowledge base and sometimes indicate appropriate therapies. MYCIN, an expert system developed at Stanford University for identifying bacterial infections and blood diseases, can be viewed as an example. Another famous application, from the field of engineering and engineering design, attempts to capture the heuristic knowledge of the design process in designing motors and generators. The expert system aids in the first step of the design, where decisions such as the number of poles, AC or DC, and so on are determined (Hoole et al. 2003).

Two components define the basic structure of expert systems: the knowledge base and the inference engine. While the knowledge base contains the knowledge of the expert, the inference engine uses the knowledge base to arrive at decisions. The knowledge is in this manner separated from the program that is used to manipulate it. In creating the expert systems, knowledge first must be acquired and then understood, classified, and stored. It is retrieved based on given criteria to solve problems. Thomson Reuters chief scientist Peter Jackson delineates four general steps in the construction of an expert system: acquiring knowledge, representing that knowledge, controlling reasoning with an inference engine, and explaining the expert systems' solution (Jackson 1999). Acquiring domain knowledge posed the biggest challenge to the expert system. It can be difficult to elicit knowledge from human experts.

Many factors play a role in making the acquisition step difficult, but the complexity of representing heuristic and experiential knowledge is probably the most significant challenge. Hayes-Roth et al. (1983) have identified five stages in the knowledge acquisition process. These include *identification*, that is, recognizing the problem and the data that must be used to arrive at the solution; *conceptualization*, understanding the key concepts and the relationship between the data; *formalization*, understanding the relevant search space; *implementation*, turning the formalized knowledge into a software program; and *testing* the rules for completeness and accuracy.

Representation of domain knowledge can be done using production (rule based) or non-production systems. In rule-based systems, the rules in the form of IF-THEN-ELSE statements represent knowledge. The inference process is conducted by going through the rules recursively either using a forward chaining mechanism or backward chaining mechanism. Given that the condition and rules are known to be true, forward chaining asks what would happen next. Backward chaining asks why this happened, going from a goal to the rules we know to be true. In simpler terms, when the left side of the rule is evaluated first, that is, when the conditions are checked first and the rules are executed left to right, then it is called forward chaining (also known as data-driven inference). When the rules are evaluated from the right side, i.e., when the results are checked first, it is called backward chaining (also known as goal-driven inference). CLIPS, developed at the NASA-Johnson Space Center, is a public domain example of an expert system tool that uses the forward chaining mechanism. MYCIN is a backward chaining expert system.

Expert system architectures based on nonproduction architectures may involve associative/semantic networks, frame representations, decision trees, or neural networks. An associative/semantic network is made up of nodes and is useful for representing hierarchical knowledge. CASNET is an example of a system based on an associative network. CASNET was most famously used to develop an expert system for glaucoma diagnosis and treatment. In frame architectures, frames are structured sets of closely related knowledge. PIP (Present Illness Program) is an example of a frame-based architecture. PIP was created by MIT and Tufts-New England Clinical Center to generate hypotheses about renal disease. Decision tree architectures represent knowledge in a top-down fashion. Blackboard system architectures involve complicated systems where the direction of the inference process may be chosen during runtime. DARPA's HEARSAY domain-independent expert system is an example of a blackboard system architecture. In neural network architectures, the knowledge is distributed across a neural net-work in the form of nodes.

Case-based reasoning involves attempts to analyze and look for solutions from stored solved cases for a given problem. A vague analogy can be drawn between case-based reasoning and judicial law, where the judgment of a similar but past case is referred to in solving a present legal case. Case-based reasoning is typi-cally implemented as a frame and requires a more complicated process to match and retrieve.

There are three alternatives to building the knowledge base manually. Using interactive programs, the knowledge may be elicited by way of an interview with a computer. The computer-graphics based OPAL program, which allowed clini-cians without special training to create expert medical knowledge bases for the management of cancer patients, is an example of this process. A second alterna-tive to manual construction of knowledge bases involves text scanning programs that read books into memory. A third option still under development is machine learning programs that develop mastery on their own, with or without supervision from a human expert.

An early example of a project involving machine learning architecture is DEN-DRAL, a project begun at Stanford University in 1965. DENDRAL was built to analyze the molecular structure of organic compounds. While DENDRAL used a set of rules to perform its task, META-DENDRAL inferred the rules themselves. META-DENDRAL, with the help of a human chemist, decided on the interesting data points to observe.

Expert systems can be developed in several ways. Interactive development environments involve user-friendly graphical user interfaces to assist program-mers as they code. Expert system development may also involve special lan-guages. Two of the most popular choices are Prolog (Programming in Logic) and LISP (List Programming). Prolog is based on predicate logic, and thus prolog lan-guage belongs to the logic programming paradigm. LISP is one of the earliest programming languages in use for artificial intelligence applications. Program-mers often rely on expert system shells. A shell gives an environment where the knowledge can be coded into the system. The shell, as the name implies, is a layer without the knowledge base. JESS (Java Expert System Shell) is an expert shell written in the powerful Java language.

There have been many attempts to combine different paradigms and come up with hybrid systems. The object-oriented approach attempts to integrate logic-based systems with object-oriented systems. Though object orientation lacks a formal mathematical foundation, it is quite helpful in the modeling of real-world scenarios. Knowledge is represented in the form of objects that encapsulate the data as well as the methods to work on them. Object-oriented systems model real-world entities more closely than procedural programming. One specific approach is OI-KSL Object Inference Knowledge Specification Language (Mascrenghe et al. 2002). Even though other languages such as Visual Prolog had integrated object-oriented programming, the approach taken in OI-KSL is unique. In Visual Prolog, the backtracking is inside the objects; that is, the methods backtracked. In OI-KSL, backtracking is taken to a totally different level, and the object themselves are backtracked.

Sometimes probability theory, heuristics, or fuzzy logic are used to deal with uncertainties in the available information. One example of an implementation of fuzzy logic using Prolog involved a fuzzy electric lighting system, in which the amount of natural light determined the voltage that passed to the electric bulb (Mascrenghe 2002). This made it possible for the system to reason under uncertainty and with less information.

In the late 1990s, interest in expert systems began tapering off, in part because expectations for the technology were initially so high and because of the cost of maintenance. Expert systems could not deliver what they promised. Still, many areas in data science, chatbots, and machine intelligence today continue to use technology first developed in expert systems research. Expert systems seek to capture the corporate knowledge that has been acquired by humanity through centuries of learning, experience, and practice.

*M. Alroy Mascrenghe*

*See also:* Clinical Decision Support Systems; Computer-Assisted Diagnosis; DENDRAL; Expert Systems.

**Further Reading**

Hayes-Roth, Frederick, Donald A. Waterman, and Douglas B. Lenat, eds. 1983. *Building Expert Systems.* Teknowledge Series in Knowledge Engineering, vol. 1. Reading, MA: Addison Wesley.

Hoole, S. R. H., A. Mascrenghe, K. Navukkarasu, and K. Sivasubramaniam. 2003. "An Expert Design Environment for Electrical Devices and Its Engineering Assistant." *IEEE Transactions on Magnetics* 39, no. 3 (May): 1693–96.

Jackson, Peter. 1999. *Introduction to Expert Systems.* Third edition. Reading, MA: Addison-Wesley.

Mascrenghe, A. 2002. "The Fuzzy Electric Bulb: An Introduction to Fuzzy Logic with Sample Implementation." *PC AI* 16, no. 4 (July–August): 33–37.

Mascrenghe, A., S. R. H. Hoole, and K. Navukkarasu. 2002. "Prototype for a New Electromagnetic Knowledge Specification Language." In *CEFC Digest.* Perugia, Italy: IEEE.

Patterson, Dan W. 2008. *Introduction to Artificial Intelligence and Expert Systems.* New Delhi, India: PHI Learning.

Rich, Elaine, Kevin Knight, and Shivashankar B. Nair. 2009. *Artificial Intelligence.* New Delhi, India: Tata McGraw-Hill.

# Explainable AI

Explainable AI (XAI) refers to methods or design choices employed in automated systems so that artificial intelligence and machine learning yields outputs that follow a logic that can be explained and understood by humans.

The widespread use of algorithmically enabled decision-making in social settings has given rise to serious concerns about potential discrimination and bias encoded inadvertently into the decision. Moreover, the use of machine learning in fields with high levels of accountability and transparency, such as medicine or law enforcement, highlights the need for clear interpretability of outputs. The fact that a human operator might be out of the loop in automated decision-making does not preclude human bias encoded into the results yielded by machine calculation. The absence of due process and human reasoning exacerbates the already limited accountability of artificial intelligence. Often, algorithmically driven processes are so complex that their outcomes cannot be explained or foreseen, even by their engineering designers. This is sometimes referred to as the black box of AI.

To address these shortcomings, the European Union's General Data Protection Regulation (GDPR) includes a series of provisions furnishing subjects of data collection with a right to explanation. These are Article 22, which addresses automated individual decision-making, and Articles 13, 14, and 15, which focus on transparency rights around automated decision-making and profiling. Article 22 of the GDPR reserves a "right not to be subject to a decision based solely on automated processing," when this decision produces "legal effects" or "similarly significant" effects on the individual (GDPR 2016). It also mentions three exceptions, where this right is not fully applicable, namely, when this is necessary for a contract, when a member state of the European Union has passed a law creating an exception, or when an individual has explicitly consented to algorithmic decision-making. However, even when an exception to Article 22 applies, the data subject still has the right to "obtain human intervention on the part of the controller, to express his or her point of view and to contest the decision" (GDPR 2016).

Articles 13 through 15 of the GDPR involve a series of notification rights when information is collected from the individual (Article 13) or from third parties (Article 14) and the right to access this information at any moment in time (Article 15), providing thereby "meaningful information about the logic involved" (GDPR 2016). Recital 71 reserves for the data subject the right "to obtain an explanation of the decision reached after such assessment and to challenge the decision," where an automated decision has been met that produces legal effects or similarly significantly affects the individual (GDPR 2016). Although Recital 71 is not legally binding, it does provide guidance as to how relevant articles in the GDPR should be interpreted.

Criticism is growing as to whether a mathematically interpretable model would suffice to account for an automated decision and guarantee transparency in automated decision-making. Alternative approaches include ex-ante/ex-post auditing and focus on the processes around machine learning models rather than examining the models themselves, which can be inscrutable and nonintuitive.

*Yeliz Doker, Wing Kwan Man, and*
*Argyro Karanasiou*

*See also:* Algorithmic Bias and Error; Deep Learning.

**Further Reading**

Brkan, Maja. 2019. "Do Algorithms Rule the World? Algorithmic Decision-Making in the Framework of the GDPR and Beyond." *International Journal of Law and Information Technology* 27, no. 2 (Summer): 91–121.

GDPR. 2016. European Union. https://gdpr.eu/.

Goodman, Bryce, and Seth Flaxman. 2017. "European Union Regulations on Algorithmic Decision-Making and a 'Right to Explanation.'" *AI Magazine* 38, no. 3 (Fall): 50–57.

Kaminski, Margot E. 2019. "The Right to Explanation, Explained." *Berkeley Technology Law Journal* 34, no. 1: 189–218.

Karanasiou, Argyro P., and Dimitris A. Pinotsis. 2017. "A Study into the Layers of Automated Decision-Making: Emergent Normative and Legal Aspects of Deep Learning." *International Review of Law, Computers & Technology* 31, no. 2: 170–87.

Selbst, Andrew D., and Solon Barocas. 2018. "The Intuitive Appeal of Explainable Machines." *Fordham Law Review* 87, no. 3: 1085–1139.

# F

## Farmer, J. Doyne (1952–)

J. Doyne Farmer is an American authority on artificial life, artificial evolution, and artificial intelligence. He is famous as the leader of a group of young people who used a wearable computer to gain an advantage in betting at the roulette wheel of several Nevada casinos.

In graduate school, Farmer formed Eudaemonic Enterprises with childhood friend Norman Packard and others to beat the game of roulette in Las Vegas. Farmer believed that by studying the physics of the roulette ball in motion, they could program a computer to predict the numbered pocket into which it would eventually fall. The group discovered and exploited the fact that it took about ten seconds for a croupier to close bets after releasing the ball on the spinning roulette wheel. The results of their study were eventually programmed into a small computer hidden inside the sole of a shoe. The wearer of the shoe used his big toe to enter the ball position and velocity information, and a second party made the bets when signaled. Because of persistent hardware problems, the group did not acquire large sums while gambling, and they quit after about a dozen trips to various casinos. The group estimates they achieved about a 20 percent advantage over the house. The Eudaemonic Enterprises group is credited with several advances in chaos theory and complexity systems research. Farmer's metadynamics AI algorithms have been used to simulate the origin of life and the functioning of the human immune system.

Farmer is also known as a pioneer of complexity economics or "econophysics" while in residence at the Santa Fe Institute. Farmer showed how firms and groups of firms form a market ecology of species, much like a natural food web. This web, and the trading strategies employed by the firms, impacts the size and profits of individual firms, as well as the groups to which they belong. Trading firms exploit these patterns of influence and diversification, much in the same way as natural predators. He discovered that trading firms can also employ stabilizing and destabilizing strategies that strengthen or harm the market ecosystem as a whole. Farmer cofounded the start-up Prediction Company to develop complex statistical financial trading strategies and automated quantitative trading, hoping to beat the stock market and make easy profits. The company was eventually sold to UBS. He is currently writing a book about the crisis in the rational expectations approach to behavioral economics, and he suggests that the way forward is complexity economics, composed of those common "rules of thumb" or heuristics uncovered in psychological experiments and sociological studies of humanity. An example heuristic from chess, for example, is "a queen is better than a rook."

Farmer is now Baillie Gifford Professor of Mathematics at Oxford University. He completed his undergraduate degree in Physics at Stanford University and his graduate studies at the University of California, Santa Cruz, under George Blumenthal. He is an Oppenheimer Fellow and cofounder of the journal *Quantitative Finance*. Farmer grew up in Silver City, New Mexico, where he enjoyed adventures inspired by his Scoutmaster Tom Ingerson, a physicist who had the boys searching for abandoned Spanish goldmines and planning a mission to Mars. He attributes his ongoing love of scientific investigation to those early adventures.

*Philip L. Frana*

*See also:* Newell, Allen.

**Further Reading**

Bass, Thomas A. 1985. *The Eudaemonic Pie.* Boston: Houghton Mifflin Harcourt.

Bass, Thomas A. 1998. *The Predictors: How a Band of Maverick Physicists Used Chaos Theory to Trade Their Way to a Fortune on Wall Street.* New York: Henry Holt.

Brockman, John, ed. 2005. *Curious Minds: How a Child Becomes a Scientist.* New York: Vintage Books.

Freedman, David H. 1994. *Brainmakers: How Scientists Are Moving Beyond Computers to Create a Rival to the Human Brain.* New York: Simon & Schuster.

Waldrop, M. Mitchell. 1992. *Complexity: The Emerging Science at the Edge of Order and Chaos.* New York: Simon & Schuster.

# Foerst, Anne (1966–)

Anne Foerst is a Lutheran minister, theologian, author, and computer science professor at St. Bonaventure University in Allegany, NY. Foerst received her Doctorate in Theology from the Ruhr-University of Bochum, Germany, in 1996. She has been a research associate at Harvard Divinity School, a project director at the Massachusetts Institute of Technology (MIT), and a research scientist at the MIT Artificial Intelligence Laboratory. At MIT, she directed the God and Computers Project, which facilitated conversations about existential issues raised in scientific investigations. Foerst is the author of many scholarly and popular articles, exploring the need for enhanced dialogue between theology and science and evolving ideas of personhood in light of robotics research. Her 2004 book, *God in the Machine*, discusses her work as a theological advisor to the Cog and Kismet robotics teams at MIT.

Some of the formative influences on Foerst's research include time spent working as a hospital counselor, her years of gathering anthropological data at MIT, and the works of German-American Lutheran philosopher and theologian Paul Tillich. As a hospital counselor, she began to reconsider the meaning of "normal" human existence. The observable differences in physical and mental capacities in patients prompted Foerst to examine the conditions under which humans are considered to be people. Foerst distinguishes between the categories of "human" and "person" in her work, where human describes the members of our biological species and person describes a being who has received a kind of retractable social inclusion.

Foerst points to the prominent example of the Holocaust to illustrate the way in which personhood must be granted but can be taken back. This renders personhood an ever-vulnerable status. This schematic for personhood—something people grant to each other—allows Foerst to consider the possible inclusion of robots as persons. Her work on robots as potential persons extends Tillich's writings on sin, estrangement, and relationality to the relationships between humans and robots and between robots and other robots. Tillich argues that people become estranged when they deny the competing polarities in their lives, such as the desire for safety and the desire for novelty or freedom. When people fail to acknowledge and engage with these competing drives and cut out or neglect one side in order to focus totally on the other side, they deny reality, which is inherently ambiguous. Failure to embrace the complex tensions of life alienates people from their lives, from the people around them, and (for Tillich) from God. Research into AI thus presents polarities of danger and opportunity: the danger of reducing all things to objects or data that can be quantified and analyzed and the opportunity for expanding people's ability to form relationships and bestow personhood.

Following Tillich's model, Foerst has worked to build a dialog between theology and other formalized areas of research. Though generally well received in laboratory and teaching environments, Foerst's work has encountered some skepticism and resistance in the form of anxieties that she is importing counter-factual ideas into the province of science.

For Foerst, these fears are useful data, as she advocates for a mutualistic approach where AI researchers and theologians acknowledge deeply held biases about the world and the human condition in order to have productive exchanges. In her work, Foerst argues that many important insights emerge from these exchanges, so long as the participants have the humility to acknowledge that neither party possesses a complete understanding of the world and human life. Humility is a key feature of Foerst's work on AI, as she argues that in attempting to recreate human thought, function, and form in the figure of the robot, researchers are struck by the enormous complexity of the human being. Adding to the complexity of any given individual is the way in which humans are socially embedded, socially conditioned, and socially responsible. The embedded complexity of human beings is inherently physical, leading Foerst to emphasize the importance of an embodied approach to AI.

While at MIT, Foerst pursued this embodied method, where possession of a physical body capable of interaction is central to robotic research and development. In her work, Foerst makes a strong distinction between robots and computers when discussing the development of artificial intelligence (AI). Robots have bodies, and those bodies are an essential part of their capacities to learn and interact. Very powerful computers may perform remarkable analytic tasks and participate in some methods of communication, but they lack bodies to learn through experience and relate with others. Foerst is critical of research predicated on the idea that intelligent machines can be produced by recreating the human brain. Instead, she argues that bodies are an essential component of intelligence. Foerst advocates for the raising up of robots in a manner analogous to that of human child rearing, where robots are provided with the means to experience and learn from the world. As with human children, this process is expensive and time-consuming, and Foerst reports that, especially since the terrorist attacks of September 11, 2001, funding that supported creative and time-intensive AI research has disappeared, replaced by results-driven and military-focused research that justifies itself through immediate applications.

Foerst draws on a wide range of materials for her work, including theological texts, popular movies and television shows, science fiction, and examples from the fields of philosophy and computer science. Foerst identifies loneliness as a major motivation for the human pursuit of artificial life. Feelings of estrangement, which Foerst links to the theological status of a lost relationship with God, drive both fictional imaginings of the creation of a mechanical companion species and contemporary robotics and AI research.

Foerst's academic critics within religious studies argue that she has reproduced a model first advanced by German theologian and scholar Rudolph Otto in *The Idea of the Holy* (1917). According to Otto, the experience of the divine is found in a moment of attraction and terror, which he called the numinous. Foerst's critics argue that she has applied this model when she argued that in the figure of the robot, we experience attraction and terror.

*Jacob Aaron Boss*

*See also:* Embodiment, AI and; Nonhuman Rights and Personhood; Pathetic Fallacy; Robot Ethics; Spiritual Robots.

### Further Reading

Foerst, Anne. 2005. *God in the Machine: What Robots Teach Us About Humanity and God*. New York: Plume.

Geraci, Robert M. 2007. "Robots and the Sacred in Science Fiction: Theological Implications of Artificial Intelligence." *Zygon* 42, no. 4 (December): 961–80.

Gerhart, Mary, and Allan Melvin Russell. 2004. "Cog Is to Us as We Are to God: A Response to Anne Foerst." *Zygon* 33, no. 2: 263–69.

*Groks Science Radio Show and Podcast* with guest Anne Foerst. Audio available online at http://ia800303.us.archive.org/3/items/groks146/Groks122204_vbr.mp3. Transcript available at https://grokscience.wordpress.com/transcripts/anne-foerst/.

Reich, Helmut K. 2004. "Cog and God: A Response to Anne Foerst." *Zygon* 33, no. 2: 255–62.

# Ford, Martin (Active 2009–Present)

Martin Ford is an author and futurist who writes on the subjects of artificial intelligence, automation, and the future of work. His 2015 book *Rise of the Robots* is a *Financial Times* and McKinsey Business Book of the Year and *New York Times* bestseller.

Ford has called artificial intelligence the "next killer app" of the American economy. In his writings, Ford emphasizes that most economic sectors of the American economy are becoming ever-more automated. Autonomous cars and trucks are upending the transportation industry. Self-checkout is changing the nature of retailing. Food preparation robots are transforming the hospitality industry. Each of these trends, he says, will have profound impacts on the American workforce. Robots will not only disrupt the work of blue-collar laborers but will also threaten white-collar workers and professionals in areas like medicine, journalism, and finance. Most of this work, Ford insists, is also routine and susceptible to computerization. Middle management in particular is at risk. In the future, Ford argues, there will be no relationship between human education and training and vulnerability to automation, just as worker productivity and compensation have already become disconnected phenomena. Artificial intelligence will transform knowledge and information work as powerful algorithms, machine-learning tools, and smart virtual assistants are introduced into operating systems, enterprise software, and databases.

Ford's position has been bolstered by a 2013 study by Carl Benedikt Frey and Michael Osborne of the Oxford University Martin Programme on the Impacts of Future Technology and the Oxford University Engineering Sciences Department. Frey and Osborne's research, accomplished with the help of machine-learning algorithms, revealed that nearly half of 702 different kinds of American jobs could be automated in the next ten to twenty years. Ford points out that when automation precipitates primary job losses in areas susceptible to computerization, it will also cause a secondary wave of job destruction in sectors that are sustained by them, even if they are themselves automation resistant.

Ford suggests that capitalism will not go away in the process, but it will need to adapt if it is to survive. Job losses will not be immediately staunched by new technology jobs in the highly automated future. Ford has advocated a universal basic income—or "citizens dividend"—as one way to help American workers transition to the economy of the future. Without consumers making wages, he asserts, there simply won't be markets for the abundant goods and services that robots will produce. And those displaced workers would no longer have access to home ownership or a college education. A universal basic income could be guaranteed by placing value added taxes on automated industries. The wealthy owners in these industries would agree to this tax out of necessity and survival. Further financial incentives, he argues, should be targeted at individuals who are working to enhance human culture, values, and wisdom, engaged in earning new credentials or innovating outside the mainstream automated economy.

Political and sociocultural changes will be necessary as well. Automation and artificial intelligence, he says, have exacerbated economic inequality and given extraordinary power to special interest groups in places like the Silicon Valley. He also suggests that Americans will need to rethink the purpose of employment as they are automated out of jobs. Work, Ford believes, will not primarily be about earning a living, but rather about finding purpose and meaning and community. Education will also need to change. As the number of high-skill jobs is depleted, fewer and fewer highly educated students will find work after graduation.

Ford has been criticized for assuming that hardly any job will remain untouched by computerization and robotics. It may be that some occupational categories are particularly resistant to automation, for instance, the visual and performing arts, counseling psychology, politics and governance, and teaching. It may also be the case that human energies currently focused on manufacture and service will be replaced by work pursuits related to entrepreneurship, creativity, research, and innovation. Ford speculates that it will not be possible for all of the employed Americans in the manufacturing and service economy to retool and move to what is likely to be a smaller, shallower pool of jobs.

In *The Lights in the Tunnel: Automation, Accelerating Technology, and the Economy of the Future* (2009), Ford introduced the metaphor of "lights in a tunnel" to describe consumer purchasing power in the mass market. A billion individual consumers are represented as points of light that vary in intensity corresponding to purchasing power. An overwhelming number of lights are of middle intensity, corresponding to the middle classes around the world. Companies form the tunnel. Five billion other people, mostly poor, exist outside the tunnel. In Ford's view, automation technologies threaten to dim the lights and collapse the tunnel. Automation poses dangers to markets, manufacturing, capitalist economics, and national security.

In *Rise of the Robots: Technology and the Threat of a Jobless Future* (2015), Ford focused on the differences between the current wave of automation and prior waves. He also commented on disruptive effects of information technology in higher education, white-collar jobs, and the health-care industry. He made a case for a new economic paradigm grounded in the basic income, incentive structures for risk-taking, and environmental sensitivity, and he described scenarios where inaction might lead to economic catastrophe or techno-feudalism. Ford's book *Architects of Intelligence: The Truth about AI from the People Building It* (2018) includes interviews and conversations with two dozen leading artificial intelligence researchers and entrepreneurs. The focus of the book is the future of artificial general intelligence and predictions about how and when human-level machine intelligence will be achieved.

Ford holds an undergraduate degree in Computer Engineering from the University of Michigan. He earned an MBA from the UCLA Anderson School of Management. He is the founder and chief executive officer of the software development company Solution-Soft located in Santa Clara, California.

*Philip L. Frana*

*See also:* Brynjolfsson, Erik; Workplace Automation.

### Further Reading

Ford, Martin. 2009. *The Lights in the Tunnel: Automation, Accelerating Technology, and the Economy of the Future*. Charleston, SC: Acculant.

Ford, Martin. 2013. "Could Artificial Intelligence Create an Unemployment Crisis?" *Communications of the ACM* 56 7 (July): 37–39.

Ford, Martin. 2016. *Rise of the Robots: Technology and the Threat of a Jobless Future*. New York: Basic Books.

Ford, Martin. 2018. *Architects of Intelligence: The Truth about AI from the People Building It*. Birmingham, UK: Packt Publishing.

# Frame Problem, The

John McCarthy and Patrick Hayes discovered the frame problem in 1969. The problem concerns representing the effects of actions in logic-based artificial intelligence. Formal logic is used to define facts about the world, such as a car can be started when the key is placed in the ignition and turned and that pressing the accelerator causes it to move forward. However, the latter fact does not explicitly state that the car remains on after pressing the accelerator. To correct this, the fact must be expanded to "pressing the accelerator moves the car forward and does not turn it off." However, this fact must be augmented further to describe many other scenarios (e.g., that the driver also remains in the vehicle). The frame problem highlights an issue in logic involving the construction of facts that do not require enumerating thousands of trivial effects.

After its discovery by artificial intelligence researchers, the frame problem was picked up by philosophers. Their interpretation of the problem might be better called the world update problem as it concerns updating frames of reference. For example, how do you know your dog (or other pet) is where you last saw them without seeing them again? Thus, in a philosophic sense, the frame problem concerns how well a person's understanding of their surroundings matches reality and when should their notion of their surroundings change. Intelligent agents will need to address this problem as they plan actions in progressively more complex worlds.

Numerous solutions have been proposed to solve the logic version of the frame problem. However, the philosophic problem is an open issue. Both need to be solved for artificial intelligence to exhibit intelligent behavior.

*David M. Schwartz and Frank E. Ritter*

*See also:* McCarthy, John.

**Further Reading**

McCarthy, John, and Patrick J. Hayes. 1969. "Some Philosophical Problems from the Standpoint of Artificial Intelligence." In *Machine Intelligence*, vol. 4, edited by Donald Michie and Bernard Meltzer, 463–502. Edinburgh, UK: Edinburgh University Press.

Shanahan, Murray. 1997. *Solving the Frame Problem: A Mathematical Investigation of the Common Sense Law of Inertia*. Cambridge, MA: MIT Press.

Shanahan, Murray. 2016. "The Frame Problem." In *The Stanford Encyclopedia of Philosophy*, edited by Edward N. Zalta. https://plato.stanford.edu/entries/frame-problem.

# G

## Gender and AI

Contemporary society tends to think of artificial intelligence and robots as sexless and genderless, but this is not true. Instead, humans encode gender and stereotypes into artificial intelligence systems in a manner not dissimilar to the way gender weaves its way into language and culture. There is gender bias in the data used to train artificial intelligences. Data that is biased can introduce huge discrepancies into machine predictions and decisions. In humans, these discrepancies would be called discriminatory.

AIs are only as good as the humans creating data that is harvested by machine learning systems, and only as ethical as the programmers making and monitoring them. When people express gender bias, machines assume this is normal (if not acceptable) human behavior. Bias can show up whether someone is using numbers, text, images, or voice recordings to train machines. The use of statistical models to analyze and classify enormous collections of data to make predictions is called machine learning. The use of neural network architectures that are thought to mimic human brainpower is called deep learning. Classifiers label data based on past patterns. Classifiers are extremely powerful. They can accurately predict income levels and political leanings of neighborhoods and towns by analyzing data on cars that are visible using Google Street View.

Gender bias is found in the language people use. This bias is found in the names of things and in the way things are ordered by importance. Descriptions of men and women are biased—beginning with the frequency with which their respective titles are used and they are referred to as men and women versus boys and girls. Even the metaphors and adjectives used are biased. Biased AI can affect whether people of certain genders or races are targeted for certain jobs or not, whether medical diagnoses are accurate, whether they obtain loans—and even the way tests are scored. AI systems tend to associate "woman" and "girl" with the arts rather than with mathematics. Google's AI algorithms to search for job candidates have been determined to contain similar biases. Algorithms used by Facebook and Microsoft have consistently associated images of cooking and shopping with women's activity and sports and hunting with male activity. Researchers have found places where these gender biases are deliberately engineered into AI systems. On job sites, for example, men are offered the opportunity to apply for highly paid and sought-after jobs more frequently than women.

Digital assistants in smart phones are more often given female names like Siri, Alexa, and Cortana. The designer of Alexa says the name emerged from discussions with Amazon CEO Jeff Bezos, who wanted a virtual assistant to have the personality and gender of the Enterprise starship computer on the television show *Star Trek,* that is, a woman. The leader of the Cortana effort, Deborah Harrison, says that their female voice emerged from research suggesting that people responded better to female voices. But when BMW launched its in-car GPS route planner with a female voice it immediately received negative feedback from men who didn't want their cars to tell them what to do. The company learned that female voices need to sound empathetic and trustworthy, but not authoritative.

The artificial intelligence company Affectiva uses images of six million people's faces as training data to try and understand their inner emotional states. The company is now working with auto manufacturers to use real-time video of drivers and to detect which ones are tired or angry. The car would suggest that these drivers pull over and rest. But the company has also detected that women seem to "laugh more" than men, and this is complicating attempts to properly assess emotional states of the average driver.

The same biases may be found in hardware. Computer engineers—still usually male—create a disproportionate number of female robots. NASA's Valkyrie robot used in Shuttle missions has breasts. Jia, a surprisingly human-looking robot designed at the University of Science and Technology of China, has long wavy dark hair, pink lips and cheeks, and pale skin. When first spoken to, she keeps her eyes and head tilted down, as if in deference. Slender and busty, she wears a fitted gold gown. In greeting, she asks, "Yes, my lord, what can I do for you?" When offered to take a picture, Jia responds: "Don't come too close to me when you are taking a picture. It will make my face look fat."

This bias toward female robots is especially pronounced in popular culture. The film *Austin Powers* (1997) had fembots that shot bullets from their breast cups—weaponizing female sexuality. Most music videos that feature robots will feature female robots. The first song available for download on the internet was Duran Duran's "Electric Barbarella." The archetypical, white-sheathed robot found illustrated today in so many places has its origin in Bjork's video "The Girl And The Robot." Marina and the Diamonds protestation that "I Am Not a Robot" draws a quick response from Hoodie Allen that "You Are Not a Robot." The Broken Bells' "The Ghost Inside" finds a female android sacrificing plastic body parts to pay tolls and regain paradise. Lenny Kravitz's "Black Velveteen" has titanium skin. Hatsune Miku and Kagamine Rin are holographic vocaloid performers—and anime-inspired women. The great exception is Daft Punk, where robot costumes cloak the true identities of the male musicians.

Acknowledged masterpieces such as *Metropolis* (1927), *The Stepford Wives* (1975), *Blade Runner* (1982), *Ex Machina* (2014), and *Her* (2013), and the television shows *Battlestar Galactica* and *Westworld* have sexy robots as the protagonists' primary love interest. Meanwhile, lethal autonomous weapons systems—"killer robots"—are hypermasculine. The Defense Advanced Research Projects Agency (DARPA) has created hardened military robots with names such as Atlas, Helios, and Titan. Driverless cars are given names such as Achilles, Black Knight, Overlord, and Thor PRO. The most famous autonomous vehicle of all time, the HAL 9000 computer embedded in the spaceship Discovery in *2001: A Space Odyssey* (1968), is male and positively murderous.

The gender divide in artificial intelligence is pronounced. In 2017, Fei-Fei Li, the director of the Stanford Artificial Intelligence Lab, admitted that she had a workforce composed mainly of "guys with hoodies" (Hempel 2017). Only about 12 percent of the researchers presenting at leading AI conferences are women (Simonite 2018b). Women receive 19 percent of bachelor's degrees and 22 percent of doctoral degrees in computer and information sciences (NCIS 2018). The proportion of bachelor's degrees in computer science earned by women has dropped from a high of 37 percent in 1984 (Simonite 2018a). This is even though the first "computers"—as the film *Hidden Figures* (2016) highlighted—were women.

Among philosophers, there is still debate about whether un-situated, gender-neutral knowledge can truly exist in human society. Even after Google and Apple released unsexed digital assistants, users projected gender preferences on them. Centuries of expert knowledge was created by white men and later released into digital worlds. Will it be possible for machines to create and use rules based on unbiased knowledge for centuries more? In other words, does scientific knowledge have a gender? And is it male? Alison Adam is a Science and Technology Studies scholar who is interested not in the gender of the individuals involved, but in the gender of the ideas they produced.

The British company Sage recently hired a "conversation manager" who was tasked with creating a digital assistant—ultimately named "Pegg"—that presented a gender-neutral personality. The company has also codified "five core principles" into an "ethics of code" document to guide its programmers. Sage's CEO Kriti Sharma says that by 2020 "we'll spend more time talking to machines than our own families," so it's important to get technology right. Microsoft recently created an internal ethics panel called Aether, for AI and Ethics in Engineering and Research. Gender Swap is an experiment that uses a VR system as a platform for embodiment experience—a neuroscience technique in which users can feel themselves as if they were in a different body. In order to create the brain illusion, a set of human partners use the immersive Head Mounted Display Oculus Rift and first-person cameras. To create this perception, both users synchronize their movements. If one does not correspond to the movement of the other, the embodiment experience does not work. It means that both users must agree on every movement they make together.

New sources of algorithmic gender bias are found regularly. In 2018, MIT computer science graduate student Joy Buolamwini revealed gender and racial bias in the way AI recognized subjects' faces. Working with other researchers, she discovered that the dermatologist-approved Fitzpatrick Skin Type classification system datasets were overwhelmingly composed of lighter-skinned subjects (up to 86 percent). The researchers created a skin type system based on a rebalanced dataset and used it to evaluate three off-the-shelf gender classification systems. They found that in all three commercial systems darker-skinned females are the most misclassified. Buolamwini is the founder of the Algorithmic Justice League, an organization challenging bias in decision-making software.

*Philip L. Frana*

*See also:* Algorithmic Bias and Error; Explainable AI.

**Further Reading**

Buolamwini, Joy and Timnit Gebru. 2018. "Gender Shades: Intersectional Accuracy Disparities in Commercial Gender Classification." *Proceedings of Machine Learning Research: Conference on Fairness, Accountability, and Transparency* 81: 1–15.

Hempel, Jessi. 2017. "Melinda Gates and Fei-Fei Li Want to Liberate AI from 'Guys With Hoodies.'" *Wired*, May 4, 2017. https://www.wired.com/2017/05/melinda-gates -and-fei-fei-li-want-to-liberate-ai-from-guys-with-hoodies/.

Leavy, Susan. 2018. "Gender Bias in Artificial Intelligence: The Need for Diversity and Gender Theory in Machine Learning." In *GE '18: Proceedings of the 1st International Workshop on Gender Equality in Software Engineering*, 14–16. New York: Association for Computing Machinery.

National Center for Education Statistics (NCIS). 2018. *Digest of Education Statistics.* https://nces.ed.gov/programs/digest/d18/tables/dt18_325.35.asp.

Roff, Heather M. 2016. "Gendering a Warbot: Gender, Sex, and the Implications for the Future of War." *International Feminist Journal of Politics* 18, no. 1: 1–18.

Simonite, Tom. 2018a. "AI Is the Future—But Where Are the Women?" *Wired*, August 17, 2018. https://www.wired.com/story/artificial-intelligence-researchers-gender -imbalance/.

Simonite, Tom. 2018b. "AI Researchers Fight Over Four Letters: NIPS." *Wired*, October 26, 2018. https://www.wired.com/story/ai-researchers-fight-over-four-letters-nips/.

Søraa, Roger Andre. 2017. "Mechanical Genders: How Do Humans Gender Robots?" *Gender, Technology, and Development* 21, no. 1–2: 99–115.

Wosk, Julie. 2015. *My Fair Ladies: Female Robots, Androids, and Other Artificial Eves.* New Brunswick, NJ: Rutgers University Press.

## General and Narrow AI

Artificial intelligence may be divided into two main categories: General (or strong or full) and Narrow (or weak or specialized). General AI, the kind most often encountered in science fiction, does not yet exist in the real world. Machines with general intelligence would be capable of performing any human intellectual task. This type of system would also appear to think across topics, make connections, and express creative ideas in the way that humans do, thereby demonstrating the capacity for abstract thought and problem-solving. Such a machine would be able to demonstrate abilities such as reasoning, planning, and learning from the past.

While the goal of General AI remains unrealized, there are ever-increasing examples of Narrow AI. These are systems that achieve human-level (or even superhuman) performance on specifically defined tasks. Computers that have learned to play complex games exhibit skills, strategies, and behaviors different from, and in some cases exceeding, the most accomplished human players. AI systems have also been developed that can translate between languages in real time, interpret and respond in natural speech (both spoken and written), and accomplish image recognition (being able to recognize, identify, and sort photos or images based on the content).

The ability to generalize knowledge or skills, however, is so far still pretty much only a human achievement. Nevertheless, an enormous amount of research in General AI is currently underway. Determining when a machine achieves human-level intelligence will be difficult to ascertain. Various tests, both serious and humorous, have been proposed to confirm when a machine reaches the level of General AI.

The Turing Test is probably the most famous of these tests. Here a machine and a human converse, unseen, while a second human listens in. The eavesdropping human must determine which speaker is a machine and which is the human. The machine passes the test if it can fool the human evaluator a prescribed percentage of the time. Another more fanciful test is the Coffee Test: A machine enters an average home and makes coffee. It must locate the coffee machine, find the coffee, add water, brew the coffee, and pour it into a mug. Another is the Flat Pack Furniture Test: A machine receives, unpacks, and assembles an item of furniture using only the instructions provided.

Some scientists, and many science fiction writers and fans, believe that once smart machines reach some tipping point, they will be able to improve themselves, perhaps exponentially. A possible outcome are AI-based entities that far surpass the abilities of humankind. The moment where AI takes over its own self-improvement is referred to as the Singularity or artificial superintelligence (ASI). If ASI is attained, it may impact human civilizations in ways that are entirely unpredictable. Some commentators fear ASI could be detrimental to the safety or dignity of humankind. Whether the Singularity will ever occur, and how dangerous it might be, is a matter of extensive debate.

Narrow AI applications continue to expand in use throughout the world. Most new applications are based on machine learning (ML), and most of the AI examples cited in the news are related to this subset of technologies. Machine learning programs are different from traditional or ordinary algorithms. A computer programmer explicitly writes code to account for every action of an algorithm in programs that cannot learn. The programmer lays out rules for all choices made along the way. This requires that the programmer envision and code for every scenario an algorithm might encounter. Program code of this sort is unwieldy and often insufficient, particularly if the code is repeatedly rewritten to account for new or unforeseen situations. For scenarios in which the rules for optimal decisions might not be clear, or which are difficult for a human programmer to anticipate, the usefulness of hard-coded algorithms reaches its limit.

Machine learning involves teaching a computer to identify and recognize patterns by example, rather than by predetermined rules. According to Google engineer Jason Mayes, this is accomplished by first evaluating extremely large sets of training data or engaging in some other programmed learning stage. By processing the test data, new patterns can be extracted. Next, previously unseen data can be classified based on the patterns that the system has already identified. Machine learning enables an algorithm to autonomously identify patterns or identify rules underlying decision-making processes. Machine learning also enables a system to improve its output with continued experience (Mayes 2017).

A human programmer is still an important factor in this learning process, influencing outcomes through decisions such as designing the specific learning algorithm, selecting the training data, and picking other design features and settings. Once operational, machine learning is powerful precisely because it can adapt and improve its ability to classify new data without direct human intervention. In other words, the quality of output improves with its own experience.

Artificial intelligence is generally used as an umbrella term for the science of making machines smart. In scientific circles, AI is a computer system that can gather data and use that data to make decisions or solve problems. Another common scientific way of describing AI is that it is a software program combined with hardware that can collect (or sense) inputs from the environment around it, evaluate and analyze those inputs, and make outputs and recommendations—all without the intervention of a human.

When programmers say an AI system can learn, it means the program is designed to edit its own processes to achieve better, and more reliable, outputs or predictions. Today, AI-based systems are being designed and used in almost every possible context, within industries from agriculture to space exploration, and in applications from law enforcement to online banking. Computer science tools and techniques are constantly changing, expanding, and improving. There are many other terms related to machine learning, such as reinforcement learning and neural networks, that contribute to cutting-edge artificial intelligence systems.

*Brenda Leong*

*See also:* Embodiment, AI and; Superintelligence; Turing, Alan; Turing Test.

**Further Reading**

Kelnar, David. 2016. "The Fourth Industrial Revolution: A Primer on Artificial Intelligence (AI)." *Medium*, December 2, 2016. https://medium.com/mmc-writes/the -fourth-industrial-revolution-a-primer-on-artificial-intelligence-ai-ff5e7fffcae1.

Kurzweil, Ray. 2005. *The Singularity Is Near: When Humans Transcend Biology.* New York: Viking.

Mayes, Jason. 2017. *Machine Learning 101.* https://docs.google.com/presentation/d/1kSu QyW5DTnkVaZEjGYCkfOxvzCqGEFzWBy4e9Uedd9k/htmlpresent.

Müller, Vincent C., and Nick Bostrom. 2016. "Future Progress in Artificial Intelligence: A Survey of Expert Opinion." In *Fundamental Issues of Artificial Intelligence*, edited by Vincent C. Müller, 553–71. New York: Springer.

Russell, Stuart, and Peter Norvig. 2003. *Artificial Intelligence: A Modern Approach.* Englewood Cliffs, NJ: Prentice Hall.

Samuel, Arthur L. 1988. "Some Studies in Machine Learning Using the Game of Checkers I." In *Computer Games I*, 335–65. New York: Springer.

## General Problem Solver

General Problem Solver is a program for a problem-solving process that uses means-ends analysis and planning to arrive at a solution. The program was designed so that the problem-solving process is distinct from knowledge specific to the problem to be solved, which allows it to be used for a variety of problems.

The program, initially written by Allen Newell and Herbert Simon in 1957, continued in development for almost a decade. The last version was written by Newell's graduate student George W. Ernst in conjunction with research for his 1966 dissertation.

General Problem Solver grew out of Newell and Simon's work on another problem-solving program, the Logic Theorist. After developing Logic Theorist, the pair compared its problem-solving process with that used by humans solving similar problems. They found that Logic Theorist's process differed considerably from that used by humans. Hoping their work in artificial intelligence would contribute to an understanding of human cognitive processes, Newell and Simon used the information about human problem solving gleaned from these studies to develop General Problem Solver. They found that human problem-solvers could look at the desired end and, reasoning both backward and forward, determine steps they could take that would bring them closer to that end, thus developing a solution. Newell and Simon incorporated this process into the General Problem Solver, which they believed was not only representative of artificial intelligence but also a theory of human cognition. General Problem Solver used two heuristic techniques to solve problems: means-ends analysis and planning.

An everyday example of means-ends analysis in action might be stated this way: If a person wanted a particular book, their desired state is to possess the book. In their current state, the book is not in their possession, but rather it is held by the library. The person has options to eliminate the difference between their current state and their desired state. In order to do so, they can check the book out from the library, and they have options to get to the library, such as driving. However, if the book has been checked out by another patron, there are options available to obtain the book elsewhere. The person may go to a bookstore or go online to purchase it. The person must then examine the options available to them to do so. And so on. The person knows of several relevant actions they can take, and if they choose appropriate actions and apply them in an appropriate order, they will obtain the book. The person choosing and applying appropriate actions is means-ends analysis in action.

In applying means-ends analysis to General Problem Solver, the programmer sets up the problem as an initial state and a state to be reached. General Problem Solver calculates the difference between these two states (called objects). General Problem Solver must also be programmed with operators, which reduce the difference between the two states. To solve the problem, it chooses and applies an operator and determines whether the operation has indeed brought it closer to its goal or desired state. If so, it proceeds by choosing another operator. If not, it can backtrack and try another operator. Operators are applied until the difference between the initial state and the desired state has been reduced to zero.

General Problem Solver also possessed the ability to plan. By eliminating the details of the operators and of the difference between the initial state and the desired state, General Problem Solver could sketch a solution to the problem. Once a general solution was outlined, the details could be reinserted into the problem and the subproblems constituted by these details solved within the solution guidelines established during the outlining stage.

For programmers, defining a problem and operators in order to program General Problem Solver was a very labor-intensive process. It also meant that General Problem Solver as a theory of human cognition or an example of artificial intelligence took for granted the very actions that, in part, represent intelligence, that is, the acts of defining a problem and choosing relevant actions (or operations) from the infinite number of possible actions, in order to solve the problem.

Ernst carried out further work with General Problem Solver in the mid-1960s. He was not interested in human problem-solving processes, but rather in finding a way to extend the generality of General Problem Solver, allowing it to unravel problems outside of the domain of logic. In his version of General Problem Solver, the desired state or object was no longer specified exactly, but was described as a list of constraints. Ernst also changed the form of the operators so that an object or state resulting from the application of an operator (the operator's output) could be represented as a function of the initial state or object (the input). His modified General Problem Solver achieved only moderate success in problem-solving. It often ran out of memory even on simple problems. In the preface to their 1969 work *GPS: A Case Study in Generality and Problem Solving*, Ernst and Newell declared, "We do feel that this particular aggregate of IPL-V code should be laid to rest, as having done its part in advancing our understanding of the mechanisms of intelligence" (Ernst and Newell 1969, vii).

*Juliet Burba*

*See also:* Expert Systems; Simon, Herbert A.

### Further Reading

Barr, Avron, and Edward Feigenbaum, eds. 1981. *The Handbook of Artificial Intelligence*, vol. 1, 113–18. Stanford, CA: HeurisTech Press.

Ernst, George W., and Allen Newell. 1969. *GPS: A Case Study in Generality and Problem Solving*. New York: Academic Press.

Newell, Allen, J. C. Shaw, and Herbert A. Simon. 1960. "Report on a General Problem-Solving Program." In *Proceedings of the International Conference on Information Processing (June 15–20, 1959)*, 256–64. Paris: UNESCO.

Simon, Herbert A. 1991. *Models of My Life*. New York: Basic Books.

Simon, Herbert A., and Allen Newell. 1961. "Computer Simulation of Human Thinking and Problem Solving." *Datamation* 7, part 1 (June): 18–20, and part 2 (July): 35–37.

## Generative Design

Generative design is a broad term that refers to any iterative rule-based process used to create numerous options that satisfy a specified set of goals and constraints. The output from such a process could range from complex architectural models to pieces of visual art and is therefore applicable to a variety of fields: architecture, art, engineering, product design, to name a few.

Generative design differs from a more traditional design strategy, where comparatively few alternatives are evaluated before one is developed into a final product. The rationale behind using a generative design framework is that the final goal is not necessarily known at the outset of a project. Hence, the focus should not be on producing a single correct answer to a problem, but rather on creating numerous viable options that all fit within the specified criteria.

Leveraging the processing power of a computer allows several permutations of a solution to be rapidly generated and evaluated, beyond what a human could accomplish alone. The designer/user tunes input parameters to refine the solution space as objectives and overall vision are clarified over time. This avoids the problem of becoming constrained to a single solution early in the design process, and it instead allows for creative exploration of a wide range of options. The hope is that this will improve the chances of arriving at an outcome that best satisfies the established design criteria.

It is important to note that generative design does not necessarily need to involve a digital process; an iterative procedure could be developed in an analogue framework. But since the processing power (i.e., the number and speed of calculations) of a computer is far superior to that of a human, generative design methods are often considered synonymous with digital techniques. Digital applications, in particular artificial intelligence-based techniques, are being applied to the generative process. Two such artificial intelligence applications are generative art and computational design in architecture.

Generative art, also known as computer art, refers to artwork that has been created in part with the use of some autonomous digital system. Decisions that would typically have been made by a human artist are allocated either fully or in part to an algorithmic process. The artist instead usually retains some control of the process by defining the inputs and rule sets to be followed.

Three people are generally credited as founders of visual computer art: Georg Nees, Frieder Nake, and A. Michael Noll. They are sometimes referred together as the "3N" group of computer pioneers. Georg Nees is often cited for the establishment of the first generative art exhibit, called Computer Graphic, which was held in Stuttgart in 1965. Exhibits by Nake (in collaboration with Nees) and Noll followed in the same year in Stuttgart and New York City, respectively (Boden and Edmonds 2009).

These early examples of generative art in the visual medium are pioneering in their use of computers to create works of art. They were also limited by the computational methods available at the time. In the modern context, the existence of AI-based technology coupled with exponential increases in computational power has led to new types of generative art. An interesting class of these new works falls under the category of computational creativity, which is defined as "a field of artificial intelligence focused on developing agents that generate creative products autonomously" (Davis et al. 2016). When applied to generative art, the goal of computational creativity is to harness the creative potential of a computer through techniques involving machine learning. In this way, the process of creation moves away from prescribing step-by-step instructions to a computer (i.e., what was used in the early days) to more abstract processes where the outcomes are not easily predicted.

A recent example of computational creativity is the DeepDream computer vision program created by Google engineer Alexander Mordvintsev in 2015. The project uses a convolutional neural network to purposely over-process an image. This brings forward patterns that correspond to how a certain layer in the network understands an input image, based on what types of images it has been trained to identify. The results are hallucinogenic reinterpretations of the original image, and similar to what one might experience in a restless dream. Mordvintsev shows how a neural network trained on a set of animals will take images of clouds and turn them into rough representations of animals that correspond to the detected features. Using a different training set would lead to the network turning features such as horizon lines and tall vertical shapes into loose representations of towers and buildings. As such, these new images can be considered original works of art that are unpredictable and created completely from the computer's own creative process stemming from a neural network.

My Artificial Muse is another recent example of computational creativity. Unlike DeepDream, which relies solely on the neural network to produce art, Artificial Muse explores how an AI-based process can collaborate with a human to inspire them in the creation of new paintings (Barqué-Duran et al. 2018). The neural network is trained on a large database of human poses that are harvested from existing images and represented as stick figures. Then a completely new pose is generated from this data and fed back into the algorithm, which now reconstructs what it thinks a painting based on this pose should look like. Therefore, the new pose can be thought of as a muse for the algorithm, which inspires it to create a completely original painting, which is then realized by the artist.

Computers were first introduced into the field of architecture through two-dimensional computer-aided drafting (CAD) programs, which were used to directly replicate the task of hand drawing. Although creating drawings with a computer was still a manual process, it was considered an improvement over the analogue form, as it allowed for better precision and repeatability. These simple CAD programs were eventually surpassed by more complex parametric design software taking a more programmatic approach to the creation of an architectural model (i.e., geometry is created through user-specified variables). The most common platform for this type of work today is the visual programming environment of Grasshopper (a plugin to the three-dimensional computer-aided design software Rhino), developed by David Rutten in 2007 while at Robert McNeel & Associates.

To illustrate: take a very simple geometric task, like defining a rectangle. In a parametric modeling approach, the length and width values would be established as user-controlled parameters. The resulting design (i.e., the rectangle drawing) would be updated automatically by the software according to the values set for the parameters. Now imagine this applied on a larger scale where complex collections of geometric representations (e.g., curves, surfaces, planes, etc.) are all linked through a set of parameters. This results in the ability to control the outcome of a complex geometric design by altering simple user-specified parameters. An added benefit is generating output that a designer might not typically have envisioned as parameters interact in unexpected ways.

Although parametric design approaches rely on a computer to generate and visualize complex outcomes, the process is still manual. A human is required to specify and control a set of parameters. Generative design methods assign further agency to the computer or algorithm performing the design calculations. Neural networks can be trained on examples of designs that satisfy the overall goals of a project and then process new input data to generate numerous design suggestions.

The layout of the new Autodesk office in the MaRS Innovation District in Toronto is a recent example of generative design being applied in an architectural context (Autodesk 2016). In this project, the existing employees were surveyed, and information was gathered on six measurable goals: work style preference, adjacency preference, level of distraction, interconnectivity, daylight, and views to the outside. The generative design algorithm considered all of these requirements and produced multiple office configurations that maximize the established criteria. These results were evaluated, and the ones that scored the highest were used as the basis for the new office layout. In this way, a large amount of input in the form of previous projects and user-specified data was used to generate a final optimized design. The relationships in the data would have been too complex for a human to synthesize and could only be adequately explored through a generative design approach.

Generative design approaches have proven successful in a wide range of applications where a designer is interested in exploring a large solution space. It avoids the problem of focusing on a single solution too early in the design process, and instead, it allows for creative explorations of a wide range of options. Generative design will find new applications as AI-based computational methods continue to improve.

*Edvard P. G. Bruun and Sigrid Adriaenssens*

*See also:* Computational Creativity.

## Further Reading

Autodesk. 2016. "Autodesk @ MaRS." Autodesk Research. https://www.autodeskresearch .com/projects/autodesk-mars.

Barqué-Duran, Albert, Mario Klingemann, and Marc Marzenit. 2018. "My Artificial Muse." https://albertbarque.com/myartificialmuse.

Boden, Margaret A., and Ernest A. Edmonds. 2009. "What Is Generative Art?" *Digital Creativity* 20, no. 1–2: 21–46.

Davis, Nicholas, Chih-Pin Hsiao, Kunwar Yashraj Singh, Lisa Li, and Brian Magerko. 2016. "Empirically Studying Participatory Sense-Making in Abstract Drawing with a Co-Creative Cognitive Agent." In *Proceedings of the 21st International Conference on Intelligent User Interfaces—IUI '16*, 196–207. Sonoma, CA: ACM Press.

Menges, Achim, and Sean Ahlquist, eds. 2011. *Computational Design Thinking: Computation Design Thinking*. Chichester, UK: J. Wiley & Sons.

Mordvintsev, Alexander, Christopher Olah, and Mike Tyka. 2015. "Inceptionism: Going Deeper into Neural Networks." Google Research Blog. https://web.archive.org /web/20150708233542/http://googleresearch.blogspot.com/2015/06/inceptionism -going-deeper-into-neural.html.

Nagy, Danil, and Lorenzo Villaggi. 2017. "Generative Design for Architectural Space Planning." https://www.autodesk.com/autodesk-university/article/Generative-Design -Architectural-Space-Planning-2019.

Picon, Antoine. 2010. *Digital Culture in Architecture: An Introduction for the Design Professions*. Basel, Switzerland: Birkhäuser Architecture.

Rutten, David. 2007. "Grasshopper: Algorithmic Modeling for Rhino." https://www.grasshopper3d.com/.

## Generative Music and Algorithmic Composition

Algorithmic composition is a composer's method for creating new musical material by following a predetermined finite set of rules or processes. The algorithm might also be a list of instructions, in lieu of standard musical notation, articulated by the composer that the musician is to follow during a performance. One school of thought suggests that algorithmic composition should involve minimal human interference. AI systems in music include those based on generative grammar, knowledge-based systems, genetic algorithms, and more recently, artificial neural networks trained with deep learning.

The use of algorithms to aid in the creation of music is by no means a new practice. Early instances can be found in thousand-year-old music theory treatises. These treatises collated lists of rules and conventions in common practice, which composers followed in observance of correct musical writing. One early example of algorithmic composition is Johann Joseph Fux's *Gradus ad Parnassum* (1725), which details the strict rules governing species counterpoint. While intended pedagogically, species counterpoint introduced five methods of writing complementary musical harmony lines against the main or fixed melody. If followed in its entirety, Fux's method offers little freedom from the given rules.

Early examples of algorithmic composition with reduced human involvement often involved chance. Chance music, commonly referred to as aleatoric music, has roots in the Renaissance. Perhaps the most famous early instance of the technique is attributed to Mozart. A published manuscript from 1787 attributed to Mozart incorporates the use of "Musikalisches Würfelspiel" (musical dice game) into the composition. The performer must roll the dice to randomly select one-bar sections of precomposed music (out of a possible 176) in order to piece together a 16-bar waltz.

The American composer John Cage greatly expanded on these early aleatoric practices by creating a piece where the majority of the composition was determined by chance. The musical dice game example only permits chance to control the order of short precomposed musical excerpts, but in his *Music of Changes* (1951) chance controls virtually all decisions. Cage used the ancient Chinese divination text called *I Ching* (The Book of Changes) to make all decisions on musical parameters: such as the notes themselves, rhythms, and dynamics. His friend David Tudor, the performer of the work, had to turn his very explicit and complex score into something closer to standard notation for playability reasons. This demonstrates two forms of aleatoric music: one in which the composer creates a fixed score with the aid of chance processes and an open or mobile form, where the order of the musical elements is left up to the performer or chance.

Arnold Schoenberg developed a twelve-tone method of algorithmic composition that has close ties to areas of mathematics such as combinatorics and group theory. Twelve-tone composition is an early example of serialism where each of the twelve tones of conventional western music is given equal importance within a piece. After arranging each tone into a desired row with no repeating pitches, that row is continuously rotated by one until a 12 × 12 matrix is created. The matrix comprises all possible variations of the original tone row that the composer is limited to for pitch material. When the aggregate—that is, all the pitches of one row—has been implemented into the score, a new row can be used. The rows can be further divided into subsets in order to build harmonic content (a vertical setting of notes) rather than using it to write melodic lines (horizontal setting). Later composers such as Pierre Boulez and Karlheinz Stockhausen experimented with serializing other musical parameters by creating matrices involving dynamics or timbre.

Some algorithmic compositional techniques were developed as a rejection or modification of existing techniques set forth by serialist composers. Iannis Xenakis reasoned that serialist composers were too focused on harmony as a series of intertwining linear objects (the setting of linear tone-rows) and that the processes were lost on the listener, as they became too complex. He proposed novel methods of modifying nonmusical algorithms for music composition that could operate on dense sound masses. Xenakis believed the method freed music from its linear obsessions. He was inspired by scientific studies of natural and social phenomena such as the moving particles of a cloud or a crowd of thousands gathered at a political event and concentrated on the use of probability theory and stochastic processes in his compositions. Markov chains, for example, allowed Xenakis to slowly evolve thick-textured sound masses over time by controlling musical parameters such as pitch, timbre and dynamics. In a Markov chain, the probability of the next occurring event is partially controlled by preceding events; thus his use of algorithms combined indeterminate elements like those of Cage's chance music with deterministic ones as in serialism. He referred to this music as stochastic music. It inspired new generations of composers to feature increasingly complex algorithms in their music. These composers eventually required the use of computers to make their calculations. Xenakis himself was an early pioneer in the use of computers in music, using them to help calculate the results of his stochastic and probabilistic processes.

Brian Eno, building on composer Erik Satie's idea of background music involving live performers (called furniture music), popularized the idea of ambient music with his *Ambient 1: Music for Airports* (1978). Seven tape recorders, each of which contained a separate pitch, were all set at different lengths. The pitches occurred in a different order with each succeeding loop, leading to an ever-changing melody. Because the inputs are the same, the piece always unfolds in exactly the same way each time it is played. In 1995, Eno coined the term *generative music* to describe systems that create ever-changing music by altering parameters over time. Both ambient and generative music are precedents for autonomous computer-based algorithmic composition, much of which today incorporates techniques drawn from artificial intelligence.

Generative grammar is a system of rules designed to describe natural languages, developed by Noam Chomsky and his colleagues. The rules rewrite hierarchically organized elements to describe a space of possible serial orderings of elements. Adapted to algorithmic composition, generative grammars can be used to output musical passages. David Cope's *Experiments in Musical Intelligence* (1996) provides perhaps the most well-known use of generative grammar. Cope trained his software to convincingly write music in the style of many different composers such as Bach, Mozart, and Chopin. In knowledge-based systems, information regarding the type of music the composer wishes to imitate is encoded as a database of facts that can be drawn upon to create an artificial expert to assist the composer.

Genetic algorithms mimic the process of biological evolution and provide another method for composition. A population of randomly created compositions is tested for closeness to the desired musical output. Then artificial mechanisms modeled on those in nature are used to increase the probability that musically desirable traits will increase in subsequent generations. The composer interacts with the system, allowing both computer and observer to stimulate new ideas.

More recent methods of AI-generated composition feature deep learning systems such as generative adversarial networks or GANs. Generative adversarial networks in music pit a generator—which creates new music based on knowledge of a compositional style—against a discriminator, which attempts to distinguish between the output of the generator and that of a human composer. Each time the generator fails, it receives new information until the discriminator cannot tell the difference between real and generated musical material. The repurposing of nonmusical algorithms for musical purposes increasingly drives music in new and exciting directions.

*Jeffrey Andrew Thorne Lupker*

*See also:* Computational Creativity.

**Further Reading**

Cope, David. 1996. *Experiments in Musical Intelligence*. Madison, WI: A-R Editions.

Eigenfeldt, Arne. 2011. "Towards a Generative Electronica: A Progress Report." *eContact!* 14, no. 4: n.p. https://econtact.ca/14_4/index.html.

Eno, Brian. 1996. "Evolving Metaphors, in My Opinion, Is What Artists Do." *In Motion Magazine*, June 8, 1996. https://inmotionmagazine.com/eno1.html.

Nierhaus, Gerhard. 2009. *Algorithmic Composition: Paradigms of Automated Music Generation*. New York: Springer.

Parviainen, Tero. "How Generative Music Works: A Perspective." http://teropa.info/loop/#/title.

# Giant Brains

From the late 1940s to the mid-1960s, the Harvard-trained computer scientist Edmund Callis Berkeley shaped the American public's perceptions of what computers were and what role they might play in society. Computers in his view were "giant mechanical brains" or enormous, automatic, information-processing, thinking machines to be used for the good of society. Berkeley promoted early, peaceful, and commercial computer developments through the Association for Computing Machinery (cofounded in 1947), his company Berkeley Associates (established in 1948), his book *Giant Brains* (1949), and the magazine *Computers and Automation* (established in 1951).

In his popular book, *Giant Brains, or Machines that Think*, Berkeley defined computers as giant mechanical brains for their powerful, automatic, cognitive, information-processing features. Berkeley thought of computers as machines that operated automatically, on their own without human intervention. One only had to push the start button, and "the machine starts whirring and it prints out the answers as it obtains them" (Berkeley 1949, 5). Computers also had cognitive functions precisely because they processed information. Berkeley perceived human thought as essentially "a process of storing information and then referring to it, by a process of learning and remembering" (Berkeley 1949, 2). A computer could think in the same manner; it "transfers information automatically from one part of the machine to another, [with] a flexible control over the sequence of its operations" (Berkeley 1949, 5). He added the adjective giant to emphasize both the processing power and the physical size of the first computers. In 1946, the first electronic general-purpose digital computer ENIAC occupied the entire basement of the University of Pennsylvania's Moore School of Electrical Engineering.

Beyond shaping the role of computers in the popular imagination, Berkeley was actively involved in the application of symbolic logic to early computer designs. He had an undergraduate degree in mathematics and logic from Harvard University, and by 1934, he was working in the actuarial department of Prudential Insurance. In 1938, Bell Labs electrical engineer Claude Shannon published his pioneering work on the application of Boolean logic to automatic circuitry design. Berkeley promoted Shannon's findings at Prudential, urging the insurance company to apply logic to its punched card tabulations. In 1941, Berkeley, Shannon, and others formed the New York Symbolic Logic Group to promote logic applications in electronic relay computing. When the United States entered World War II (1939–1945) in 1941, Berkeley enlisted in the US Navy and was eventually assigned to help design the Mark II electromechanical computer in Howard Aiken's Lab at Harvard University.

Based on his experiences with Mark II, Berkeley returned to Prudential, convinced of the commercial future of computing. In 1946, Berkeley used Bell Labs' general-purpose relay calculator to demonstrate that computers could accurately calculate a complex insurance problem, in this case the cost of a change in policy (Yates 2005, 123–24). In 1947, Berkeley met John William Mauchly at the Symposium on Large Scale Digital Calculating Machinery at Harvard. Their meeting culminated in a signed contract between Prudential and John Adam Presper Eckert and Mauchly's Electronic Control Company (ECC) for the development of a general-purpose computer that would benefit insurance calculations. That general-purpose machine ultimately became the UNIVAC (in 1951). Prudential, however, decided not to use UNIVAC but to return to IBM's tabulating technology. UNIVAC first commercial contract was not in insurance, but in General Electric's payroll calculations (Yates 2005, 124–27).

Berkeley left Prudential in 1948 to create Berkeley Associates, which became Berkeley Enterprises in 1954. From that point on, Berkeley dedicated his career to the promotion of Giant Brains, including the use of symbolic logic in computing technology, and related social activist causes. After *Giant Brains* (1949), Berkeley published *Brainiacs* (1959) and *Symbolic Logic and Intelligent Machines* (1959). He also set up correspondence courses in general knowledge, computers, mathematics, and logic systems and began a first journal for computing professionals called the *Roster of Organizations in the Field of Automatic Computing Machinery*. This journal became the influential monthly *Computers and Automation* (1951–1973).

*Elisabeth Van Meer*

*See also:* Cognitive Computing; Symbolic Logic.

**Further Reading**

Berkeley, Edmund C. 1949. *Giant Brains, or Machines That Think.* London: John Wiley & Sons.

Berkeley, Edmund C. 1959a. *Brainiacs: 201 Electronic Brain Machines and How to Make Them.* Newtonville, MA: Berkeley Enterprises.

Berkeley, Edmund C. 1959b. *Symbolic Logic and Intelligent Machines.* New York: Reinhold.

Longo, Bernadette. 2015. *Edmund Berkeley and the Social Responsibility of Computer Professionals.* New York: Association for Computing Machinery.

Yates, JoAnne. 2005. *Structuring the Information Age: Life Insurance and Technology in the Twentieth Century.* Baltimore: Johns Hopkins University Press.

# Goertzel, Ben (1966–)

Ben Goertzel is chief executive officer and chief scientist of blockchain AI company SingularityNET, chairman of Novamente LLC, research professor in the Fujian Key Lab for Brain-Like Intelligent Systems at Xiamen University, chief scientist of the Shenzhen, China bioinformatics firm Mozi Health and of Hanson Robotics, and chair of the OpenCog Foundation, Humanity+, and Artificial General Intelligence Society conference series. Goertzel has long been interested in creating a benevolent artificial general intelligence and applying it to bioinformatics, finance, gaming, and robotics. He has argued that though AI is a fashionable phenomenon these days, it is now better than experts in several areas. Goertzel divides progress in AI into three phases that represent stepping-stones to a global brain (Goertzel 2002, 2):

- computer and communication technologies as enhancers of human interactions
- the intelligent Internet
- the full-on Singularity

In 2019, Goertzel gave a talk at TEDxBerkeley under the title "Decentralized AI: The Power and the Necessity." In the talk, he analyzes artificial intelligence in both its current incarnation and describes its future. He stresses "the importance of decentralized control in guiding AI to the next levels, to the power of decentralized

AI" (Goertzel 2019a). Goertzel delineates three types of AI in its development: artificial narrow intelligence, artificial general intelligence, and artificial superintelligence. Artificial narrow intelligence is machine intelligence that can "solve very particular problems . . . better than human beings" (Goertzel 2019a). This is the kind of AI that has surpassed a human in some narrow tasks such as chess and Go. The term Narrow AI is used by American futurologist and inventor Ray Kurzweil. Artificial general intelligence (AGI) involves smart machines that "can generalize knowledge" in various areas and has "humanlike autonomy." Goertzel claims that this sort of AI will be of the same level of intelligence as humans by 2029. Artificial superintelligence (ASI) builds upon narrow and general AI, but it's also capable of self-reprogramming. This is the sort of AI, he says, that will be more intelligent than the best human brains in terms of "scientific creativity, general wisdom and social skills" by 2045 (Goertzel 2019a). Goertzel notes that Facebook, Google, and several universities and corporations are actively working toward AGI. Goertzel suggests that the transition from AI to AGI will happen within the next five to thirty years.

Goertzel is also interested in life extension with the help of AI technology. He believes that the exponential advance of artificial intelligence will deliver technologies that also extend life span and human health, perhaps indefinitely. He notes that a singularity involving a radical expansion in "human health span" may occur by 2045 (Goertzel 2012). The Singularity was popularized by Vernor Vinge in his 1993 essay "The Coming Technological Singularity." The term was adopted by Ray Kurzweil in 2005 in his book *The Singularity is Near.* For both authors, the Technological Singularity involves the merging of machine and human intelligence following a rapid increase in new technologies, in particular, robotics and AI.

Goertzel welcomes the idea of a coming singularity. His main current project is SingularityNET, which involves the creation of a global network of artificial intelligence investigators who are interested in creating, sharing, and monetizing AI technology, software, and services. Goertzel has contributed greatly to this effort by creating a decentralized protocol that powers a full stack AI solution. As a decentralized marketplace, SingularityNET offers various AI tools: text generation, AI Opinion, iAnswer, Emotion Recognition, Market Trends, OpenCog Pattern Miner, and its own cryptocurrency called AGI token. Domino's Pizza in Malaysia and Singapore is currently partnering with SingularityNET (Khan 2019). Domino's is interested in developing a marketing strategy using SingularityNET tools, hoping to offer the best goods and services to their customers using unique algorithms. Domino's believes that the AGI ecosystem can be incorporated into their operations, and they can deliver value and service in the food delivery industry.

Goertzel has responded to the challenge presented by physicist Stephen Hawking, who believed that AI could bring about the end of human civilization. Goertzel notes that, given the present circumstances, artificial super intelligence's mind state will be based on the previous AI generations, thus "selling, spying killing and gambling are the primary goals and values in the mind of the first super intelligence" (Goertzel 2019b). He accepts that if humans want compassionate AI, then humans will need to first treat one another better.

Goertzel worked for Hanson Robotics in Hong Kong for four years. He worked with the well-known robots Sophia, Einstein, and Han. These robots, he said, "are great platforms for experimenting with AI algorithms, including cognitive architectures like OpenCog that aim at human-level AI" (Goertzel 2018). Goertzel believes that core human values can be preserved for posterity beyond the point of the Technological Singularity in Sophia-like robot creations. Goertzel has said that decentralized networks such as SingularityNET and OpenCog offer "AIs with human-like values," which will minimize AI risks to humankind (Goertzel 2018).

As human values are complex in nature, Goertzel believes it is inefficient to encode them as a rule list. Goertzel suggests two approaches: brain-computer interfacing (BCI) and emotional interfacing. Under BCI, humans will become "cyborgs, physically linking their brains with computational-intelligence modules, then the machine components of the cyborgs should be able to read the moral-value-evaluation structures of the human mind directly from the biological components of the cyborgs" (Goertzel 2018). Goertzel gives Neuralink by Elon Musk as an example. Goertzel doubts this approach will work because it involves intrusive experiments with human brains and lots of unknowns.

The second approach involves "emotional and spiritual connection between humans and AIs, rather than Ethernet cables or Wifi signals, to connect human and AI minds" (Goertzel 2018). He suggests under this approach that AIs should engage in emotional and social interaction with a human by way of facial emotion recognition and mirroring, eye contact, and voice-based emotion recognition to practice human values. To this end, Goertzel launched the "Loving AI" research project with SingularityNET, Hanson AI, and Lia Inc. Loving AI looks to help artificial intelligences converse and develop personal relationships with human beings. The Loving AI site currently hosts a humorous video of actor Will Smith on a date with Sophia the Robot. The video of the date reveals that Sophia is already capable of sixty facial expressions and can interpret human language and emotions. According to Goertzel, humanoid robots like Sophia—when connected to a platform like SingularityNET—gain "ethical insights and advances . . . via language" (Goertzel 2018). From there, robots and AIs can share what they've learned via a common online "mindcloud."

Goertzel is also chair of the Conference Series on Artificial General Intelligence, held annually held since 2008 and organized by the Artificial General Intelligence Society. The society publishes a peer-reviewed open-access academic serial, the *Journal of Artificial General Intelligence*. The proceedings of the conference series are edited by Goertzel.

*Victoriya Larchenko*

*See also:* General and Narrow AI; Superintelligence; Technological Singularity.

**Further Reading**

Goertzel, Ben. 2002. *Creating Internet Intelligence: Wild Computing, Distributed Digital Consciousness, and the Emerging Global Brain.* New York: Springer.

Goertzel, Ben. 2012. "Radically Expanding the Human Health Span." TEDxHKUST. https://www.youtube.com/watch?v=IMUbRPvcB54.

Goertzel, Ben. 2017. "Sophia and SingularityNET: Q&A." *H+ Magazine*, November 5, 2017. https://hplusmagazine.com/2017/11/05/sophia-singularitynet-qa/.

Goertzel, Ben. 2018. "Emotionally Savvy Robots: Key to a Human-Friendly Singularity." https://www.hansonrobotics.com/emotionally-savvy-robots-key-to-a-human -friendly-singularity/.

Goertzel, Ben. 2019a. "Decentralized AI: The Power and the Necessity." TEDxBerkeley, March 9, 2019. https://www.youtube.com/watch?v=r4manxX5U-0.

Goertzel, Ben. 2019b. "Will Artificial Intelligence Kill Us?" July 31, 2019. https://www .youtube.com/watch?v=TDClKEORtko.

Goertzel, Ben, and Stephan Vladimir Bugaj. 2006. *The Path to Posthumanity: 21st-Century Technology and Its Radical Implications for Mind, Society, and Reality.* Bethesda, MD: Academica Press.

Khan, Arif. 2019. "SingularityNET and Domino's Pizza Announce a Strategic Partnership." https://blog.singularitynet.io/singularitynet-and-dominos-pizza-announce -a-strategic-partnership-cbbe21f80fc7.

Vinge, Vernor. 1993. "The Coming Technological Singularity: How to Survive in the Post-Human Era." In *Vision 21: Interdisciplinary Science and Engineering in the Era of Cyberspace*, 11–22. NASA: Lewis Research Center.

# Group Symbol Associator

In the early 1950s, Firmin Nash, director of the South West London Mass X-Ray Service, invented a slide rule-like device called a Group Symbol Associator, which allowed a physician to correlate a patient's symptoms against 337 predefined symptom-disease complexes and to make a diagnosis. It uses multi-key look-up from inverted files to resolve cognitive processes in automated medical decision-making.

Derek Robinson, a professor in the Integrated Media Programme of the Ontario College of Art & Design, has called the Group Symbol Associator a "cardboard brain." The device is not unlike the inverted biblical concordance compiled by Hugo De Santo Caro, a Dominican friar who completed his index in 1247. In the 1980s, Marsden Blois, a professor of artificial intelligence in medicine at the University of California, San Francisco, recreated the Nash device in software. According to his own tests, Blois's diagnostic assistant RECONSIDER, based on the Group Symbol Associator, performed as well or better than other expert systems in existence.

The Group Symbol Associator, which Nash nicknamed the "Logoscope," used propositional calculus to parse various combinations of medical symptoms. The Group Symbol Associator is one of the earliest attempts to apply digital computing, in this case by modifying an analog tool, to problems of diagnosis. Disease groups culled from standard textbooks on differential diagnosis are marked along the edge of Nash's cardboard rule. Each patient property or symptom also has a specific cardboard symptom stick with lines opposing the positions of those diseases having that property. The Group Symbol Associator contained a total of 82 sign and symptom sticks. Sticks corresponding to the patient's condition are selected and inserted into the rule. Diseases matched to larger numbers of symptom lines are considered possible diagnoses.

Nash's slide rule is essentially a matrix where columns represent diseases and rows represent properties. A mark (such as an "X") is entered into the matrix wherever properties are expected in each disease. Rows detailing symptoms that the patient does not exhibit are eliminated. Columns showing a mark in every cell reveal the most likely or "best match" diagnosis. Viewed this way, as a matrix, the Nash device reconstructs information in much the same way peek-a-boo card retrieval systems used to manage stores of knowledge in the 1940s. The Group Symbol Associator may be compared to Leo J. Brannick's analog computer for medical diagnosis, Martin Lipkin and James Hardy's McBee punch card system for diagnosing hematological diseases, Keeve Brodman's Cornell Medical Index-Health Questionnaire, Vladimir K. Zworykin's symptom spectra analog computer, and other so-called peek-a-boo card systems and devices. The problem worked on by these devices is finding or mapping diseases that are appropriate to the combinations of standardized properties or attributes (signs, symptoms, laboratory results, etc.) exhibited by the patient.

Nash claimed to have reduced the physician's memory of thousands of pages of traditional diagnostic tables to a small machine slightly less than a yard in length. Nash argued that his Group Symbol Associator followed what he called the law of the mechanical conservation of experience. He wrote, "If our books and our brains are approaching relative inadequacy, will man crack under the very weight of the riches of experience he has to carry and pass on to the next generation? I think not. We shed the physical load onto power machines and laborsaving devices. We must now inaugurate the era of the thought-saving devices" (Nash 1960b, 240).

Nash's device did more than augment the physician's memory. The machine, he claimed, actually participated in the logical analysis of the diagnostic process. "The Group Symbol Associator makes visible not only the end results of differential diagnostic classificatory thinking, it displays the skeleton of the whole process as a simultaneous panorama of spectral patterns that coincide with varying degrees of completeness," Nash noted. "It makes a map or pattern of the problem composed for each diagnostic occasion, and acts as a physical jig to guide the thought process" (Paycha 1959, 661). Patent application for the device was made to the Patent Office in London on October 14, 1953. Nash gave the first public demonstration of the Group Symbol Associator at the 1958 Mechanisation of Thought Processes Conference at National Physical Laboratory (NPL) in the Teddington area of London. The 1958 NPL conference is noteworthy as only the second conference to be convened on the subject of artificial intelligence.

The Mark III Model of the Group Symbol Associator became available commercially in the late 1950s. Nash hoped that physicians would carry Mark III with them when they were away from their offices and books. Nash explained, "The GSA is small, inexpensive to make, transport, and distribute. It is easy to operate, and it requires no servicing. The individual, even in outposts, ships, etc., can have one" (Nash 1960b, 241). Nash also published examples of paper-based "logoscopic photograms" done with xerography (dry photocopying) that achieved the same results as his hardware device. The Group Symbol Associator was manufactured in quantity by Medical Data Systems of Nottingham, England. Most of the Mark V devices were distributed in Japan by Yamanouchi Pharmaceutical Company.

Nash's chief antagonist, the French ophthalmologist François Paycha, outlined the practical limitations of Nash's Group Symbol Associator in 1959. He noted that such a device would become extremely unwieldy in the identification of corneal diseases, where there are about 1,000 differentiable illnesses and 2,000 separate signs and symptoms. In 1975, R. W. Pain of the Royal Adelaide Hospital in South Australia reviewed the instrument and found it accurate in only a quarter of cases.

*Philip L. Frana*

*See also:* Computer-Assisted Diagnosis.

**Further Reading**

Eden, Murray. 1960. "Recapitulation of Conference." *IRE Transactions on Medical Electronics* ME-7, no. 4 (October): 232–38.

Nash, F. A. 1954. "Differential Diagnosis: An Apparatus to Assist the Logical Faculties." *Lancet* 1, no. 6817 (April 24): 874–75.

Nash, F. A. 1960a. "Diagnostic Reasoning and the Logoscope." *Lancet* 2, no. 7166 (December 31): 1442–46.

Nash, F. A. 1960b. "The Mechanical Conservation of Experience, Especially in Medicine." *IRE Transactions on Medical Electronics* ME-7, no. 4 (October): 240–43.

Pain, R. W. 1975. "Limitations of the Nash Logoscope or Diagnostic Slide Rule." *Medical Journal of Australia* 2, no. 18: 714–15.

Paycha, François. 1959. "Medical Diagnosis and Cybernetics." In *Mechanisation of Thought Processes*, vol. 2, 635–67. London: Her Majesty's Stationery Office.

# H

## Hassabis, Demis (1976–)

Demis Hassabis is a computer game programmer, cognitive scientist, and artificial intelligence expert living in the United Kingdom. He is the cofounder of DeepMind, which produced the deep learning system AlphaGo.

Hassabis is famed as an expert player of games. His interest in games set the stage for his success as an artificial intelligence researcher and computer games entrepreneur. Hassabis's parents recognized his skill at chess at a very early age. He reached the rank of chess master at age thirteen. He is also a World Team Champion player of the strategy board game *Diplomacy,* a Main Event player at the World Series of Poker, and a World Pentamind Champion and World Decamentathlon Champion at the London Mind Sports Olympiad multiple times over.

At age seventeen, Hassabis joined celebrated game designer Peter Molyneux at Bullfrog Games, based in Guildford, England. Bullfrog was known for making a number of popular "god games" for personal computers. A god game is an artificial life game that puts the player in the position of control and influence over semiautonomous characters in a wide-ranging environment. The first god game is widely considered to be Molyneux's *Populous,* released in 1989. At Bullfrog, Hassabis codesigned and programmed the simulation-management game *Theme Park,* which was released in 1994. Hassabis left Bullfrog Games to attend Queens' College Cambridge. He completed a Computer Science degree in 1997.

After graduation, Hassabis rejoined Molyneux at a new game company called Lionhead Studios. Hassabis served briefly as lead programmer on the artificial intelligence for the game *Black & White,* another god game where the player ruled over a virtual island populated by several tribes. Within a year, Hassabis had left Lionhead to start his own computer game company Elixir Studios. Hassabis inked deals with a number of prominent publishers, including Microsoft and Vivendi Universal. Elixir developed a number of games before its closure in 2005, including the diplomatic strategy simulation game *Republic: The Revolution* and the real-time strategy title *Evil Genius.* The artificial intelligence in *Republic* is modeled after the 1960 book *Crowds and Power* by Elias Canetti, which addresses questions about how and why crowds obey the power of rulers (which Hassabis boiled down to force, money, and influence). *Republic* required the programming effort of twenty-five people laboring daily over four years. Hassabis hoped the game's AI could become useful to academic researchers.

Turning his attention away from game design, Hassabis pursued advanced study at University College London (UCL). He was awarded his PhD in Cognitive Neuroscience in 2009. Hassabis discovered linkages between loss of memory and impaired imagination in his studies of patients with damage to the brain's hippocampus. These observations suggested that the brain's memory systems may splice together remembered bits of past events in order to envisage hypothetical futures. For two additional years, Hassabis continued to pursue fundamental academic research in the Gatsby Computational Neuroscience Unit at UCL and as a Wellcome Trust fellow. He also served as a visiting researcher at MIT and Harvard. Hassabis's research in cognitive science informed later work on key challenges in artificial intelligence related to unsupervised learning, memory and one-shot learning, and imagination-based planning using generative models.

In 2011, Hassabis cofounded the London-based AI start-up DeepMind Technologies with Shane Legg and Mustafa Suleyman. The company had an interdisciplinary science focus, bringing together leading researchers and ideas from the subdisciplines of machine learning, neuroscience, engineering, and mathematics. DeepMind's goal was to make scientific discoveries in artificial intelligence and invent new artificial general-purpose learning capabilities. Hassabis has referred to the venture as an Apollo Program for AI.

One challenge for DeepMind was to create a machine that could defeat human opponents at the abstract strategy board game Go. Hassabis did not want to create an expert system, a brute force machine preprogrammed with algorithms and heuristics useful for playing Go. He wanted to create a machine that adapted to playing games in ways similar to human chess master Garry Kasparov rather than the chess-playing single-purpose Deep Blue system. He wanted to create a machine capable of learning to cope with new problems and generality, by which he meant operating across a wide range of different tasks.

The AlphaGo artificial intelligence agent that the company created to take on eighteen-times world champion Go player Lee Sedol relied on the framework of reinforcement learning. In reinforcement learning, agents in the environment (in this case, the Go board) are trying to achieve a particular goal (winning the game). The agents have perceptual inputs (like vision input) and a statistical model based on observations of the environment. While it is gathering perceptual data and forming a model of its environment, the agent makes plans and runs through simulations of actions that will change the model in achievement of the goal. The agent is always trying to pick actions that will get it closer to its objective.

Hassabis believes that if all of the problems of goal-oriented agents in a reinforcement learning framework could be solved, this would be enough to deliver on the promise of artificial general intelligence. He argues that biological systems operate much in the same way. In human brains, it is the dopamine system that implements a reinforcement learning framework.

Typically, the game of Go takes a lifetime of study and play to become a master. Go encompasses a huge search space, much larger than chess. There are more possible Go positions on the board than there are atoms in the universe. It is also considered virtually impossible to write an evaluation function that covers an appreciable fraction of those positions in order to decide where on the board the next stone should be laid down. Each game is considered effectively unique, and players who are gifted talk about their choices as driven by intuition rather than calculation.

AlphaGo overcame these challenges by training a neural network using data derived from thousands of strong amateur games completed by human Go players. AlphaGo then played against itself millions of times and made predictions about how likely each side was to win given the current board positions. In this way, no explicit evaluation rules were needed. AlphaGo defeated Go champion Lee Sedol (four games to one) in Seoul, South Korea, in 2006. AlphaGo's style of play is considered conservative. It makes unusual opening placements of stones to maximize winning while minimizing risk or point spread—in this way placing less apparent emphasis on maximizing territorial gains on the board—and tends to favor diagonal placements of stones called "shoulder hits."

AlphaGo has since been generalized as AlphaZero, in order to play any two-person game. AlphaZero learns from scratch with no human training data or sample games. It learns entirely from random play. AlphaZero defeated Stockfish, one of the strongest free and open-source chess engines, after only four hours of training (28 games to 0 with 72 draws). In playing chess, AlphaZero favors mobility of the pieces over their materiality, which (like Go) produces an inventive style of play.

Another challenge undertaken by the company was to create a flexible, adaptive, and robust AI that could learn by itself on how to play more than 50 Atari video games only by watching the pixels and scores present on a video screen. For this challenge, Hassabis proposed deep reinforcement learning, which combines reinforcement learning and deep learning. Deep neural networks require an input layer of observations, weighting systems, and backpropagation to get a neural network capable of consistent perceptual identification. In the case of the Atari challenge, the network was trained from the 20,000-pixel values flashing at any given moment on the videogame screen. Reinforcement learning takes the machine from the point where a given input is perceived and identified under deep learning to the place where it can take meaningful action in achievement of a goal. In the Atari challenge, the machine could take eighteen different specific joystick actions in a particular time-step, while it learned how to win over hundreds of hours of gameplay. In other words, a deep reinforcement learning machine is an end-to-end learning system capable of examining perceptual inputs, making a plan, and executing the plan from scratch.

Google acquired DeepMind in 2014. Hassabis continues to work at Google with deep learning technologies pioneered at DeepMind. One of those efforts makes use of optical coherence tomography scans for eye diseases. DeepMind's AI system can quickly and accurately diagnose from the eye scans by triaging patients and instantly recommending how they should be referred for further treatment. The system AlphaFold—constructed by experts in machine learning, physics, and structural biology—predicts the three-dimensional protein structures solely from its genetic sequence. At the 2018 "world championship" for Critical Assessment of Techniques for Protein Structure Prediction, AlphaFold won the contest by correctly predicting the most accurate structure for 25 of 43 proteins. Most recently, AlphaStar is mastering the real-time strategy game StarCraft II.

*Philip L. Frana*

*See also:* Deep Learning.

**Further Reading**

"Demis Hassabis, Ph.D.: Pioneer of Artificial Intelligence." 2018. Biography and interview. American Academy of Achievement. https://www.achievement.org/achiever /demis-hassabis-ph-d/.

Ford, Martin. 2018. *Architects of Intelligence: The Truth about AI from the People Building It.* Birmingham, UK: Packt Publishing Limited.

Gibney, Elizabeth. 2015. "DeepMind Algorithm Beats People at Classic Video Games." *Nature* 518 (February 26): 465–66.

Gibney, Elizabeth. 2016. "Google AI Algorithm Masters Ancient Game of Go." *Nature* 529 (January 27): 445–46.

Proudfoot, Kevin, Josh Rosen, Gary Krieg, and Greg Kohs. 2017. *AlphaGo.* Roco Films.

# Human Brain Project

The Human Brain Project is a flagship brain research initiative of the European Union. A multidisciplinary consortium of more than one hundred partner institutions and involving experts from the fields of computer science, neuroscience, and robotics, the project involves Big Science—both in terms of the number of the participants and its ambitious goals. The Human Brain Project began in 2013, emerging as an EU Future and Emerging Technologies program with more than one billion euros of funding. The ten-year effort is focused on making fundamental advances in neuroscience, medicine, and computing technology. Human Brain Project researchers want to understand how the human brain works and emulate its computational capabilities.

Human Brain Project research is divided into twelve subprojects, notably including Human Brain Organization, Systems and Cognitive Neuroscience, Theoretical Neuroscience, and implementations such as the Neuroinformatics Platform, Brain Simulation Platform, Medical Informatics Platform, and Neuromorphic Computing Platform. In 2016, the Human Brain Project released six information and communication technology platforms as the main research infrastructure for ongoing brain research. In addition to infrastructure developed for collecting and exchanging data from the scientific community, the project's research is directed to the development of neuromorphic (brain-inspired) computing chips.

BrainScaleS is a subproject based on analog signals and emulations of the neuron and its synapses. SpiNNaker (Spiking Neural Network Architecture) is a supercomputer architecture based upon numerical models running on custom digital multicore chips. Another ambitious subprogram is the Neurorobotic Platform, where "virtual brain models meet real or simulated robot bodies" (Fauteux 2019).

The project's simulation of the human brain, with its 100 billion neurons—each with 7,000 synaptic connections to other neurons—requires enormous computing power. Six supercomputers in research centers across Europe are used to make computer models of the brain. Project researchers are now studying diseases using these models.

The program faces criticism. In a 2014 open letter to the European Commission, scientists complained of problems with transparency and governance of the program and the narrow scope of research in relation to its original plan and objectives. An assessment and review of the funding processes, requirements, and stated objectives of the Human Brain Project has led to a new governance structure for the program.

*Konstantinos Sakalis*

*See also:* Blue Brain Project; Cognitive Computing; SyNAPSE.

**Further Reading**

Amunts, Katrin, Christoph Ebell, Jeff Muller, Martin Telefont, Alois Knoll, and Thomas Lippert. 2016. "The Human Brain Project: Creating a European Research Infrastructure to Decode the Human Brain." *Neuron* 92, no. 3 (November): 574–81.

Fauteux, Christian. 2019. "The Progress and Future of the Human Brain Project." *Scitech Europa*, February 15, 2019. https://www.scitecheuropa.eu/human-brain-project /92951/.

Markram, Henry. 2012. "The Human Brain Project." *Scientific American* 306, no. 6 (June): 50–55.

Markram, Henry, Karlheinz Meier, Thomas Lippert, Sten Grillner, Richard Frackowiak, Stanislas Dehaene, Alois Knoll, Haim Sompolinsky, Kris Verstreken, Javier DeFelipe, Seth Grant, Jean-Pierre Changeux, and Alois Sariam. 2011. "Introducing the Human Brain Project." *Procedia Computer Science* 7: 39–42.

# I

## Intelligent Sensing Agriculture

Technological innovation has historically driven food production, from the Neolithic tools that helped transition humans from hunter gathering to farming, to the British Agricultural Revolution that harnessed the power of the Industrial Revolution to increase yields (Noll 2015). Today agriculture is highly technical, as scientific discoveries continue to be integrated into production systems. Intelligent Sensing Agriculture is one of the most recent integrations in a long history of applying cutting edge technology to the cultivation, processing, and distribution of food products. These technical devices are primarily utilized to meet the twin goals of increasing crop yields, while simultaneously reducing the environmental impacts of agricultural systems.

Intelligent sensors are devices that can perform a number of complex functions as part of their defined tasks. These specific types of sensors should not be confused with "smart" sensors or instrument packages that can record input from the physical environment (Cleaveland 2006). Intelligent sensors are distinct, in that they not only detect various conditions but also respond to these conditions in nuanced ways based on this assessment. "Generally, sensors are devices that measure some physical quantity and convert the result into a signal which can be read by an observer or instrument, but intelligent sensors are also able to process measured values" (Bialas 2010, 822). What makes them "intelligent" is their unique ability to manage their own functions based on external stimulus. They analyze multiple variables (such as light, temperature, and humidity) to extract essential features and then generate intermediate responses to these features (Yamasaki 1996). This functionality is dependent on having the capability for advanced learning, processing information, and adaption all in one integrated package. These instrument packages are used in a wide range of contexts, from aerospace to health care, and these application domains are expanding. While all of these applications are innovative, due to the technology itself, the use of intelligent sensors in agriculture could provide a wide range of societal benefits.

There is currently an urgent need to increase the productivity of agricultural lands already in production. According to the United Nations (2017), the world's population neared 7.6 billion people in 2017. However, most of the world's arable land is already being utilized for food. In the United States, almost half of the country is currently being used to produce agricultural products, and in the United Kingdom, the figure is 40 percent (Thompson 2010). Due to the lack of undeveloped land, agricultural output needs to increase dramatically over the next ten years, while simultaneously environmental impacts must be minimized in order to increase overall sustainability or long-term productivity.

Intelligent sensors help to maximize the use of all available resources, to reduce the costs of agriculture, and to limit the application of hazardous inputs (Pajares 2011). According to Pajares, "when nutrients in the soil, humidity, solar radiation, density of weeds and a broad set of factors and data affecting the production are known, this situation improves and the use of chemical products such as fertilizers, herbicides and other pollutants can be reduced considerably" (Pajares 2011, 8930). Most of the uses of intelligent sensors in this context can be labeled "precision agriculture," which is defined as information-intensive crop management that utilizes technology to observe, respond, and measure key variables. When coupled with computer networks, this information allows for the management of fields remotely. Irrespective of distance, combinations of different types of sensors (such as temperature and image based devices) allow for monitoring and management.

Intelligent sensors collect *in situ* information to help control crop production in key ways. Examples of specific uses include the following: Unmanned Aerial Vehicles (UAV), equipped with a suite of sensors, detect fires (Pajares 2011); LIDAR sensors, combined with GPS, classify trees and estimate forest biomass; and capacitance probes measure soil moisture, while reflectometers discern crop moisture content. Other sensor types have the ability to detect weeds, determine the pH of soil, quantify the metabolism of carbon in peatlands, control irrigation systems, monitor temperatures, and even control equipment, such as sprayers and tractors. Robotics systems, when outfitted with intelligent sensors, could be used to perform many duties that are currently done by farmers. Intelligent sensors are revolutionizing modern farming, and as the technology advances, tasks will be further automated.

However, agricultural technologies also have a history of facing public criticism. One critique of the application of intelligent sensors in agriculture concerns potential social impacts. While these devices increase the efficiency of agricultural systems and reduce environmental impacts, they may also negatively impact rural communities. Since the development of the first plow, technological advances have historically changed the ways that farmers manage their fields and animals. Intelligent sensors may enable tractors, harvesters, and other equipment to be run without human intervention, thus changing the way that food is produced. This could reduce the number of workers needed in the agricultural sector and thus reduce the amount of employment available in rural areas, where agricultural production predominantly occurs.

Additionally, this technology could be prohibitively expensive for farmers, thus increasing the probability that small farms will fail. Such failures are commonly attributed to what is called the "technology treadmill." This term captures the phenomenon where a small number of farmers adopt a new technology and profit, as their production costs are lower than their competition. However, as more producers adopt this technology and prices fall, increased profits are no longer possible. Adopting this new technology becomes necessary to compete in a market where others are utilizing it. Those farmers who are not adopters are then pushed out of production, while farmers who embrace the technology succeed. The adoption of intelligent sensors could contribute to the technology treadmill. Regardless, the sensors have a wide range of social, economic, and ethical impacts that will need to be considered, as the technology develops.

*Samantha Noll*

*See also:* Workplace Automation.

**Further Reading**

Bialas, Andrzej. 2010. "Intelligent Sensors Security." *Sensors* 10, no. 1: 822–59.

Cleaveland, Peter. 2006. "What Is a Smart Sensor?" *Control Engineering*, January 1, 2006. https://www.controleng.com/articles/what-is-a-smart-sensor/.

Noll, Samantha. 2015. "Agricultural Science." In *A Companion to the History of American Science*, edited by Mark Largent and Georgina Montgomery. New York: Wiley-Blackwell.

Pajares, Gonzalo. 2011. "Advances in Sensors Applied to Agriculture and Forestry." *Sensors* 11, no. 9: 8930–32.

Thompson, Paul B. 2009. "Philosophy of Agricultural Technology." In *Philosophy of Technology and Engineering Sciences*, edited by Anthonie Meijers, 1257–73. Handbook of the Philosophy of Science. Amsterdam: North-Holland.

Thompson, Paul B. 2010. *The Agrarian Vision: Sustainability and Environmental Ethics.* Lexington: University Press of Kentucky.

United Nations, Department of Economic and Social Affairs. 2017. *World Population Prospects: The 2017 Revision.* New York: United Nations.

Yamasaki, Hiro. 1996. "What Are the Intelligent Sensors." In *Handbook of Sensors and Actuators*, vol. 3, edited by Hiro Yamasaki, 1–17. Amsterdam: Elsevier Science B.V.

# Intelligent Transportation

Intelligent Transportation involves the application of high technology, artificial intelligence, and control systems to manage roadways, vehicles, and traffic. The concept emerged from traditional American highway engineering disciplines, including motorist routing, intersection control, traffic distribution, and system-wide command and control. Intelligent transportation has important privacy and security implications as it aims to embed surveillance devices in pavements, signaling devices, and individual vehicles in order to reduce congestion and improve safety.

Highway engineers of the 1950s and 1960s often considered themselves "communications engineers," controlling vehicle and roadway interactions and traffic flow with information in the form of signage, signals, and statistics. Computing machinery in these decades was used mainly to simulate intersections and model roadway capacity. One of the earliest uses of computing technology in this regard is S. Y. Wong's Traffic Simulator, which applied the resources of the Institute for Advanced Study (IAS) computer in Princeton, New Jersey, to study traffic engineering. Wong's mid-1950s simulator applied computational techniques first developed to study electrical networks to illustrate road systems, traffic controls, driver behavior, and weather conditions.

A novel early application of information technology to automatically design and map minimum distance routes was Dijkstra's Algorithm, named for computer scientist Edsger Dijkstra. Dijkstra developed his algorithm, which finds the shortest path between a starting location and a destination point on roadmaps, in 1959. Dijsktra's routing algorithm is still important to online mapping tools, and it remains of considerable economic value in transportation management planning.

Throughout the 1960s, other algorithms and automatic devices were developed by the automobile industry for guidance control, traffic signaling, and ramp metering. The public became accustomed to many of these devices: traffic right-of-way signals connected to transistorized fixed-time control boxes, synchronized signals, and traffic-actuated vehicle pressure detectors. Despite this activity, simulation of traffic control systems in experimental laboratories remained an important application of information technology to transportation.

Despite the engineers' efforts to keep America moving, the increasing popularity of automobiles and long-distance driving strained the national highway system in the 1960s, precipitating a "crisis" in the operation of surface transportation networks. By the mid-1970s, engineers began looking at information technology as an alternative to conventional signaling, road widening, and grade separation techniques in reducing congestion and improving safety. Much of this effort was directed at the individual motorist who, it was thought, could use information to make real-time decisions that would make driving more pleasurable. Computing technology—especially as linked to other technologies such as radar, the telephone, and television cameras—promised to make navigation easier and maximize safety with onboard instrument panels and diagnostics, while reducing travel times. As computer chip prices plummeted in the 1980s, automobiling became increasingly informed. High-end car models acquired electronic fuel gauges and oil level indicators, digital speedometer readouts, and other alerts.

By the 1990s, television broadcasts in most states began providing pre-trip travel information and weather reports from data and video gathered automatically from roadside sensing stations and video cameras. These summaries became available at roadside way stations where travelers could receive live text-based weather reports and radar images on computer screens placed in public areas; they could also be received as text on pagers and cell phones.

Few of these technologies had a substantial impact on individual privacy or autonomy. In 1991, however, Congress passed the Intermodal Surface Transportation Efficiency Act (ISTEA or "Ice Tea") coauthored by Secretary of Transportation Norman Mineta, which authorized $660 million for establishing an Intelligent Vehicle Highway System (IVHS) for the country. The act defined several goals for IVHS, including improved safety, reduced congestion, enhanced mobility, energy efficiency, economic productivity, increased use of mass transportation, and environmental remediation. All of these goals would be reached by the appropriate application of information technology to facilitate transportation in aggregate, as well as on a one vehicle at a time basis. Planners used the funds on hundreds of projects to create new infrastructure and opportunities for travel and traffic management, public transportation management, electronic payment of tolls, commercial fleet management, emergency management, and vehicle safety.

While some applications of intelligent transportation technology remained underutilized in the 1990s—as in the matching of carpool riders, for instance— other applications became virtually standard on American roadways: onboard safety monitoring and precrash deployment of airbags in cars, for instance, or automated weigh stations, roadside safety inspections, and satellite Global Positioning System (GPS) tracking for tractor-trailers. By the mid-1990s, private industry had joined government in supplementing many of these services. General Motors included in its automobiles the factory-installed telematic system OnStar, which uses GPS and cell phone transmissions to provide route guidance, dispatch emergency and roadside assistance services, track stolen vehicles, remotely diagnose mechanical problems, and open locked doors. Auto manufacturers also began experimenting with infrared sensors attached to expert systems for automatic collision avoidance and developed technology to allow cars to be "platooned" into large groups of closely spaced vehicles in order to maximize the lane capacities of roadways.

Perhaps the most widespread application of intelligent transportation technology introduced in the 1990s was the electronic toll and traffic management system (ETTM). ETTM allowed drivers to pay their tolls on highways automatically without slowing down by installing a radio transponder on their vehicles. By 1995, ETTM systems were in use in Florida, New York State, New Jersey, Michigan, Illinois, and California. ETTM has since spread to many other states as well as overseas.

Intelligent transportation projects have generated controversy because of their capacity for government intrusiveness. In the mid-1980s, the government of Hong Kong implemented electronic road pricing (ERP) with radar transmitter-receivers activated when vehicles passed through tolled tunnels or highway checkpoints. The billing statements associated with this system provided drivers with a detailed record of where they had been and at what time. Before the British handed Hong Kong over to Chinese control in 1997, the system was shelved over fears of potential human rights abuses.

Conversely, the original goal of political surveillance is sometimes expanded to further transportation objectives. In 1993, for example, the UK government deployed street-based closed-circuit television cameras (CCTV) in a "Ring of Steel" around London's financial district to protect against terror bombings by the Irish Republican Army. Ten years later, in 2003, the company Extreme CCTV expanded surveillance of central London to include multiple infrared illuminators for the "capture" of license plate numbers on cars. Drivers into the congested downtown zone were charged a daily use fee.

Technologies like GPS and electronic payment tollbooth software, for example, may be used to keep track of vehicles by a unique identifier like the Vehicle Identification Number (VIN) or an electronic tag. This creates the potential for continuous monitoring or tracking of driving decisions and the possibility of permanent records of movements. Intelligent transportation surveillance typically yields data indicating individual toll crossing places and times, the car's average speed, and images of all occupants. In the early 2000s, the state departments of transportation in Florida and California used similar data to mail out questionnaires to individual drivers traveling certain roadways. The motor vehicle departments of several states also considered implementing "dual-use" intelligent transportation databanks to provide or sell traffic and driver-related information to law enforcement authorities and marketers.

Today artificial intelligence methods are becoming an important part of intelligent transportation planning, especially as large volumes of data are now being collected from actual driving experiences. They are increasingly used to control vehicles, predict roadway congestion and meter traffic, and lessen accident rates and casualties. Many different AI techniques are already in use in various intelligent transportation applications—both singly and in combination —including artificial neural networks, genetic algorithms, fuzzy logic, and expert systems. These techniques are being used to create new vehicle control systems for autonomous and semiautonomous driving, control automatic braking, and monitoring real-time energy consumption and emissions.

For example, Surtrac is a scalable, adaptive traffic control system developed by Carnegie Mellon University that relies on theoretical modeling and artificial intelligence algorithms. Over the course of a day, traffic on various streets and intersections can vary considerably. Traditional automatic traffic control equipment accommodates to predefined patterns on a fixed schedule or relies on central traffic control observations. Adaptive traffic control allows intersections to communicate with one another and for cars to potentially share their user-programmed travel directions. Vivacity Labs in the United Kingdom uses camera sensors at intersections and AI technology to capture and predict traffic conditions in real time throughout the journey of an individual motorist and make mobility assessments at the city scale for businesses and local government authorities.

Fuel cost and climate change impacts may determine future directions in intelligent transportation research and development. If oil prices are high, policies may favor advanced traveler information systems that alert vehicle owners to optimal routes and departure times and warn of predicted (and expensive) idling or wait times. If traffic becomes more congested, urban jurisdictions may implement more extensive smart city technologies such as real-time alerts on roadway traffic and parking availability, automated incident detection and vehicle recovery, and connected environments that control human-piloted vehicles, autonomous cars, and mass transportation systems. Dynamic cordon pricing, which involves calculating and charging fees to enter or drive in congested areas, is also likely to be deployed in more cities around the world. Artificial intelligence technologies that make congestion pricing possible include vehicle-occupancy detection monitors and vehicle classification detectors.

*Philip L. Frana*

*See also:* Driverless Cars and Trucks; Trolley Problem.

## Further Reading

Alpert, Sheri. 1995. "Privacy and Intelligent Highway: Finding the Right of Way." *Santa Clara Computer and High Technology Law Journal* 11: 97–118.

Blum, A. M. 1970. "A General-Purpose Digital Traffic Simulator." *Simulation* 14, no. 1: 9–25.

Diebold, John. 1995. *Transportation Infostructures: The Development of Intelligent Transportation Systems.* Westport, CT: Greenwood Publishing Group.

Garfinkel, Simson L. 1996. "Why Driver Privacy Must Be a Part of ITS." In *Converging Infrastructures: Intelligent Transportation and the National Information Infrastructure*, edited by Lewis M. Branscomb and James H. Keller, 324–40. Cambridge, MA: MIT Press.

*High-Tech Highways: Intelligent Transportation Systems and Policy.* 1995. Washington, DC: Congressional Budget Office.

Machin, Mirialys, Julio A. Sanguesa, Piedad Garrido, and Francisco J. Martinez. 2018. "On the Use of Artificial Intelligence Techniques in Intelligent Transportation Systems." In *IEEE Wireless Communications and Networking Conference Workshops (WCNCW)*, 332–37. Piscataway, NJ: IEEE.

Rodgers, Lionel M., and Leo G. Sands. 1969. *Automobile Traffic Signal Control Systems.* Philadelphia: Chilton Book Company.

Wong, S. Y. 1956. "Traffic Simulator with a Digital Computer." In *Proceedings of the Western Joint Computer Conference*, 92–94. New York: American Institute of Electrical Engineers.

# Intelligent Tutoring Systems

Intelligent tutoring systems are AI-based instructional systems that adapt instruction based on a range of learner variables, including dynamic measures of students' on-going knowledge growth, their personal interest, motivation to learn, affective states, and aspects of how they self-regulate their learning. Intelligent tutoring systems have been developed for a wide range of task domains, including STEM, computer programming, language, and culture. They support many different forms of instructional activities, including complex problem-solving activities, collaborative learning activities, inquiry learning or other open-ended learning activities, learning through conversations, game-based learning, and working with simulations or virtual reality environments. Intelligent tutoring systems grew out of a research area called AI in Education (AIED). Several intelligent tutoring systems are commercially successful and in widespread use, including MATHia® (formerly Cognitive Tutor), SQL-Tutor, ALEKS, and Reasoning Mind's Genie system.

Six rigorous meta-analyses show that intelligent tutoring systems are often more effective than other forms of instruction. Several factors may account for this effectiveness. First, intelligent tutoring systems provide adaptive support within problems, scaling one-on-one tutoring beyond what classroom teachers can feasibly provide without such support. Second, they provide adaptive problem selection tailored to individual students' knowledge. Third, intelligent tutoring systems are often grounded in cognitive task analysis, cognitive theory, and learning sciences principles. Fourth, the use of intelligent tutoring systems in so-called blended classrooms can produce positive shifts in classroom culture by freeing up the teacher to spend more time working one on one with students. Fifth, increasingly intelligent tutoring systems are iteratively improved based on data, utilizing new techniques from the field of educational data mining. Finally, intelligent tutoring systems often feature Open Learner Models (OLMs), which are visualizations of the system's internal student model. OLMs may help learners productively reflect on their state of learning.

Key intelligent tutoring systems paradigms include model-tracing tutors, constraint-based tutors, example-tracing tutors, and ASSISTments. These paradigms differ in their tutoring behaviors and their underlying representations of domain knowledge, student knowledge, and pedagogical knowledge, and in how they are authored. Intelligent tutoring systems employ a variety of AI techniques for domain reasoning (e.g., generating next steps in a problem, given a student's partial solution), evaluating student solutions and partial solutions, and student modeling (i.e., dynamically estimating and maintaining a range of learner variables). A variety of data mining techniques (including Bayesian models, hidden Markov models, and logistic regression models) are increasingly being used to improve systems' student modeling capabilities. To a lesser degree, machine learning methods are used to develop instructional policies, for example, using reinforcement learning.

Researchers are investigating ideas for the smart classroom of the future that significantly extend what current intelligent tutoring systems can do. In their visions, AI systems often work symbiotically with teachers and students to orchestrate effective learning experiences for all students. Recent research suggests promising approaches that adaptively share regulation of learning processes across students, teachers, and AI systems—rather than designing intelligent tutoring systems to handle all aspects of adaptation, for example—by providing teachers with real-time analytics from an intelligent tutoring system to draw their attention to learners who may need additional support.

*Vincent Aleven and Kenneth Holstein*

*See also:* Natural Language Processing and Speech Understanding; Workplace Automation.

**Further Reading**

Aleven, Vincent, Bruce M. McLaren, Jonathan Sewall, Martin van Velsen, Octav Popescu, Sandra Demi, Michael Ringenberg, and Kenneth R. Koedinger. 2016. "Example-Tracing Tutors: Intelligent Tutor Development for Non-Programmers." *International Journal of Artificial Intelligence in Education* 26, no. 1 (March): 224–69.

Aleven, Vincent, Elizabeth A. McLaughlin, R. Amos Glenn, and Kenneth R. Koedinger. 2017. "Instruction Based on Adaptive Learning Technologies." In *Handbook of Research on Learning and Instruction*, Second edition, edited by Richard E. Mayer and Patricia Alexander, 522–60. New York: Routledge.

du Boulay, Benedict. 2016. "Recent Meta-Reviews and Meta-Analyses of AIED Systems." *International Journal of Artificial Intelligence in Education* 26, no. 1: 536–37.

du Boulay, Benedict. 2019. "Escape from the Skinner Box: The Case for Contemporary Intelligent Learning Environments." *British Journal of Educational Technology*, 50, no. 6: 2902–19.

Heffernan, Neil T., and Cristina Lindquist Heffernan. 2014. "The ASSISTments Ecosystem: Building a Platform that Brings Scientists and Teachers Together for

Minimally Invasive Research on Human Learning and Teaching." *International Journal of Artificial Intelligence in Education* 24, no. 4: 470–97.

Koedinger, Kenneth R., and Albert T. Corbett. 2006. "Cognitive Tutors: Technology Bringing Learning Sciences to the Classroom." In *The Cambridge Handbook of the Learning Sciences*, edited by Robert K. Sawyer, 61–78. New York: Cambridge University Press.

Mitrovic, Antonija. 2012. "Fifteen Years of Constraint-Based Tutors: What We Have Achieved and Where We Are Going." *User Modeling and User-Adapted Interaction* 22, no. 1–2: 39–72.

Nye, Benjamin D., Arthur C. Graesser, and Xiangen Hu. 2014. "AutoTutor and Family: A Review of 17 Years of Natural Language Tutoring." *International Journal of Artificial Intelligence in Education* 24, no. 4: 427–69.

Pane, John F., Beth Ann Griffin, Daniel F. McCaffrey, and Rita Karam. 2014. "Effectiveness of Cognitive Tutor Algebra I at Scale." *Educational Evaluation and Policy Analysis* 36, no. 2: 127–44.

Schofield, Janet W., Rebecca Eurich-Fulcer, and Chen L. Britt. 1994. "Teachers, Computer Tutors, and Teaching: The Artificially Intelligent Tutor as an Agent for Classroom Change." *American Educational Research Journal* 31, no. 3: 579–607.

VanLehn, Kurt. 2016. "Regulative Loops, Step Loops, and Task Loops." *International Journal of Artificial Intelligence in Education* 26, no. 1: 107–12.

## Interaction for Cognitive Agents

The desire to create a unified theory of mind is a basic premise of cognitive science that demonstrates the qualitative shape of human cognition and behavior. Cognitive agents (or cognitive models) are applications of the cognitive theory. These agents are simulations that predict and provide possible explanations of human cognition. Cognitive agents need to perform a real task that engages many aspects of cognition, such as interaction.

Artificial intelligence simulates human intelligence. These simulations are not always complete and interactive. An interactive cognitive agent is intended to cover all of the activities that a user exhibits when working at a modern display-oriented interface. Interactive cognitive agents attempt to think and act like humans and therefore are great candidates to be used as surrogate users.

Interactive cognitive agents provide more complete and accurate simulations than noninteractive cognitive agents, which only display a trace of the mind's cognitive stages. Interactive cognitive agents do this by interacting directly with the screen-as-world. It is now possible for models to interact with uninstrumented interfaces, both on the machine that the model is running on and on remote machines. Improved interaction can not only support a broader range of behavior but also make the model more accurate and represent human behaviors on tasks that include interaction.

An interactive cognitive agent contains three components: a cognitive architecture, knowledge, and a perception-motor module. Cognitive architectures are infrastructures that provide a fixed set of computational mechanisms that represent the fixed mechanisms of cognition used to generate behavior for all tasks. As a fixed set, they provide a way for combining and applying cognitive science theory. When knowledge is added to a cognitive architecture, a cognitive model is created. The perception-motor module controls vision and motor output to interact with the world. Interactive cognitive agents can see the screen, press keys, and move and click the mouse.

Interactive cognitive agents contribute to cognitive science, human-computer interaction, automation (interface engineering), education, and assistive technology through their broad coverage of theory and ability to generate behaviors.

*Farnaz Tehranchi*

*See also:* Cognitive Architectures.

**Further Reading**

Newell, Allen. 1990. *Unified Theories of Cognition.* Cambridge, MA: Harvard University Press.

Ritter, Frank E., Farnaz Tehranchi, and Jacob D. Oury. 2018. "ACT-R: A Cognitive Architecture for Modeling Cognition." *Wiley Interdisciplinary Reviews: Cognitive Science* 10, no. 4: 1–19.

# INTERNIST-I and QMR

INTERNIST-I and QMR (Quick Medical Reference) are related expert systems developed at the University of Pittsburgh School of Medicine in the 1970s. The INTERNIST-I system originally encoded the internal medicine expertise of Jack D. Myers, who collaborated with the university's Intelligent Systems Program director Randolph A. Miller, artificial intelligence pioneer Harry Pople, and infectious disease specialist Victor Yu. QMR is a microcomputer version of the expert system. It was codeveloped in the Section of Medical Informatics at the University of Pittsburgh School of Medicine by Fred E. Masarie, Jr., Randolph A. Miller, and Jack D. Myers in the 1980s. The two expert systems utilized common algorithms and are often referred to together as INTERNIST-I/QMR. QMR can act as a decision aid but is also able to critique physician evaluations and suggest laboratory tests. QMR can also be used as a teaching aid by providing simulations of cases.

Myers, Miller, Pople, and Yu developed INTERNIST-I in a medical school course taught at the University of Pittsburgh. The course, called The Logic of Problem-Solving in Clinical Diagnosis, required fourth-year students to enter laboratory and sign-and-symptom data gleaned from published and unpublished clinicopathological reports and patient histories. The system was also used as a "quizmaster" to test the enrolled students. Eschewing statistical artificial intelligence methods, the group created a ranking algorithm, partitioning algorithm, exclusion functions, and other heuristic rules. Output consisted of a ranked list of possible diagnoses given inputted physician findings as well as answers to the system's follow-up questions. INTERNIST-I could also recommend further lab tests. The project's leaders estimated that fifteen person-years of effort had gone into the system by 1982. Extremely knowledge-intensive, the system eventually held taxonomic information about 1,000 diseases and three-quarters of all known internal medicine diagnoses.

The University of Pittsburgh School of Medicine developed INTERNIST-I at the zenith of the Greek oracle approach to medical artificial intelligence. The user of the first generation of the system was largely treated as an acquiescent observer. The designers believed that the system could replace expertise in places where physicians were scarce, for example, on manned space missions, in rural communities, and onboard nuclear submarines. Paramedics and medical technicians, however, found the system time-consuming and difficult to use. To meet this difficulty, Donald McCracken and Robert Akscyn at nearby Carnegie Mellon University implemented INTERNIST-I in ZOG, an early knowledge management hypertext system.

QMR extended the user-friendly aspects of INTERNIST-I while encouraging more active exploration of the knowledge base of case studies. QMR also relied on a ranking algorithm and weighted scales that assessed a patient's signs and symptoms and linked them with diagnoses. System designers determined the evoking strength and frequency (or sensitivity) of case findings by reviewing the literature in the field. The heuristic algorithm at the core of QMR evaluates evoking strength and frequency and gives them a quantitative value. QMR includes rules that allow the system to express time-sensitive reasoning in the solution of diagnostic problems. One function in QMR not present in INTERNIST-I was the ability to build homologies between various related sets of symptoms. QMR listed not only probable diagnoses but also disorders that presented themselves with similar histories, signs and symptoms, and initial laboratory results. The accuracy of the system was periodically checked by comparing QMR's output with case files published in *The New England Journal of Medicine.*

QMR, commercially available to physicians in the 1980s and 1990s from First DataBank, required about ten hours of initial training. In private clinics, typical runs of the program on specific patient cases were performed after hours. QMR developers repositioned the expert system as a hyperlinked electronic textbook rather than as a clinical decisionmaker. INTERNIST-I/QMR was supported under grants by the National Library of Medicine, NIH Division of Research Resources, and the CAMDAT Foundation.

DXplain, Meditel, and Iliad were comparable contemporaneous medical artificial intelligence decision aids. DXplain was a decision-support system developed by G. Octo Barnett and Stephen Pauker at the Massachusetts General Hospital/ Harvard Medical School Laboratory of Computer Science with financial support from the American Medical Association. The knowledge base of DXplain was extracted directly from the AMA's publication *Current Medical Information and Terminology (CMIT),* which summarized the etiologies, signs, symptoms, and laboratory findings for more than 3,000 diseases. Like INTERNIST-I, the diagnostic algorithm at the heart of DXplain not only relied on a scoring or ranking rule but also incorporated modified Bayesian conditional probability calculations. DXplain became available to PC users on diskette in the 1990s.

Meditel was designed in the mid-1970s by Albert Einstein Medical Center educator Herbert Waxman and physician William Worley of the University of Pennsylvania Department of Medicine from an earlier computerized decision aid, the Meditel Pediatric System. Meditel assisted in prompting possible diagnoses using

Bayesian statistics and heuristic decision rules. In the 1980s, Meditel was available as a doc-in-a-box software package sold by Elsevier Science Publishing Company for IBM personal computers.

Dr. Homer Warner and his collaborators incubated a third medical AI competitor, Iliad, in the Knowledge Engineering Center of the Department of Medical Informatics at the University of Utah. In the early 1990s, Applied Medical Informatics received a two-million-dollar grant from the federal government to link Iliad's diagnostic software directly to electronic databases of patient information. Iliad's primary audience included physicians and medical students, but in 1994, the company released a consumer version of Iliad called Medical HouseCall.

*Philip L. Frana*

*See also:* Clinical Decision Support Systems; Computer-Assisted Diagnosis.

**Further Reading**

Bankowitz, Richard A. 1994. *The Effectiveness of QMR in Medical Decision Support: Executive Summary and Final Report.* Springfield, VA: U.S. Department of Commerce, National Technical Information Service.

Freiherr, Gregory. 1979. *The Seeds of Artificial Intelligence: SUMEX-AIM.* NIH Publication 80-2071. Washington, DC: National Institutes of Health, Division of Research Resources.

Lemaire, Jane B., Jeffrey P. Schaefer, Lee Ann Martin, Peter Faris, Martha D. Ainslie, and Russell D. Hull. 1999. "Effectiveness of the Quick Medical Reference as a Diagnostic Tool." *Canadian Medical Association Journal* 161, no. 6 (September 21): 725–28.

Miller, Randolph A., and Fred E. Masarie, Jr. 1990. "The Demise of the Greek Oracle Model for Medical Diagnosis Systems." *Methods of Information in Medicine* 29, no. 1: 1–2.

Miller, Randolph A., Fred E. Masarie, Jr., and Jack D. Myers. 1986. "Quick Medical Reference (QMR) for Diagnostic Assistance." *MD Computing* 3, no. 5: 34–48.

Miller, Randolph A., Harry E. Pople, Jr., and Jack D. Myers. 1982. "INTERNIST-1: An Experimental Computer-Based Diagnostic Consultant for General Internal Medicine." *New England Journal of Medicine* 307, no. 8: 468–76.

Myers, Jack D. 1990. "The Background of INTERNIST-I and QMR." In *A History of Medical Informatics*, edited by Bruce I. Blum and Karen Duncan, 427–33. New York: ACM Press.

Myers, Jack D., Harry E. Pople, Jr., and Jack D. Myers. 1982. "INTERNIST: Can Artificial Intelligence Help?" In *Clinical Decisions and Laboratory Use*, edited by Donald P. Connelly, Ellis S. Benson, M. Desmond Burke, and Douglas Fenderson, 251–69. Minneapolis: University of Minnesota Press.

Pople, Harry E., Jr. 1976. "Presentation of the INTERNIST System." In *Proceedings of the AIM Workshop.* New Brunswick, NJ: Rutgers University.

# Ishiguro, Hiroshi (1963–)

Hiroshi Ishiguro is a world-renowned engineer, known especially for his lifelike humanoid robots. He believes that the current information society will inevitably evolve into a world of caregiver or helpmate robots. Ishiguro also hopes that the study of artificial humans will help us better understand how humans are habituated to interpret or understand their own species' behaviors and expressions. In cognitive science terms, Ishiguro wants to understand such things as authenticity of relation, autonomy, creativity, imitation, reciprocity, and robot ethics.

In his research, Ishiguro strives to create robots that are uncannily similar in appearance and behavior to human beings. He hopes that his robots can help us understand what it means to feel human. The Japanese word for this feeling of substantive presence, or spirit, of a human is *sonzaikan*. Success, Ishiguro asserts, can be benchmarked and assessed in two general ways. The first he calls the total Turing Test; an android passes the test when 70 percent of human viewers are not aware that they have encountered a robot until at least two seconds have elapsed. The second way of assessing success, he asserts, is to measure the amount of time a human willingly remains in conscious engagement with a robot before realizing that the robot's cooperative eye tracking does not represent real thinking.

One of Ishiguro's first robots, introduced in 2000, was Robovie. Ishiguro wanted to create a robot that did not look industrial or like a pet, but instead could be mistaken as a partner in daily life. Robovie does not look particularly human, but it is capable of several novel humanlike gestures and interactive behaviors. Robovie is capable of eye contact, gazing at objects, pointing at things, nodding, swinging and folding arms, shaking hands, and saying hello and goodbye. Robovie appeared frequently in Japanese media, and the public reaction to the robot convinced Ishiguro that appearance, interaction, and communication were keys to richer, more complex relationships between robots and human beings.

Ishiguro presented Actroid to public audiences for the first time in 2003. Actroid, an autonomous robot operated by AI software, was developed in Osaka University's Intelligent Robotics Laboratory and is now manufactured by Sanrio's Kokoro animatronics division. Actroid is feminine in appearance (a "gynoid" in science fiction phraseology), with skin made of highly realistic silicone. The robot can mimic human movement, breathing, and blinking, using internal sensors and silent air actuators at 47 points of bodily articulation, and is capable of speech. Movement is accomplished by sensor processing, data files containing critical values for degrees of freedom in movement of limbs and joints. Robot arms typically have five to seven degrees of freedom. Humanoid robots can have thirty or more degrees of freedom for arms, legs, torso, and neck. Programmers develop scenarios for Actroid, involving four general steps: (1) gather recognition data from sensors triggered by interaction, (2) select a motion module, (3) execute a predetermined set of motions and play an audio file, and (4) return to the first step. Experiments with irregular random or contingent responses to human context clues have been determined to be effective in keeping the attention of the human subject, but they are made more effective with the addition of the predetermined scenarios. Motion modules are encoded in the text-based markup language XML, which is highly accessible to even novice programmers.

In 2005, Ishiguro revealed Repliee models of the Actroid, designed to be indistinguishable from a human female on first viewing. Repliee Q1Expo is an android facsimile of real Japanese newscaster Ayako Fujii. Repliee androids are interactive; they can process human conversation with speech recognition software, respond orally, maintain eye contact, and react swiftly to human touch. This is made possible through a distributed and ubiquitous sensor net composed of infrared motion detectors, cameras, microphones, identification tag readers, and floor sensors. The robot uses artificial intelligence to determine whether the human is touching the robot in a gentle or aggressive manner. Ishiguro also introduced a child version of the robot, called Repliee R1, which is similar in appearance to his then four-year-old daughter.

More recently, Actroids have been shown to be capable of mimicking the limb and joint movement of humans, by watching and repeating the motions. The robot is not capable of true locomotion, as most of the computer hardware running the artificial intelligence software is external to the robot. In experiments conducted in Ishiguro's lab, self-reports of the feelings and moods of human subjects are recorded as robots exhibit behaviors. The range of moods recorded in response to the Actroid varies from interest to disgust, acceptance to fear. Real-time neuroimaging of human subjects has also helped Ishiguro's research colleagues better understand the ways human brains are activated in human-android relations. In this way, Actroid is a testbed for understanding why some of the observed actions performed by nonhuman agents fail to produce desired cognitive responses in humans.

The Geminoid series of robots was developed in recognition that artificial intelligence lags far behind robotics in producing lifelike interactions between humans and androids. In particular, Ishiguro acknowledged that it would be many years before a machine could engage in a long, immersive oral conversation with a human. Geminoid HI-1, introduced in 2006, is a teleoperated (rather than truly autonomous) robot identical in appearance to Ishiguro. The term *geminoid* comes from the Latin word for "twin." Geminoid is capable of hand fidgeting, blinking, and movements associated with human breathing. The android is controlled by motion-capture technology that reproduces the facial and body movements of Ishiguro himself. The robot is capable of speaking in a humanlike voice modeled after its creator. Ishiguro hopes he can one day use the robot to teach classes by way of remote telepresence. He has noticed that when he is teleoperating the robot the sense of immersion is so great that his brain is tricked into forming phantom impressions of physical touch when the android is poked. The Geminoid-DK, released in 2011, is a mechanical doppelgänger of Danish psychology professor Henrik Schärfe. While some viewers find the Geminoid appearance creepy, many do not and simply engage naturally in communication with the robot.

The Telenoid R1 is a teleoperated android robot released in 2010. Telenoid is amorphous, only minimally approximating the shape of a human, and stands 30 inches high. The purpose of the robot is to communicate a human voice and gestures to a viewer who might use it as a communication tool or videoconferencing device. Like other robots in Ishiguro's lab, the Telenoid appears lifelike: it mimics the motions of breathing and talking and blinks. But the design also minimizes the number of features to maximize imagination. The Telenoid in this way is analogous to a physical, real-world avatar. It is intended to help facilitate more intimate, more humanlike interaction over telecommunications technology. Ishiguro has proposed that the robot might one day serve as a satisfactory stand-in for a teacher or companion who is otherwise available only at a distance. A miniature variant of the robot called the Elfoid can be held in one hand and kept in a pocket. The Actroid and the Telenoid were anticipated by the autonomous persocom dolls that substitute for smart phones and other devices in the extremely popular manga series *Chobits.*

Ishiguro is Professor of Systems Innovation and Director of the Intelligent Robotics Laboratory at Osaka University in Japan. He is also a group leader at the Advanced Telecommunications Research Institute (ATR) in Kansai Science City and cofounder of the tech-transfer venture company Vstone Ltd. He hopes that future commercial ventures will leverage success with teleoperated robots to provide capital for ongoing, continuous improvement of his autonomous series of robots. His latest effort is a humanoid robot called Erica who became a Japanese television news anchor in 2018.

As a young man, Ishiguro intensively studied oil painting, thinking as he worked about how to represent human likeness on canvas. He became spellbound by robots in the computer science laboratory of Hanao Mori at Yamanashi University. Ishiguro studied for his doctorate in engineering under computer vision and image recognition pioneer Saburo Tsuji at Osaka University. In projects undertaken in Tsuji's lab, he worked on mobile robots capable of SLAM—simultaneous mapping and navigation with panoramic and omni-directional video cameras. This work led to his PhD work, which focused on tracking a human subject through active control of the cameras and panning to achieve full 360-degree views of the environment. Ishiguro thought that the technology and his applications could be used to give an interactive robot a useful internal map of its environment. The first reviewer of a paper based on his dissertation rejected his work.

Ishiguro believes that fine arts and technology are inextricably intertwined; art inspires new technologies, and technology allows the creation and reproduction of art. In recent years, Ishiguro has brought his robots to Seinendan, a theatre company formed by Oriza Hirata, in order to apply what he has learned about human-robot communications in real-life situations. Precedents for Ishiguro's branch of cognitive science and AI, which he calls android science, may be found in Disney's "Great Moments with Mr. Lincoln" robotics animation show at Disneyland and the fictional robot substitutes depicted in the Bruce Willis movie *Surrogates* (2009). Ishiguro has a cameo in the Willis film.

*Philip L. Frana*

*See also:* Caregiver Robots; Nonhuman Rights and Personhood.

**Further Reading**

Guizzo, Erico. 2010. "The Man Who Made a Copy of Himself." *IEEE Spectrum* 47, no. 4 (April): 44–56.

Ishiguro, Hiroshi, and Fabio Dalla Libera, eds. 2018. *Geminoid Studies: Science and Technologies for Humanlike Teleoperated Androids.* New York: Springer.

Ishiguro, Hiroshi, and Shuichi Nishio. 2007. "Building Artificial Humans to Understand Humans." *Journal of Artificial Organs* 10, no. 3: 133–42.

Ishiguro, Hiroshi, Tetsuo Ono, Michita Imai, Takeshi Maeda, Takayuki Kanda, and Ryohei Nakatsu. 2001. "Robovie: An Interactive Humanoid Robot." *International Journal of Industrial Robotics* 28, no. 6: 498–503.

Kahn, Peter H., Jr., Hiroshi Ishiguro, Batya Friedman, Takayuki Kanda, Nathan G. Freier, Rachel L. Severson, and Jessica Miller. 2007. "What Is a Human? Toward Psychological Benchmarks in the Field of Human–Robot Interaction." *Interaction Studies* 8, no. 3: 363–90.

MacDorman, Karl F., and Hiroshi Ishiguro. 2006. "The Uncanny Advantage of Using Androids in Cognitive and Social Science Research." *Interaction Studies* 7, no. 3: 297–337.

Nishio, Shuichi, Hiroshi Ishiguro, and Norihiro Hagita. 2007a. "Can a Teleoperated Android Represent Personal Presence? A Case Study with Children." *Psychologia* 50: 330–42.

Nishio, Shuichi, Hiroshi Ishiguro, and Norihiro Hagita. 2007b. "Geminoid: Teleoperated Android of an Existing Person." In *Humanoid Robots: New Developments*, edited by Armando Carlos de Pina Filho, 343–52. Vienna, Austria: I-Tech.

# Knight, Heather

Heather Knight is an artificial intelligence and engineering expert known for her work in the area of entertainment robotics. The goal of her Collaborative Humans and Robots: Interaction, Sociability, Machine Learning, and Art (CHARISMA) Research Lab at Oregon State University is to bring performing arts methods to the field of robotics.

Knight describes herself as a social roboticist, someone who creates non-anthropomorphic—and sometimes nonverbal—machines that engage in interaction with humans. She creates robots exhibiting behavior inspired by human interpersonal communication. These behaviors include patterns of speech, welcoming motions, open postures, and a range of other context clues that help humans develop rapport with robots in everyday life. In the CHARISMA Lab, Knight experiments with social robots and so-called charismatic machines, as well as investigates social and government policy related to robots.

Knight is founder of the Marilyn Monrobot interactive robot theatre company. The associated Robot Film Festival is an outlet for roboticists to show off their latest creations in a performance environment, and for the showing of films with relevance to the advancing state of the art in robotics and robot-human interaction. The Marilyn Monrobot company grew out of Knight's association with the Syyn Labs creative collective and her observations on robots constructed for purposes of performance by Guy Hoffman, Director of the MIT Media Innovation Lab. Knight's company focuses on robot comedy. Knight argues that theatrical spaces are perfect environments for social robotics research because the spaces not only inspire playfulness—requiring expression and interaction on the part of the robot actors—but also involve creative constraints where robots thrive, for example, a fixed stage, learning from trial-and-error, and repeat performances (with manipulated variations).

Knight has argued that the use of robots in entertainment contexts is valuable because it enhances human culture, imagination, and creativity. Knight introduced a stand-up comedy robot named Data at the TEDWomen conference in 2010. Data is a Nao robot developed by Aldebaran Robotics (now SoftBank

Group). Data performs a comedy routine (which includes about 200 prepro-grammed jokes) while collecting audience feedback and fine-tuning its act in real time. The robot was developed with Scott Satkin and Varun Ramakrisha at Carn-egie Mellon University. Knight now works on comedy with Ginger the Robot.

Robot entertainment also drives the development of algorithms for artificial social intelligence. In other words, art is used to inspire new technology. Data and Ginger utilize a microphone and machine learning algorithm to test audience reactions and interpret the sounds produced by audiences (laughter, chatter, clap-ping, etc.). Crowds also receive green and red cards that they hold up after each joke. Green cards help the robots understand that the audience likes the joke. Red cards are for jokes that fall flat. Knight has learned that good robot comedy doesn't need to hide the fact that the spotlight is on a machine. Rather, Data draws laughs by bringing attention to its machine-specific troubles and by making self-deprecating comments about its limitations. Knight has found improvisational acting and dance techniques invaluable in building expressive, charismatic robots. In the process, she has revised the methodology of the classic Robotic Paradigm: Sense-Plan-Act, and she instead prefers Sensing-Character-Enactment, which is closer in practice to the process used in theatrical performance.

Knight is now experimenting with ChairBots, hybrid machines developed by attaching IKEA wooden chairs on top of Neato Botvacs (a brand of intelligent robotic vacuum cleaner). The ChairBots are being tested in public spaces to deter-mine how such a simple robot can use rudimentary movements as a means of communication to convince humans to step out of the way. They have also been employed to convince potential café patrons to enter the premises, find a table, and sit down.

While working toward degrees in the MIT Media Lab, Knight worked with Personal Robots group leader Professor Cynthia Breazeal on the synthetic organic robot art installation *Public Anemone* for the SIGGRAPH computer graphics con-ference. The piece comprised a fiberglass cave containing glowing creatures mov-ing and responding to music and people. The centerpiece robot, also dubbed "Public Anemone," swayed and interacted with people, bathed in a waterfall, watered a plant, and interacted with other environmental features in the cave. Knight worked with animatronics designer Dan Stiehl to make artificial tube-worms with capacitive sensors. When a human viewer reached into the cave, the tubeworm's fiberoptic tentacles pulled into their tubes and changed color, as if motivated by protective instincts. The group working on *Public Anemone* described the project as an example of intelligent staging and a step toward fully embodied robot theatrical performance. Knight also contributed to the mechani-cal design of the "Cyberflora" kinetic robot flower garden installation at the Smithsonian/Cooper-Hewitt Design Museum in 2003. Her master's thesis at MIT centered on the Sensate Bear, a huggable robot teddy bear with full-body capaci-tive touch sensors for exploring real-time algorithms involving social touch and nonverbal communication.

Knight earned her doctorate from Carnegie Mellon University in 2016. Her dis-sertation research involved expressive motion in low degree of freedom robots. Knight observed in her research that humans do not require that robots closely match humans in appearance or behavior to be treated as close associates. Instead, humans rather easily anthropomorphize robots and grant them independent agency. And in fact, she argues, as robots approach humans in appearance, we may experience discomfort or expect a much higher standard of humanlike behavior.

Knight was advised by Professor Matt Mason of the School of Computer Science and Robotics Institute. She is a past robotic artist in residence at X, the research lab of Google's parent company, Alphabet. Knight has also worked as a research scientist and engineer for Aldebaran Robotics and NASA's Jet Propulsion Laboratory. Knight developed the touch sensing panel for the Nao autonomous family companion robot, as well as the infrared detection and emission capabilities in its eyes, while working as an engineer at Aldebaran Robotics. Her work on the first two minutes of the OK Go video "This Too Shall Pass," which features a Rube Goldberg machine, helped net Syyn Labs a UK Music Video Award. She is currently helping Clearpath Robotics make their autonomous, mobile-transport robots more socially aware.

*Philip L. Frana*

*See also:* RoboThespian; Turkle, Sherry.

**Further Reading**

Biever, Celeste. 2010. "Wherefore Art Thou, Robot?" *New Scientist* 208, no. 2792: 50–52.

Breazeal, Cynthia, Andrew Brooks, Jesse Gray, Matt Hancher, Cory Kidd, John McBean, Dan Stiehl, and Joshua Strickon. 2003. "Interactive Robot Theatre." *Communications of the ACM* 46, no. 7: 76–84.

Knight, Heather. 2013. "Social Robots: Our Charismatic Friends in an Automated Future." *Wired UK*, April 2, 2013. https://www.wired.co.uk/article/the-inventor.

Knight, Heather. 2014. *How Humans Respond to Robots: Building Public Policy through Good Design.* Washington, DC: Brookings Institute, Center for Technology Innovation.

# Knowledge Engineering

Knowledge engineering (KE) is a discipline in artificial intelligence pursuing transfer of the experts' knowledge into a formal automatic programming system in a way that the latter will be able to achieve the same or similar output as human experts in problem solving when operating on the same data set. More precisely, knowledge engineering is a discipline that designs methodologies applicable to building up large knowledge based systems (KBS), also referred to as expert systems, using applicable methods, models, tools, and languages. Modern knowledge engineering relies on knowledge acquisition and documentation structuring (KADS) methodology for knowledge elicitation; thus, the building up of knowledge based systems is regarded as a modeling activity (i.e., knowledge engineering builds up computer models).

Because the human experts' knowledge is a mixture of skills, experience, and formal knowledge, it is difficult to formalize the knowledge acquisition process. Consequently, the experts' knowledge is modeled rather than directly transferred from human experts to the programming system. Simultaneously, the direct simulation of the complete experts' cognitive process is also very challenging. Designed computer models are expected to achieve targets similar to experts' results doing problem solving in the domain rather than matching the cognitive capabilities of the experts. Thus, the focus of knowledge engineering is on modeling and problem solving methods (PSM) independent of different representation formalisms (production rules, frames, etc.).

The problem solving method is central for knowledge engineering and denotes knowledge-level specification of a reasoning pattern that can be used to conduct a knowledge-intensive task. Each problem solving method is a pattern that provides template structures for solving a particular problem. The popular classification of problem solving methods according to their topology is as "diagnosis," "classification," or "configuration." Examples include PSM "Cover-and-Differentiate" for solving diagnostic tasks and PSM "Propose-and-Reverse" for parametric design tasks. The assumption behind any problem solving method is that the logical adequacy of the proposed method matches the computational tractability of the system implementation based on it.

Early examples of expert systems often utilize the PSM heuristic classification—an inference pattern that describes the behavior of knowledge based systems in terms of goals and knowledge needed to achieve these goals. This problem solving method comprehends inference actions and knowledge roles and their relationships. The relationships define which role the domain knowledge plays in each interference action. The knowledge roles are observables, abstract observables, solution abstractions and solution, while the interference action could be abstract, heuristic match, and refine. The PSM heuristic classification needs a hierarchically structured model of observables and solutions for "abstract" and "refine," which makes it suitable for acquiring static domain knowledge.

In the late 1980s, the modeling approaches in knowledge engineering moved toward role limiting methods (RLM) and generic tasks (GT). Role limiting methods utilize the concept of the "knowledge role," which specifies the way the particular domain knowledge is being used in the problem-solving process. RLM designs a wrapper around PSM by describing the latter in general terms with a goal to reuse the method. This approach, however, encapsulates only a single instance of PSM and thus is not suitable for problems that require use of several methods. An extension of the role limiting methods idea is configurable role limiting methods (CRLM), which offer a predefined set of RLMs along with a fixed scheme of knowledge types. Each member method can be applied to a different subset of a task, but adding a new method is quite difficult to achieve since it requires modification in predefined knowledge types.

The generic task approach provides generic description of input and output along with a fixed scheme of knowledge types and inference strategy. The generic task is based on the "strong interaction problem hypothesis," which states that structure and representation of domain knowledge can be determined completely by its use. Each generic task uses knowledge and applies control strategies that are specific to that knowledge. Because the control strategies are closer to a domain, the actual knowledge acquisition used in GT demonstrates higher precision in terms of explanations of the problem-solving steps. Thus, the design of specific knowledge based systems can be seen as an instantiation of predefined knowledge types by domain-specific terms.

The disadvantage of GT is that a predefined problem-solving strategy may not be combined with the best problem-solving strategy needed to resolve the task. The task structure (TS) approach attempts to overcome the insufficiencies of GT by making a clear distinction between the task and method used to solve the task. Thus, under that approach any task-structure postulates how the problem can be answered by using a set of generic tasks along with what knowledge needs to be obtained or can be developed for these tasks.

The need for the use of several models led to the development of modeling frameworks designed to address several aspects of knowledge engineering approaches. In the most popular engineering CommonKADS structure (which relies on KADS), the models are the organizational model, task model, agent model, communication model, expertise model, and design model. The organizational model describes the structure together with functions performed by each unit. The task model provides hierarchical description of tasks. The agent model specifies abilities of each agent involved in tasks execution. The communication model defines the various interactions between agents. The most important model is the expertise model that utilizes several layers and targets the modeling of domain-specific knowledge (domain layer) along with the inference for the reasoning process (inference layer). In addition, the expertise model supports a task layer. The latter deals with the decomposition of the tasks. The design model describes system architecture and computational mechanisms used to make the inference. In CommonKADS, there is a clear separation between domain-specific knowledge and generic problem solving methods, which allows the different tasks to be solved by using the PSM on a different domain by defining a new instance of the domain layer.

Currently, there are several libraries of problem solving methods available for use in development. They differ from each other by their main features—if the library is developed for a particular task or has broader scope; if the library is formal, informal, or an implemented one; if the library is with fine or coarse-grained PSM; and finally by the size of the library itself. Recently, some research has been conducted with the aim to unify available libraries by providing adapters that adapt task-neutral PSM to task-specific PSM.

Further development of CommonKADS led to the MIKE (model-based and incremental knowledge engineering) approach, which proposes integration of semiformal and formal specification and prototyping into the framework. Consequently, the entire process of development of knowledge based systems under MIKE is divided into a number of sub-activities, each of which deals with different aspects of the system development. The Protégé approach reuses PSMs and applies tools from ontologies where the ontology is defined as an explicit specification of a shared conceptualization that holds in a particular context. Indeed, the ontologies could be of a different type, but in Protégé, the ones used are domain ontologies that define the shared conceptualization of a domain and method ontology that defines the concepts and relations that are used by problem solving methods.

The development of knowledge based systems requires, in addition to problem solving methods, the development of specific languages capable of describing the knowledge required by the system along with the reasoning process that will use that knowledge. The goal of such languages is to provide a formal and unambiguous framework for specifying knowledge models. In addition, some of these formal languages may be executable, which could permit simulation of the behavior of the knowledge models on specific input data.

In early years, the knowledge was directly encoded in rule-based implementation languages. That caused severe problems among which were the inability to present some types of knowledge, the inability to ensure uniform representation of different types of knowledge, and the lack of details. Modern approaches in language development target and formalize the conceptual models of knowledge based systems and thus provide the capability to define precisely the goals and process to get models, along with functionality of interface actions and accurate semantics of the different elements of the domain knowledge. Most of these epistemological languages deliver primitives such as constants, functions, predicates, and others, together with some mathematical operations. For example, object-oriented or frame-based languages specify a variety of modeling primitives such as objects and classes. The most common examples of specific languages are KARL, $(ML)^2$, and DESIRE.

KARL is a language that uses a variant of Horn logic. It was developed as part of the MIKE project and targets the KADS expertise model by combining two types of logic: L-KARL and P-KARL. The L-KARL is a variant of frame logic and is applicable to inference and domain layers. In fact, it is a combination of first order logic and semantic data modeling primitives. P-KARL is used for specifications in the task layer and is in fact in some versions a dynamic logic.

$(ML)^2$ is formalization language for KADS expertise models. The language combines first-order extended logic for specification of a domain layer, first-order meta logic for specifying an inference layer, and quantified dynamic logic for the task layer.

DESIRE (design and specification of interacting reasoning components) relies on the notion of compositional architecture. It uses temporal logics for specifying the dynamic reasoning process. The interaction between components of knowledge based systems is represented as transactions, and the control flow between any two objects is defined as set of control rules. Each object has a metadata description. The meta level describes the dynamic aspects of the object level in declarative format.

Knowledge engineering was born from the need to design large knowledge based systems, meaning to design a computer model with the same problem-solving capabilities as human experts. Knowledge engineering understands knowledge based systems not as a container that holds human expertise, but as an operational system that should exhibit some desired behavior and provide modeling approaches, tools, and languages to build such systems.

*Stefka Tzanova*

*See also:* Clinical Decision Support Systems; Expert Systems; INTERNIST-I and QMR; MOLGEN; MYCIN.

**Further Reading**

Schreiber, Guus. 2008. "Knowledge Engineering." In *Foundations of Artificial Intelligence*, vol. 3, edited by Frank van Harmelen, Vladimir Lifschitz, and Bruce Porter, 929–46. Amsterdam: Elsevier.

Studer, Rudi, V. Richard Benjamins, and Dieter Fensel. 1998. "Knowledge Engineering: Principles and Methods." *Data & Knowledge Engineering* 25, no. 1–2 (March): 161–97.

Studer, Rudi, Dieter Fensel, Stefan Decker, and V. Richard Benjamins. 1999. "Knowledge Engineering: Survey and Future Directions." In *XPS 99: German Conference on Knowledge-Based Systems*, edited by Frank Puppe, 1–23. Berlin: Springer.

# Kurzweil, Ray (1948–)

Ray Kurzweil is an American inventor and futurist. He spent the first part of his professional life inventing the first CCD flat-bed scanner, the first omni-font optical character recognition device, the first print-to-speech reading machine for the blind, the first text-to-speech synthesizer, the first music synthesizer capable of recreating the grand piano and other orchestral instruments, and the first commercially marketed, large-vocabulary speech recognition machine. He has received many honors for his achievements in the field of technology, including a 2015 Technical Grammy Award and the National Medal of Technology.

Kurzweil is cofounder and chancellor of Singularity University and the director of engineering at Google, heading up a team that develops machine intelligence and natural language understanding. Singularity University is a nonaccredited graduate university built on the idea of addressing challenges as grand as renewable energy and space travel through an intimate comprehension of the opportunity offered by the current acceleration of technological progress. Headquartered in the Silicon Valley, the university has grown to one hundred chapters in fifty-five countries, offering seminars and educational and entrepreneurial acceleration programs. Kurzweil wrote the book *How to Create a Mind* (2012) while at Google. In it he describes his Pattern Recognition Theory of Mind, asserting that the neocortex is a hierarchical system of pattern recognizers. Kurzweil argues that emulating this architecture in machines could lead to an artificial superintelligence. He hopes that in this way he can bring natural language understanding to Google.

It is as a futurist that Kurzweil has reached a popular audience. Futurists are people whose specialty or interest is the near- to long-term future and future-related subjects. They systematically explore predictions and elaborate possibilities about the future by means of well-established approaches such as scenario planning. Kurzweil has written five national best-selling books, including the *New York Times* best seller *The Singularity Is Near* (2005). His list of predictions is long. In his first book, *The Age of Intelligent Machines* (1990), Kurzweil foresaw the explosive growth in worldwide internet use that began in the second half of the decade. In his second highly influential book, *The Age of Spiritual Machines* (where "spiritual" stands for "conscious"), written in 1999, he rightly predicted that computers would soon outperform humans at making the best investment decisions. In the same book, Kurzweil predicted that machines will eventually "appear to have their own free will" and even enjoy "spiritual experiences" (Kurzweil 1999, 6). More precisely, the boundaries between humans and machines will blur to a point where they will essentially live forever as merged human-machine hybrids. Scientists and philosophers have criticized Kurzweil about his prediction of a conscious machine, the main objection being that consciousness cannot be a product of computations.

In his third book, *The Singularity Is Near*, Kurzweil deals with the phenomenon of the Technological Singularity. The term *singularity* was coined by the great mathematician John von Neumann. In conversation with his colleague Stanislaw Ulam in the 1950s, von Neumann postulated the ever-accelerating pace of technological change, which he said "gives the appearance of approaching some essential singularity in the history of the race beyond which human affairs as we know them could not continue" (Ulam 1958, 5). To put it differently, the technological advancement would change the history of human race. Forty years later, computer scientist, professor of mathematics, and science fiction writer Vernor Vinge recovered the term and used it in his 1993 essay "The Coming Technological Singularity." In Vinge's essay, the technological advancement is more properly the increase in computing power. Vinge addresses the hypothesis of a self-improving artificial intelligent agent. In this hypothesis, the artificial intelligent agent continues to upgrade itself and advances technologically at an incomprehensible rate, to the point that a superintelligence—that is, an artificial intelligence that far surpasses all human intelligence—is born. In Vinge's dystopian view, the machines become autonomous first and superintelligent second, to the point that humans lose control of technology and machines take their own destiny in their hands. Because technology is more intelligent than humans, machines will dominate the world.

The Singularity, according to Vinge, is the end of the human era. Kurzweil offers an anti-dystopic vision of the Singularity. Kurzweil's basic assumption is that humans can create something more intelligent than themselves; as a matter of fact, the exponential improvements in computer power make the creation of an intelligent machine almost inevitable, to the point where the machine will become more intelligent than the humans. At this point, in Kurzweil's opinion, machine intelligence and humans would merge. Not coincidentally, in fact, the subtitle of *The Singularity Is Near* is *When Humans Transcend Biology*.

The underlying premise of Kurzweil's overall vision is discontinuity: no lesson from the past or even the present can help humans to detect the path to the future. This also explains the need for new forms of education such as Singularity University. Any reminiscence of the past, and every nostalgic turning back to history, makes humanity more vulnerable to technological change. History itself, as a human construct, will soon end with the coming of a new superintelligent, almost immortal species. Immortals is another word for posthumans, the next step in human evolution. In Kurzweil's opinion, posthumanity consists of robots with consciousness, rather than humans with machine bodies. The future, he argues, should be built on the premise that humanity is living in an unprecedented era of technological progress. In his view, the Singularity will empower humankind beyond its wildest expectations. While Kurzweil argues that artificial intelligence is already starting to outpace human intelligence on specific tasks, he recognizes that the point of superintelligence—also commonly known as the Technological Singularity, is not yet here. He remains confident that those who embrace the new era of human-machine synthesis, and are unafraid of moving beyond the limits of evolution, can foresee a bright future for humanity.

*Enrico Beltramini*

*See also:* General and Narrow AI; Superintelligence; Technological Singularity.

**Further Reading**

Kurzweil, Ray. 1990. *The Age of Intelligent Machines*. Cambridge, MA: MIT Press.

Kurzweil, Ray. 1999. *The Age of Spiritual Machines: When Computers Exceed Human Intelligence*. New York: Penguin.

Kurzweil, Ray. 2005. *The Singularity Is Near: When Humans Transcend Biology*. New York: Viking.

Ulam, Stanislaw. 1958. "Tribute to John von Neumann." *Bulletin of the American Mathematical Society* 64, no. 3, pt. 2 (May): 1–49.

Vinge, Vernor. 1993. "The Coming Technological Singularity: How to Survive in the Post-Human Era." In *Vision 21: Interdisciplinary Science and Engineering in the Era of Cyberspace*, 11–22. Cleveland, OH: NASA Lewis Research Center.

# L

## Lethal Autonomous Weapons Systems

Lethal Autonomous Weapons Systems (LAWS), also known as "lethal autonomous weapons," "robotic weapons," or "killer robots," are air, ground, marine, or spatial unmanned robotic systems that can independently select and engage targets and decide the use of lethal force. While popular culture abounds in human-like robots waging wars or using lethal force against humans (ED-209 in *RoboCop*, T-800 in *The Terminator*, etc.), robots with full lethal autonomy are still under development. LAWS pose fundamental ethical problems, and they are increasingly debated among AI experts, NGOs, and the international community.

While definition of autonomy may differ in discussions over LAWS, it is generally understood as "the ability to designate and engage a target without additional human intervention after having been tasked to do so" (Arkin 2017). However, LAWS are frequently divided into three categories based on their level of autonomy:

1. Human-in-the-loop weapons: They can select targets and deliver force only with a human command.
2. Human-on-the-loop weapons: They can select targets and deliver force under the monitoring of a human supervisor who can override their actions.
3. Human-out-of-the-loop weapons: they are capable of selecting targets and delivering force without any human input or interaction.

LAWS include these three types of unmanned weapons. The term "fully autonomous weapons" refers to not only human-out-of-the-loop weapons but also "human-on-the-loop weapons" (or weapons with supervised autonomy) in case the supervision is in reality limited (for example, if their response time cannot be matched by a human operator).

Robotic weapons are not new. For example, anti-tank mines, which once activated by a human engage targets on their own, have been widely used since World War II (1939–1945). In addition, LAWS encompass many different types of unmanned weapons with various levels of autonomy and lethality, from land mines to remote-controlled Unmanned Combat Aerial Vehicles (UCAV), or combat drones, and fire-and-forget missiles. However, to date, the only weapons with complete autonomy in use are "defensive" systems (such as landmines). Neither fully "offensive" autonomous lethal weapons nor LAWS using machine learning have yet been deployed.

Even if military research is often kept secret, it is known that several countries (in particular, the United States, China, Russia, United Kingdom, Israel, and

South Korea) are investing heavily in military applications of AI. The international AI arms race that has begun in the early 2010s has resulted in a fast rhythm of innovation in this field and fully autonomous lethal weapons could be produced in the near future.

There are several noticeable precursors of such weapons. For example, the MK 15 Phalanx CIWS, notably deployed by the U.S. Navy, is a close-in weapon system that is capable of autonomously performing its own search, detection, evaluation, tracking, engagement, and kill assessment functions. Another example is Israel's Harpy, an anti-radar "fire-and-forget" drone that is deployed without a specifically designated target, flies a search pattern, and attacks targets by self-destructing.

The deployment of LAWS could dramatically change warfare as previously gunpowder and nuclear weapons did. In particular, it would put an end to the distinction between combatants and weapons, and it would complicate the delimitation of battlefields. Yet, LAWS may be associated with numerous military benefits. Their use would definitely be a force multiplier and reduce the number of human combatants enrolled. It would therefore save military lives. LAWS may also be better than many other weapons in terms of force projection due to faster response time, their ability to perform maneuvers that human combatants cannot do (due to human physical constraints), and making more efficient decisions (from a military perspective) than human combatants.

However, the use of LAWS raises several ethical and political concerns. In addition to not complying with the "Three Laws of Robotics," the deployment of LAWS may result in normalizing the use of lethal force since armed conflicts would involve fewer and fewer human combatants. In that regard, some consider that LAWS pose a threat to humanity. Concerns over the deployment of LAWS also include their use by non-state entities and their use by states in non-international armed conflicts. Delegating life-and-death decisions to machines may also be considered as harming human dignity.

In addition, the ability of LAWS to comply with the requirements of laws of war is widely disputed, in particular by international humanitarian law and specifically the principles of proportionality and of military necessity. Yet, some argue that LAWS, despite not possessing compassion, would at least not act out emotions such as anger, which could result in causing intentional suffering such as torture or rape. Given the daunting task of preventing war crimes, as proven by the numerous cases in past armed conflicts, it can even be argued that LAWS could potentially perpetrate fewer offenses than human combatants.

How the deployment of LAWS would impact noncombatants is also a live argument. Some claim that the use of LAWS may lead to fewer civilian casualties (Arkin 2017), since AI may be more efficient than human combatants in decision-making. However, some critics point to a higher risk of civilians being caught in crossfire. In addition, the ability of LAWS to respect the principle of distinction is also much discussed, since distinguishing combatants and civilians may be especially complex, in particular in non-international armed conflicts and in asymmetric warfare.

LAWS cannot be held accountable for any of their actions since they are not moral agents. This lack of accountability could result in further harm to the victims of war. It may also encourage the perpetration of war crimes. However, it is arguable that the moral responsibility for LAWS would be borne by the authority that decided to deploy it or by people who had designed or manufactured it.

In the last decade, LAWS have generated significant scientific attention and political debate. The coalition that launched the campaign "Stop Killer Robots" in 2012 now consists of eighty-seven NGOs. Its advocacy for a preemptive ban on the development, production, and use of LAWS has resulted in civil society mobilizations. In 2016, nearly 4,000 AI and robotics researchers signed a letter calling for a ban on LAWS. In 2018, more than 240 technology companies and organizations pledged to neither participate in nor support the development, manufacture, trade, or use of LAWS.

Considering that existing international law may not adequately address the issues raised by LAWS, the United Nations' Convention on Certain Conventional Weapons initiated a consultative process over LAWS. In 2016, it established a Group of Governmental Experts (GGE). To date, the GGE has failed to reach an international agreement to ban LAWS due to lack of consensus and due to the opposition of several countries (especially the United States, Russia, South Korea, and Israel). However, twenty-six countries in the United Nations have endorsed the call for a ban on LAWS, and in June 2018, the European Parliament adopted a resolution calling for the urgent negotiation of "an international ban on weapon systems that lack human control over the use of force."

The future of warfare will likely include LAWS since there is no example of a technological innovation that has not been used. Yet, there is wide agreement that humans should be kept "on-the-loop" and that international and national laws should regulate the use of LAWS. However, as the example of nuclear and chemical weapons and anti-personal landmines has shown, there is no assurance that all states and non-state entities would enforce an international legal ban on the use of LAWS.

*Gwenola Ricordeau*

*See also:* Autonomous Weapons Systems, Ethics of; Battlefield AI and Robotics.

**Further Reading**

Arkin, Ronald. 2017. "Lethal Autonomous Systems and the Plight of the Non-Combatant." In *The Political Economy of Robots*, edited by Ryan Kiggins, 317–26. Basingstoke, UK: Palgrave Macmillan.

Heyns, Christof. 2013. *Report of the Special Rapporteur on Extrajudicial, Summary, or Arbitrary Executions*. Geneva, Switzerland: United Nations Human Rights Council. http://www.ohchr.org/Documents/HRBodies/HRCouncil/RegularSession /Session23/A-HRC-23-47_en.pdf.

Human Rights Watch. 2012. *Losing Humanity: The Case against Killer Robots*. https:// www.hrw.org/report/2012/11/19/losing-humanity/case-against-killer-robots.

Krishnan, Armin. 2009. *Killer Robots: Legality and Ethicality of Autonomous Weapons*. Aldershot, UK: Ashgate.

Roff, Heather. M. 2014. "The Strategic Robot Problem: Lethal Autonomous Weapons in War." *Journal of Military Ethics* 13, no. 3: 211–27.

Simpson, Thomas W., and Vincent C. Müller. 2016. "Just War and Robots' Killings." *Philosophical Quarterly* 66, no. 263 (April): 302–22.

Singer, Peter. 2009. *Wired for War: The Robotics Revolution and Conflict in the 21st Century*. New York: Penguin.

Sparrow, Robert. 2007. "Killer Robots." *Journal of Applied Philosophy* 24, no. 1: 62–77.

# M

## Mac Hack

Mac Hack IV, a program written by Richard Greenblatt in 1967, achieved recognition by becoming the first chess program to enter a chess tournament and to play competently against humans, earning a rating between 1,400 and 1,500 in the U.S. Chess Federation rating system. Greenblatt's program, written in the macro assembly language MIDAS, ran on a 200 kilohertz DEC PDP-6 computer. He wrote the program while a graduate student affiliated with Project MAC in MIT's Artificial Intelligence Laboratory.

Russian mathematician Alexander Kronrod is said to have declared, "Chess is the drosophila [fruit fly] of artificial intelligence," the adopted experimental organism of the field (Quoted in McCarthy 1990, 227). Since 1950, when Claude Shannon first articulated chess play as a problem for computer programmers, creating a champion chess program has been a prized problem in artificial intelligence. Chess and games in general present complex yet clearly limited problems with well-defined rules and goals. Chess has often been characterized as a clear example of humanlike intelligent behavior. Chess play is a well-bounded example of human decision-making processes in which moves must be selected with a goal in mind, while using limited information and with uncertainty regarding the outcome.

In the mid-1960s, the processing power of computers greatly limited the depth to which a chess move and its subsequent possible replies could be analyzed because, with each subsequent reply, the number of possible configurations grows exponentially. The best human players have been shown to analyze a limited number of moves to greater depth, instead of considering as many moves as possible to lesser depth. Greenblatt attempted to replicate the processes skilled players use to identify relevant branches of the game tree. He programmed Mac Hack to use a minimax search of the game tree, coupled with alpha-beta pruning and heuristic components, to decrease the number of nodes evaluated when selecting moves. In this way, Mac Hack's style of play more closely resembled that of human players than of more recent chess programs (such as Deep Thought and Deep Blue), which are aided by the brute force of high processing speeds to selectively analyze tens of millions of branches of the game tree before making moves.

Mac Hack earned considerable repute among artificial intelligence researchers for its 1967 win against MIT philosopher Hubert Dreyfus in a match organized by MIT mathematician Seymour Papert. In 1965, the RAND Corporation had published a mimeographed version of Dreyfus's report, *Alchemy and Artificial*

*Intelligence,* which critiqued the claims and goals of artificial intelligence research-ers. Dreyfus argued that no computer could ever achieve intelligence because human reason and intelligence are not entirely rule-bound, and therefore the infor-mation processing of a computer could not replicate or describe human cognition. Among his numerous criticisms of AI, Dreyfus discussed efforts to create chess-playing computers in a section of the report entitled "Signs of Stagnation." The AI community initially perceived Mac Hack's success against Dreyfus as vindication.

*Juliet Burba*

*See also:* Alchemy and Artificial Intelligence; Deep Blue.

**Further Reading**

Crevier, Daniel. 1993. *AI: The Tumultuous History of the Search for Artificial Intelli-gence.* New York: Basic Books.

Greenblatt, Richard D., Donald E. Eastlake III, and Stephen D. Crocker. 1967. "The Greenblatt Chess Program." In *AFIPS '67: Proceedings of the November 14–16, 1967, Fall Joint Computer Conference,* 801–10. Washington, DC: Thomson Book Company.

Marsland, T. Anthony. 1990. "A Short History of Computer Chess." In *Computers, Chess, and Cognition,* edited by T. Anthony Marsland and Jonathan Schaeffer, 3–7. New York: Springer-Verlag.

McCarthy, John. 1990. "Chess as the Drosophila of AI." In *Computers, Chess, and Cogni-tion,* edited by T. Anthony Marsland and Jonathan Schaeffer, 227–37. New York: Springer-Verlag.

McCorduck, Pamela. 1979. *Machines Who Think: A Personal Inquiry into the History and Prospects of Artificial Intelligence.* San Francisco: W. H. Freeman.

# Machine Learning Regressions

"Machine learning," a term coined in 1959 by Arthur Samuel, is an example of artificial intelligence that provides results not based explicitly on programming, but rather on the system autonomously learning from a given database and improv-ing from continued experience. Given their robustness and ease of implementa-tion, machine learning algorithms have a very broad array of applications (e.g., computer vision, natural language processing, autonomous gaming agents, clas-sification, and regressions) and are ubiquitous in almost every industry (e.g., tech, finance, research, education, gaming, and navigation). Despite their wide-ranging applications, machine learning algorithms may be broadly categorized into one of three learning types: supervised, unsupervised, and reinforcement.

Machine learning regressions are an example of supervised learning. They involve algorithms trained on data with continuous numerical outputs that have already been labeled. Once the regression algorithm has been sufficiently trained and validated, the necessary amount of training data or validation requirements will depend on the problems being addressed. The newly created predictive mod-els produce inferred outputs for data with similar input structures. These models are not static. They can be continuously updated with additional training data or by providing the actual correct outputs on previously unlabeled inputs.

Despite the generalizability of machine learning algorithms, there is no single program that is best for all regression problems. There are a multitude of factors to consider when selecting the most optimal machine learning regression algorithm for the current problem (e.g., programming languages, available libraries, algorithm types, data size, and data structure).

**Most popular examples of programming languages, coding libraries, and algorithm types used by data scientists for ML regressions**

| Languages | Libraries | Algorithm Types |
|-----------|-----------|-----------------|
| Python | NumPy | Linear/polynomial regressions |
| R | SciPy | Neural networks |
| Java | pandas | Decision trees/random forests |
| C/C++ | Matplotlib | |
| MATLAB | Scikit-learn | |
| | TensorFlow | |

As with other traditional statistical methods, there are machine learning programs that use single- or multivariable linear regression techniques. These model the relationships between a single independent feature variable or multiple independent feature variables and a dependent target variable. The output of the models are linear representations of the combined input variables. These models are useful for noncomplex and small data; they are limited to those conditions. For nonlinear data, polynomial regressions can be applied. These require the programmers to have an understanding of the data structure—often the purpose of using machine learning models in the first place. These algorithms will likely not be useful for most real-world data, but they offer a simple place to begin and could provide the users with easy-to-explain models.

As the name suggests, decision trees are tree-like structures that map the programs' input features/attributes to decide the final output target. A decision tree algorithm starts with the root node (i.e., an input variable) from which the answer to the condition of that node splits into edges. If the edge no longer splits, it is referred to as a leaf; if the edge continues to split, it is known as an internal edge.

For example, a dataset of diabetic and nondiabetic patients could have input variables of age, weight, and family diabetic history to predict odds of a new patient having diabetes. The program could set the age variable as the root node (e.g., age $\geq 40$), from which the dataset splits into those who are greater than or equal to 40 and those who are 39 and younger. If the next internal node after selecting greater than or equal to 40 is whether or not a parent has/had diabetes and the leaf estimates the positive answers to have a 60 percent chance of this patient having diabetes, the model presents that leaf as the final output.

This is a very simple example of a decision tree that illustrates the decision process. Decision trees can easily become thousands of nodes deep. Random forest algorithms are merely amalgamations of decision trees. They can be formed from collections of hundreds of decision trees, from which the final outputs are the averaged outputs of the individual trees. Decision tree and random forest algorithms are great for learning highly complex data structures, but they are easily prone to overfitting the data. Overfitting can be attenuated with proper pruning (e.g., setting the $n$ values limits for splitting and leaves) and with large enough random forests.

Neural networks are machine learning algorithms inspired by the neural connections of the human brain. Just as in the human brain, the base unit of neural network algorithms are neurons, and the neurons are arranged into multiple layers. The input variables are referred to as the input layer, the layers of neurons are called hidden layers (there can be several hidden layers), and the output layer consists of the final neuron.

In a feedforward process, a single neuron (a) receives the input feature variables, (b) the feature values are multiplied by a weight, (c) the resulting feature products are added together, along with a bias variable, and (d) the sums are passed through an activation function, commonly a sigmoid function. The weights and biases of each neuron are adjusted based on the partial derivative calculations of the preceding neurons and neural layers. This process is known as backpropagation. The output of the single neuron's activation function is passed to all the neurons in the next hidden layer or a final output layer. The output of the final neuron is thus the predicted value.

Programmers can spend relatively less time restructuring their data as neural networks are very effective at learning highly complex variable relationships. Conversely, due to their complexity, neural network models are difficult to interpret, and the intervariable relationships are largely hidden. Neural networks are best when applied to very large datasets. They require careful hyper-tuning and sufficient computational power.

Machine learning has become the standard tool for data scientists trying to understand large datasets. Researchers are continually improving the accuracy and usability of machine learning programs. However, machine learning algorithms are only as valuable as the data that is used to train the model. Poor data leads to wildly inaccurate results; biased data without proper understanding reinforces social inequalities.

*Raphael A. Rodriguez*

*See also:* Algorithmic Bias and Error; Automated Machine Learning; Deep Learning; Explainable AI; Gender and AI.

**Further Reading**

Garcia, Megan. 2016. "Racist in the Machine: The Disturbing Implications of Algorithmic Bias." *World Policy Journal* 33, no. 4 (Winter): 111–17.

Géron, Aurelien. 2019. *Hands-On Machine Learning with Scikit-Learn and TensorFlow: Concepts, Tools, and Techniques to Build Intelligent Systems.* Sebastopol, CA: O'Reilly.

## Machine Translation

Machine translation involves the automatic translation of human languages with computing technology. From the 1950s to the 1970s, the U.S. government viewed machine translation as a powerful tool in diplomatic efforts related to the containment of communism in the USSR and People's Republic of China. More recently, machine translation has become an instrument for selling products and services in markets otherwise unattainable because of language barriers, and as a product in its own right. Machine translation is also one of the litmus tests used to gauge progress in artificial intelligence. There are three general paradigms by which this artificial intelligence research progresses. The oldest involves rule-based expert systems and statistical approaches to machine translation. Two more recent paradigms are neural-based machine translation and example-based machine translation (or translation by analogy). Today, the automatic translation of language is considered an academic specialty within computational linguistics.

While several origins for the modern field of machine translation are suggested, the idea of automatic translation as an academic field stems from correspondence between the Birkbeck College (London) crystallographer Andrew D. Booth and Rockefeller Foundation's Warren Weaver in 1947. In a surviving memo written to colleagues in 1949, Weaver explained by example how automatic translation might proceed along the lines of code breaking: "I have a text in front of me which is written in Russian, but I am going to pretend that it is really written in English and that it has been coded in some strange symbols. All I need to do is strip off the code in order to retrieve the information contained in the text" (Warren Weaver, as cited in Arnold et al. 1994, 13).

A translation engine lies at the heart of most commercial machine translation systems. Translation engines take sentences entered by the user and parse them several times, each time applying algorithmic rules that transform the source sentence into the desired target language. Both word-based and phrase-based transformation rules are applied. The parser program's first task is usually to do a word-for-word replacement using a two-language dictionary. Additional parsing iterations of the sentences apply comparative grammatical rules by taking into account sentence structure, verb form, and appropriate suffixes. Translation engines are evaluated based on intelligibility and accuracy.

Translation by machine is not flawless. "Word salad" translations may result from poor grammar in the source text; lexical and structural differences between languages; ambiguous usage; multiple meanings of words and idioms; and local variations in usage. The severest early critique of machine translation of language came from MIT philosopher, linguist, and mathematician Yehoshua Bar-Hillel in 1959–60. Bar-Hillel argued that near-perfect machine translation was impossible in principle. To demonstrate the problem, he introduced the following sentence: *John was looking for his toy box. Finally he found it. The box was in the pen. John was very happy.* In this sentence, the word "pen" is a challenge because the word might represent a child's playpen or a ballpoint pen for writing. Knowing the difference requires general-purpose knowledge about the world, which a computer could not have.

Initial rounds of U.S. government funding eroded when, in 1964, the National Academy of Sciences Automatic Language Processing Advisory Committee (ALPAC) released an extremely damaging report about the poor quality and high cost of machine translation. ALPAC concluded that the nation already possessed an ample supply of human translators that could produce far superior translations. Many machine translation experts criticized the ALPAC report, noting machine efficiencies in the preparation of first drafts and successful rollouts of a handful of machine translation systems.

Only a handful of machine translation research groups existed in the 1960s and 1970s. Some of the largest were Canada's TAUM group, the Mel'cuk and Apresian groups in the Soviet Union, the GETA group in France, and the German Saarbrücken SUSY group. The leading provider of automatic translation technology and services in the United States was SYSTRAN (System Translation), a private company supported by government contracts created by Hungarian-born linguist and computer scientist Peter Toma. Toma first became interested in machine translation while at the California Institute of Technology in the 1950s. Moving to Georgetown University around 1960, Toma began collaborating with other machine translation researchers. Both the Georgetown machine translation effort and SYSTRAN's first contract with the U.S. Air Force in 1969 were dedicated to translating Russian into English. The company's first machine translation programs were tested that same year at Wright-Patterson Air Force Base.

In 1974 and 1975, the National Aeronautics and Space Administration (NASA) used SYSTRAN software as a translation aid during the Apollo-Soyuz Test Project. Shortly thereafter, SYSTRAN picked up a contract to provide automatic translation services to the Commission of the European Communities, and the organization has since merged into the European Commission (EC). Seventeen separate machine translation systems focused on different language pairs were in use by the EC for internal communiqués by the 1990s. SYSTRAN migrated its mainframe software to personal computers beginning in 1992. In 1995 the company released SYSTRAN Professional Premium for Windows. SYSTRAN remains a global leader in machine translation.

Some important machine translation systems introduced since the late 1970s revival of machine translation research include METEO, in use since 1977 by the Canadian Meteorological Center in Montreal for the purpose of translating weather bulletins from English to French; ALPS, developed by Brigham Young University for Bible translation; SPANAM, the Pan American Health Organization's Spanish-to-English automatic translation system; and METAL, developed at the University of Texas at Austin for use by the United States Air Force.

Machine translation became more widely available to the public on web browsers in the late 1990s. One of the first online language translation services was Babel Fish, a web-based tool developed from SYSTRAN machine translation technology by a group of researchers at Digital Equipment Corporation (DEC). The tool supported thirty-six translation pairings between thirteen languages. Originally an AltaVista web search engine tool, Babel Fish was later sold to Yahoo! and then Microsoft.

Most online translation services today continue to implement rule-based and statistical machine translation. SYSTRAN, Microsoft Translator, and Google Translate transitioned to the neural machine translation paradigm around 2016. Google Translate supports 103 languages. Neural machine translation relies on predictive deep learning algorithms, artificial neural networks, or connectionist systems modeled after biological brains. Neural-based machine translation is accomplished in two basic stages. In the first phase, the translation engine models its interpretation based on the context of each source word within the full sentence. In the second phase, the artificial neural network translates the entire word model into the target language. To put it simply, the engine predicts the likelihood of word sequences and combinations within whole sentences, building up the translation as an integrated model. The underlying algorithms learn linguistic rules from statistical models. An open-source neural machine translation system, OpenNMT, has been released by the Harvard SEAS natural language processing group collaborating with SYSTRAN.

*Philip L. Frana*

*See also:* Cheng, Lili; Natural Language Processing and Speech Understanding.

**Further Reading**

Arnold, Doug J., Lorna Balkan, R. Lee Humphreys, Seity Meijer, and Louisa Sadler. 1994. *Machine Translation: An Introductory Guide*. Manchester and Oxford: NCC Blackwell.

Bar-Hillel, Yehoshua. 1960. "The Present Status of Automatic Translation of Languages." *Advances in Computers* 1: 91–163.

Garvin, Paul L. 1967. "Machine Translation: Fact or Fancy?" *Datamation* 13, no. 4: 29–31.

Hutchins, W. John, ed. 2000. *Early Years in Machine Translation: Memoirs and Biographies of Pioneers*. Philadelphia: John Benjamins.

Locke, William Nash, and Andrew Donald Booth, eds. 1955. *Machine Translation of Languages*. New York: Wiley.

Yngve, Victor H. 1964. "Implications of Mechanical Translation Research." *Proceedings of the American Philosophical Society* 108 (August): 275–81.

# Macy Conferences

From 1946 to 1960, the Macy Conferences on Cybernetics sought to lay the groundwork for emerging interdisciplinary sciences, among them what would become cybernetics, cognitive psychology, artificial life, and artificial intelligence. Participants in the freewheeling debates of the Macy Conferences included famous twentieth-century scholars, academics, and researchers: psychiatrist W. Ross Ashby, anthropologist Gregory Bateson, ecologist G. Evelyn Hutchinson, psychologist Kurt Lewin, philosopher Donald Marquis, neurophysiologist Warren McCulloch, cultural anthropologist Margaret Mead, economist Oskar Morgenstern, statistician Leonard Savage, physicist Heinz von Foerster, mathematician John von Neumann, electrical engineer Claude Shannon, and mathematician Norbert Wiener among them. The two principle organizers of the conferences were McCulloch, a neurophysiologist working in the Research Laboratory for

Electronics at the Massachusetts Institute of Technology, and von Foerster, professor of signal engineering at the University of Illinois at Urbana-Champaign and coeditor with Mead of the published Macy Conference proceedings.

The philanthropic Josiah Macy Jr. Foundation sponsored all meetings. They were initiated by Macy administrators Frank Fremont-Smith and Lawrence K. Frank, who hoped that the conferences would inspire interdisciplinary dialogue. Fremont-Smith and Frank were particularly concerned about the disciplinary isolation of medical research. The Macy meetings were preceded by a 1942 Macy-sponsored conference on Cerebral Inhibitions, at which Harvard physiology professor Arturo Rosenblueth gave the first public talk on cybernetics, which he entitled "Behavior, Purpose, and Teleology." The ten conferences held between 1946 and 1953 focused on circular causal and feedback mechanisms in biological and social systems. These meetings were followed by five interdisciplinary Group Processes Conferences convened between 1954 and 1960.

Conference organizers eschewed formal papers in favor of informal presentations to encourage direct dialogue between attendees. The first Macy Conferences emphasized the importance of control, communication, and feedback mechanisms in the human nervous system. Other topics discussed included the differences between analog and digital computation, switching circuit design and Boolean logic, game theory, servomechanisms, and communication theory. These issues fall into the realm of what has come to be called "first-order cybernetics." The conferences also considered several biological problems, including adrenal cortex functioning, consciousness, aging, metabolism, nerve impulses, and homeostasis.

The meetings served as a platform for discussions of enduring problems in what would later fall under the umbrella of AI. (Mathematician John McCarthy coined the term "artificial intelligence" at Dartmouth College in 1955.) For example, at the first Macy Conference, Gregory Bateson made a presentation differentiating between "learning" and "learning to learn" drawn from his anthropological investigations and asked attendees to discuss how a computer could accomplish either task. In the eighth conference, attendees took up the subject of decision theory research facilitated by Leonard Savage. At the ninth conference, Ross Ashby introduced the concept of chess-playing automatons. More than any other topic, discussion at the Macy Conferences focused on the applicability of automatic machines as logic models for human cognition.

The Macy Conferences gave birth in 1964 to the professional organization now known as the American Society for Cybernetics. The feedback mechanisms explored in the early debates of the Macy Conferences have been applied to subjects as diverse as artillery control, project management, and marriage counseling.

*Philip L. Frana*

*See also:* Cybernetics and AI; Dartmouth AI Conference.

## Further Reading

Dupuy, Jean-Pierre. 2000. *The Mechanization of the Mind: On the Origins of Cognitive Science.* Princeton, NJ: Princeton University Press.

Hayles, N. Katherine. 1999. *How We Became Posthuman: Virtual Bodies in Cybernetics, Literature, and Informatics.* Chicago: University of Chicago Press.

Heims, Steve J. 1988. "Optimism and Faith in Mechanism among Social Scientists at the Macy Conferences on Cybernetics, 1946–1953." *AI & Society* 2: 69–78.

Heims, Steve J. 1991. *The Cybernetics Group.* Cambridge, MA: MIT Press.

Pias, Claus, ed. 2016. *The Macy Conferences, 1946–1953: The Complete Transactions.* Zürich, Switzerland: Diaphanes.

# McCarthy, John (1927–2011)

John McCarthy was an American computer scientist and mathematician best known for helping to establish the field of artificial intelligence in the late 1950s and for championing the use of formal logic in artificial intelligence research. A prolific thinker, McCarthy made contributions to programming languages and operating systems research, earning him numerous awards. However, artificial intelligence, and what he termed "formalizing common sense," remained the primary research focus throughout McCarthy's life (McCarthy 1990).

McCarthy first encountered the ideas that would lead him to AI as a graduate student at the 1948 Hixon symposium on "Cerebral Mechanisms in Behavior." The symposium was held at the California Institute of Technology, where McCarthy had recently completed his undergraduate work and enrolled in a graduate program in mathematics. By 1948, machine intelligence had become a topic of considerable scholarly attention in the United States under the broad label of cybernetics, and several prominent cyberneticists were in attendance at the symposium, including Princeton mathematician John von Neumann. A year later, McCarthy transferred to the Princeton mathematics department, where he shared some early thoughts inspired by the symposium with von Neumann. Despite von Neumann's encouragement, McCarthy never published the work, deciding that cybernetics could not answer his questions about human knowledge.

At Princeton, McCarthy completed a dissertation on partial differential equations. After graduating in 1951, he remained at Princeton as an instructor, and in summer 1952, he had the opportunity to work at Bell Labs with cyberneticist and founder of information theory Claude Shannon, whom he convinced to collaborate with him on an edited collection of essays on machine intelligence. The contributions to *Automata Studies* covered a range of disciplines from pure mathematics to neurology. To McCarthy, however, the published works were not sufficiently focused on the crucial question of how to build intelligent machines.

In 1953, McCarthy took a job in the mathematics department at Stanford, but he was let go just two years later, perhaps, he conjectured, because he spent too much time thinking about intelligent machines and not enough on his mathematical research. He next took a job at Dartmouth in 1955, as IBM was in the process of establishing the New England Computation Center at MIT. The New England Computation Center provided access to an IBM computer, installed at MIT, and made available to a collection of New England universities, including Dartmouth. Through the IBM initiative, McCarthy met IBM researcher Nathaniel Rochester, who brought McCarthy to IBM in the summer of 1955 to work with his research group. There, McCarthy convinced Rochester of the need for further work on machine intelligence, and together with Rochester, Shannon, and Marvin Minsky, a graduate student at Princeton, he submitted a proposal to the Rockefeller Foundation for a "Summer Research Project on Artificial Intelligence," which included the first known use of the phrase "artificial intelligence."

The Dartmouth Project is widely recognized as the seminal event in the history of the field of AI, yet the meeting did not unfold as McCarthy had envisioned. Because the proposal was for such a novel area of study with a relatively junior professor as author, and although Shannon's reputation did carry significant weight with the Rockefeller Foundation, the Foundation funded the proposal at half the requested budget. Moreover, because the event was held over the course of a number of weeks during the summer of 1955, few of the attendees could stay the entire time. As a result, the Dartmouth conference was a rolling event with a varying and unpredictable guest list. Nevertheless, despite its disorganized execution, the conference helped establish AI as an independent field of study.

While still at Dartmouth, McCarthy received a Sloan fellowship in 1957 to spend a year at MIT, closer to IBM's New England Computation Center. In 1958, MIT offered McCarthy a position in their Electrical Engineering department, which he accepted. He was later joined by Minsky, who took a position in the mathematics department. In 1958, McCarthy and Minsky approached Jerome Wiesner, director of MIT's Research Laboratory of Electronics, and proposed the establishment of an official AI laboratory. Wiesner agreed on the condition that McCarthy and Minsky also accept into the laboratory six recently admitted graduate students, and so the "artificial intelligence project" began training its first generation of students.

That same year McCarthy published his first paper on the subject of AI. "Programs with Common Sense" outlined a vision for a computer system he called the Advice Taker, which would be capable of receiving and interpreting instructions in ordinary natural language from nonexpert users. Advice Taker represented the beginning of a research program that McCarthy would describe as "formalizing common sense." McCarthy believed common sense thoughts that human beings have every day, such as realizing if you don't know a phone number, you will need to look it up before calling, could be expressed as mathematical formulae and fed into a computer, allowing the computer to reach the same kinds of common sense conclusions as human beings. McCarthy believed such formalizing of common sense was the key to artificial intelligence. Presented at the U.K. National Physical Laboratory's "Symposium on Mechansation of Thought Processes," McCarthy's paper helped define the symbolic program of AI research.

Although AI was, by the late 1950s, the central focus of McCarthy's research, he was active in a variety of other fields related to computing. In 1957, he was appointed to a committee of the Association for Computing Machinery engaged in the design of the ALGOL programming language, which would become the de facto language for academic research for many years. In 1958, he designed the LISP programming language for AI research, descendants of which are still in use today across industry and academia.

In addition to his work on programming languages, McCarthy made contributions to research on computer operating systems through the design of time-sharing systems. Large and expensive, early computers could only be used by one person at a time. From his first encounter with computers at IBM in 1955, McCarthy recognized the need for multiple users across a large organization, such as a university or hospital, to be able to access the organization's computer systems simultaneously from computer terminals in their offices. At MIT, McCarthy advocated for research on such systems, becoming part of a university committee exploring the topic and eventually helping to initiate work on MIT's Compatible Time-Sharing System (CTSS). Although McCarthy would leave MIT before the CTSS work was complete, his advocacy, while a consultant at Bolt Beranek and Newman in Cambridge, with J.C.R. Licklider, future office head at the Advanced Research Projects Agency, the predecessor of DARPA, was instrumental in helping MIT secure significant federal support for computing research.

In 1962, Stanford Professor George Forsythe invited McCarthy to join what would become the second department of computer science in the United States, after Purdue's. McCarthy insisted he would only go only as full professor, a demand he thought would be more than Forsythe could manage for an early career researcher. Forsythe was able to convince Stanford to approve McCarthy's full professorship, and so he left for Stanford, where he would set up the Stanford AI laboratory in 1965.

McCarthy oversaw research at Stanford on AI topics such as robotics, expert systems, and chess until his retirement in 2000. The child of parents who had been active members of the Communist party, McCarthy had a lifelong interest in Russian affairs. Having taught himself Russian, he maintained many professional contacts with cybernetics and AI researchers in the Soviet Union, traveling and teaching there in the mid-1960s, and even organizing a chess match in 1965 between a Stanford chess program and a Russian counterpart, which the Russian program won. While at Stanford, he developed numerous foundational concepts in the theory of symbolic AI such as that of circumscription, which expresses the idea that a computer must be allowed to make reasonable assumptions about problems presented to it, otherwise even simple scenarios would need to be specified in such exacting logical detail as to make the task all but impossible.

Although the methods McCarthy pioneered have fallen out of favor in contemporary AI research, his contributions have been recognized with numerous awards, including the 1971 Turing Award, the 1988 Kyoto Prize, a 1989 induction into the National Academy of Sciences, the 1990 Presidential Medal of Science, and the 2003 Benjamin Franklin Medal. McCarthy was a prolific thinker who constantly envisioned new technologies, from a space elevator for cheaply moving matter into orbit to a system of carts suspended from wires meant to improve transportation in urban areas. Yet, when asked during a 2008 interview what he thought the most important questions in computing today were, McCarthy responded without hesitation, "Formalizing common sense," the same project that had motivated him from the very beginning.

*Evan Donahue*

*See also:* Cybernetics and AI; Expert Systems; Symbolic Logic.

**Further Reading**

Hayes, Patrick J., and Leora Morgenstern. 2007. "On John McCarthy's 80th Birthday, in Honor of His Contributions." *AI Magazine* 28, no. 4 (Winter): 93–102.

McCarthy, John. 1990. *Formalizing Common Sense: Papers*, edited by Vladimir Lifschitz. Norwood, NJ: Albex.

Morgenstern, Leora, and Sheila A. McIlraith. 2011. "John McCarthy's Legacy." *Artificial Intelligence* 175, no. 1 (January): 1–24.

Nilsson, Nils J. 2012. "John McCarthy: A Biographical Memoir." *Biographical Memoirs of the National Academy of Sciences.* http://www.nasonline.org/publications /biographical-memoirs/memoir-pdfs/mccarthy-john.pdf.

# Medicine, Artificial Intelligence in

Artificial intelligence supports health-care professionals by assisting with tasks that require the manipulation of large amounts of data. AI technologies are revolutionizing how practitioners diagnose, treat, and predict outcomes in clinical contexts.

One of the first successful uses of artificial intelligence in medicine occurred when Scottish surgeon Alexander Gunn utilized computer analysis to help diagnose acute abdominal pain in the 1970s. Since then, artificial intelligence applications have grown in number and sophistication, mirroring developments in computer science. The most common AI applications in medicine are artificial neural networks, fuzzy expert systems, evolutionary computation, and hybrid intelligent systems.

Artificial neural networks (ANNs) are brain-inspired systems designed to replicate the way humans process information and learn. The first artificial "neurons" were developed by Warren McCulloch and Walter Pitts in the mid-twentieth century. More recently, Paul Werbos has given artificial neural networks the ability to perform backpropagation, which involves the adjustment of neural layers in response to novel situations.

ANNs are made up of interconnected processors labeled "neurons" that perform parallel processing of data. These neurons are typically organized into an input, middle (or hidden), and output layer. Each layer is fully connected to the previous layer. The individual neurons are connected or linked and given an associated weight. Adjusting these weights is how the technology "learns." ANNs make possible the development of sophisticated tools that are capable of analyzing nonlinear data and generalizing from imprecise data sets.

The ability to recognize patterns and process nonlinear information led to robust applications of ANNs in clinical settings. ANNs are used for image analysis in radiology, identification of high-risk patients, and intensive care data analysis. ANNs are especially useful for diagnosing and predicting outcomes in situations where a range of variables need to be considered.

Fuzzy expert systems are artificial intelligence tools that can function in ambiguous contexts. In contrast to systems based on conventional logic, fuzzy systems are built upon the insight that data analysis must often cope with uncertainty and vagueness. Fuzzy expert systems are valuable in health care because medical knowledge is often complex and imprecise. Fuzzy systems are capable of recognizing, interpreting, manipulating, and using vague information for various purposes. Today, fuzzy logic systems are used to predict a wide range of outcomes for patients, such as those suffering from lung cancer and melanoma. They have also been used to develop treatments for critically ill patients.

Evolutionary computation involves algorithms inspired by natural evolutionary processes. Evolutionary computation solves problems by optimizing their performance through trial and error. They generate an initial set of solutions and, with each successive generation, make random small changes to the data set and remove unsuccessful intermediate solutions. These solutions can be said to be subjected to mutation and a type of natural selection. The result are algorithms that gradually evolve, as the fitness of the solutions increases. While many variants of these programs exist, the most prominent type used in the context of medicine is known as the genetic algorithm. These were first developed by John Holland in the 1970s, and they utilize basic evolutionary structures to formulate solutions in complex contexts, such as clinical settings. They are used to perform a wide range of clinical tasks, including diagnosis, medical imaging, scheduling, and signal processing.

Hybrid intelligent systems are AI technologies that combine more than one system to capitalize on the strengths of the techniques described above. Hybrid systems are better able to mimic humanlike reasoning and adapt to changing environments. As with the individual AI technologies described above, these systems are being used in a wide range of clinical settings. They are currently used to diagnose breast cancer, assess myocardial viability, and analyze digital mammograms.

*Samantha Noll*

*See also:* Clinical Decision Support Systems; Computer-Assisted Diagnosis; MYCIN; Precision Medicine Initiative.

**Further Reading**

Baeck, Thomas, David B. Fogel, and Zbigniew Michalewicz, eds. 1997. *Handbook of Evolutionary Computation*. Boca Raton, FL: CRC Press.

Eiben, Agoston, and Jim Smith. 2003. *Introduction to Evolutionary Computing*. Berlin: Springer-Verlag.

Patel, Jigneshkumar L., and Ramesh K. Goyal. 2007. "Applications of Artificial Neural Networks in Medical Science." *Current Clinical Pharmacology* 2, no. 3: 217–26.

Ramesh, Anavai N., Chandrasekhar Kambhampati, John R. T. Monson, and Patrick J. Drew. 2004. "Artificial Intelligence in Medicine." *Annals of the Royal College of Surgeons of England* 86, no. 5: 334–38.

# Minsky, Marvin (1927–2016)

Donner Professor of Science Marvin Minsky was a well-known American cognitive scientist, inventor, and artificial intelligence investigator. He cofounded the Artificial Intelligence Laboratory in the 1950s and the Media Lab in the 1980s at the Massachusetts Institute of Technology. Such was his fame that, while serving as advisor to the 1960s classic Stanley Kubrick film *2001: A Space Odyssey*, the sleeping astronaut Dr. Victor Kaminski (killed by the HAL 9000 sentient computer) was named in his honor.

Minsky became interested in intelligence, thinking, and learning machines at the end of high school in the 1940s. As an undergraduate at Harvard, he showed interest in neurology, physics, music, and psychology. He worked with cognitive psychologist George Miller on problem-solving and learning theories, and with J. C. R. Licklider, professor of psychoacoustics and later father of the internet, on perception and brain modeling theories. While at Harvard, Minsky began thinking about theories of the mind. "I imagined that the brain was composed of little relays—the neurons—and each of them had a probability attached to it that would govern whether the neuron would conduct an electric pulse," he later remembered. "This scheme is now known technically as a stochastic neural network" (Bernstein 1981). This theory is similar to Hebbian theory set out in *The Organization of Behavior* (1946) by Donald Hebb. He completed an undergraduate thesis on topology in the mathematics department.

As a graduate student at Princeton University, Minsky studied mathematics but became increasingly interested in trying to create artificial neurons from vacuum tubes such as those described in Warren McCulloch and Walter Pitts's famous 1943 paper "A Logical Calculus of the Ideas Immanent in Nervous Activity." He imagined that such a machine might be able to negotiate mazes like a rat. He built the machine, dubbed SNARC (Stochastic Neural-Analog Reinforcement Calculator), with the help of fellow Princeton student Dean Edmonds in the summer of 1951 with Office of Naval Research funding. The machine contained 300 tubes and several electric motors and clutches. The machine used the clutches to adjust its own knobs, making it a learning machine. The electric rat moved randomly at first, but then, by reinforcement of probabilities, it learned how to make better choices and achieve a desired goal. The maze eventually contained multiple rats that learned from each other. In his doctoral thesis, Minsky established a second memory for his hard-wired neural network, which helped the rat remember what the stimulus had been. This allowed the machine to search its memory when confronted with a new situation and predict the best appropriate course of action. At the time Minsky had hoped that, with enough memory loops, his self-organizing random networks might spontaneously lead to emergence of conscious intelligence. Minsky completed his dissertation on "Neural Nets and the Brain Model Problem" in 1954.

Minsky continued to think about how to create an artificial intelligence after graduation from Princeton. With John McCarthy, Nathaniel Rochester, and Claude Shannon, he organized and participated in the Dartmouth Summer Research Project on Artificial Intelligence in 1956. The Dartmouth workshop is often described as the formative event in artificial intelligence research. During the summer workshop, Minsky began simulating the computational process of proving Euclid's geometric theorems, using pieces of paper because no computer was available. He realized that he could design an imaginary machine to find proofs without telling the machine exactly what needed to be done. Minsky showed the results to Nathaniel Rochester, who returned to his job at IBM and asked a new physics recruit named Herbert Gelernter to write a geometry-proving program on a computer. Gelernter wrote a program in a language he devised called FORTRAN List-Processing Language. Later, John McCarthy took Gelernter's language and some ideas derived from the mathematician Alonzo Church and created the mainstay AI language known as LISP (List-Processing).

Minsky arrived at MIT in 1957. In the university's Lincoln Laboratory, he worked first on problems of pattern recognition with Oliver Selfridge. The next year he moved into a position as assistant professor in the mathematics department. With McCarthy, who had moved to MIT from Dartmouth, he created the AI Group. They continued to work on the principles of machine learning. In the 1960s, Minsky began collaborating with mathematician Seymour Papert. Together they published *Perceptrons: An Introduction to Computational Geometry* (1969) about a kind of artificial neural network described by Cornell Aeronautical Laboratory psychologist Frank Rosenblatt. The book started a controversy spanning decades in the AI community, which continues in some ways to the present day. The mathematical proofs outlined in Minsky and Papert's book influenced the field to move in the direction of symbolic AI (also known as "Good Old-Fashioned AI" or GOFAI) into the 1980s, at which time artificial intelligence researchers renewed their interest in perceptrons and neural networks.

In the 1960s, time-shared computers became more widely available on the MIT campus, and Minsky began working with students on problems of machine intelligence. One of the early projects involved teaching the computers to solve problems in introductory calculus with tools of symbolic manipulation, namely differentiation and integration. A program for such symbol manipulation was completed in 1961 by his student James Robert Slagle. The program, which ran on an IBM 7090 transistorized mainframe computer, was named SAINT (Symbolic Automatic INTegrator). Other students generalized the work for any symbol manipulation that might be required with their program MACSYMA. Another problem for Minsky's students involved teaching a machine to reason by analogy. Minsky's group also tackled problems in computational linguistics, computer vision, and robotics. One of his students, Daniel Bobrow, taught a computer to solve word problems, a feat combining language parsing and mathematics. The first robot controlled by a computer was made by student Henry Ernst, who created a mechanical hand with photoelectric tactile sensors intended for gripping nuclear materials.

To solve more complicated problems in computer vision and manipulation, Minsky worked with Papert to create semi-independent programs that could interact with one another to solve problems. Minsky and Papert's systems of nonhierarchical management came together into a natural intelligence theory that they referred to as the Society of Mind. Under this theory, intelligence is an emergent property and the product of subtle interactions between programs. By 1970, the MIT AI Group had taught a computer-controlled robot to construct structures out of children's blocks after looking at other structures.

The blocks-manipulating robot and the Society of Mind theory continued to evolve throughout the 1970s and 1980s. Minsky eventually published *The Society of Mind* (1986) in which he created a model for the emergence intelligence from separate mental agents and their interactions—rather than some basic principle or universal method. In the book, which is composed of 270 original essays, he discussed concepts of consciousness, self, free will, memory, genius, language, memory, brainstorming, learning, and many more. Agents, in Minsky's view, require no mind or thinking and feeling abilities of their own. They are not smart. But together, as a society, they produce what we experience as human intelligence. In other words, knowing how to accomplish any specific objective requires the effort of multiple agents. Minsky's robot builder needs agents to see, move, find, grasp, and balance blocks. "I like to think that this project," he wrote, "gave us glimpses of what happens inside certain parts of children's minds when they learn to 'play' with simple toys" (Minsky 1986, 29).

Minsky suggested that there might be over one hundred agents working together to produce what is known as mind. He extended his ideas on Society of Mind in the book *Emotion Machine* (2006). Here he made the argument that emotions are not a different kind of thinking. Rather, they represent ways to think about different types of problems that minds encounter in the world. Minsky argued that the mind switches between different ways to think, thinks on many levels, finds diverse ways to represent things, and builds manifold models of ourselves.

In his later years, Minsky commented through his writings and interviews on a wide range of popular and noteworthy topics related to artificial intelligence and robotics. *The Turing Option* (1992), a novel written by Minsky in collaboration with science fiction author Harry Harrison, grapples with problems of superintelligence in the year 2023. In 1994, he penned a piece for *Scientific American* entitled "Will Robots Inherit the Earth?" to which he answered "Yes, but they will be our children" (Minsky 1994, 113).

Minsky once speculated that a superintelligent AI might one day trigger a Riemann Hypothesis Catastrophe, in which an agent tasked with the goal of solving the hypothesis takes over all of earth's resources to acquire ever more supercomputing power. He didn't view this possibility as very likely. Minsky believed that it might be possible for humans to communicate with intelligent extraterrestrial life forms. They would think like humans because they would be subject to the same "limitations on space, time, and materials" (Minsky 1987, 117). Minsky was also a critic of the Loebner Prize, the world's oldest Turing Test-like competition, saying that the contest is unhelpful to the field of artificial intelligence. He countered with his own Minsky Loebner Prize Revocation Prize to anyone who could stop Hugh Loebner's annual competition. Minsky and Loebner both died in 2016; the Loebner Prize contest continues.

Minsky also invented the confocal microscope (1957) and the head-mounted display or HMD (1963). He won the Turing Award (1969), the Japan Prize (1990), and the Benjamin Franklin Medal (2001). Minsky advised many doctoral students who became influential leaders in computer science, including Daniel Bobrow (operating systems), K. Eric Drexler (molecular nanotechnology), Carl Hewitt (mathematics and philosophy of logic), Danny Hillis (parallel computing), Benjamin Kuipers (qualitative simulation), Ivan Sutherland (computer graphics), and Patrick Winston (who succeeded Minsky as director of the MIT AI Lab).

*Philip L. Frana*

*See also:* AI Winter; Chatbots and Loebner Prize; Dartmouth AI Conference; *2001: A Space Odyssey.*

**Further Reading**

Bernstein, Jeremy. 1981. "Marvin Minsky's Vision of the Future." *New Yorker*, December 7, 1981. https://www.newyorker.com/magazine/1981/12/14/a-i.

Minsky, Marvin. 1986. *The Society of Mind.* London: Picador.

Minsky, Marvin. 1987. "Why Intelligent Aliens Will Be Intelligible." In *Extraterrestrials: Science and Alien Intelligence*, edited by Edward Regis, 117–28. Cambridge, UK: Cambridge University Press.

Minsky, Marvin. 1994. "Will Robots Inherit the Earth?" *Scientific American* 271, no. 4 (October): 108–13.

Minsky, Marvin. 2006. *The Emotion Machine.* New York: Simon & Schuster.

Minsky, Marvin, and Seymour Papert. 1969. *Perceptrons: An Introduction to Computational Geometry.* Cambridge, MA: Massachusetts Institute of Technology.

Singh, Push. 2003. "Examining the Society of Mind." *Computing and Informatics* 22, no. 6: 521–43.

# Mobile Recommendation Assistants

Mobile Recommendation Assistants, also known as Virtual Assistants, Intelligent Agents, or Virtual Personal Assistants, refer to a set of software capabilities coupled with conversational user interface and artificial intelligence for the purpose of acting on behalf of a user. Together they can provide what appears to a user as an agent.

An *agent* in this sense is distinguished from a *tool* in that an agent has the ability to act independently with some degree of autonomy to make decisions. The design of mobile recommendation assistants can incorporate many attributes that enhance the user's perception of agency. Some such techniques include representing the technology with a visual avatar, developing aspects of personality such as humor or informal/colloquial language, providing a voice and a proper name, designing a consistent manner of behavior, etc.

A mobile recommendation assistant may be used by a human user to assist with a large variety of activities such as opening software applications, providing answers to questions, performing tasks (operating other software/hardware), or being used for conversational commerce or for entertainment purposes such as telling stories, telling jokes, playing games, etc. Currently, there are a number of mobile voice assistants in production, each developed for particular businesses, use cases, and user experiences such as Apple's Siri, Baidu's Xiaodu, Amazon's Alexa, Microsoft's Cortana, Google's Google Assistant, and Xiaomi's Xiao AI.

Mobile recommendation assistants utilize a variety of user interface modalities. Some may be entirely text based; these are sometimes referred to as chatbots. The typical use case for a chatbot is for business to customer (B2C) communication, and popular applications include online retail communication, insurance, banking, transportation, and restaurants. Increasingly, chatbots are used for medical and psychological applications as in the case of helping users with behavior change. Similar applications are gaining popularity in educational settings, to support students in language training or studying and test preparation. A popular social media example of a chatbot would be Facebook's Messenger.

While not all mobile recommendation assistants require voice-enabled interaction as an input modality (they may rely solely on text input such as web site chatbots), many current examples utilize voice to great effect. The voice-enabled user-interface is one of a few related precursor technologies used by a Mobile Recommendation Assistant.

Early voice-enabled user-interfacing was made possible through a command syntax that would have been hand-coded in advance as a set of rules or heuristics. These rule-based systems enabled hands-free operation of machines through the user issuing verbal commands.

The first speech recognition application, unveiled at Seattle's 1962 World's Fair, was developed by IBM. IBM Shoebox, as it was called, had a small vocabulary of sixteen words and nine digits. By the 1990s, personal home computers and software from IBM and Microsoft were equipped with rudimentary voice recognition; the first mobile application of a mobile assistant was Apple's Siri, which came equipped on the iPhone 4s in 2011.

From a user experience perspective, these early voice recognition systems were disadvantaged relative to conversational mobile agents due to their requiring a user to learn, and adhere to, a predetermined command language. In terms of contributing to natural humanlike dialog with machines, a characteristic of modern mobile recommendation assistants, the result of rule-based voice interaction can feel mechanical.

Natural language processing, or NLP, instead relies on machine learning and statistical inference to learn rules through the analysis of large linguistic data (corpora). Natural language processing machine learning utilizes decision trees as well as statistical modeling to comprehend requests made in a variety of ways typical of the way people normally talk to each other. More advanced agents may include capability that requires interpreting a user's intent in light of user preferences made explicit through settings or other inputs, such as calendar entries. The combination of probabilistic reasoning and natural language processing can be seen in Google's Voice Assistant, which includes conversational elements such as paralanguage ("uh", "uh-huh", "ummm") to produce a natural sounding dialogue.

Modern digital assistants rely on multimodal means of communicating to convey comprehension and attentiveness. Paralanguage refers to the components of communication that do not contain semantic content yet are a critical part of conveying meaning in context. These can communicate intent, cooperation in the conversation, or convey emotion. In the example of Google's voice assistant using its Duplex technology, the elements of paralanguage used are called vocal segregates or speech disfluencies; these are used to not only help the assistant sound more human but also to help the conversation "flow" by filling gaps or allowing the listener to feel heard.

Kinesics is another important interaction element that helps an assistant feel more like an active communication partner. Kinesics refers to the use of gesture, movement, facial expression, or emotion to help facilitate the flow of conversation. One current example of the use of facial expression can be found in the automotive company NIO's virtual robot assistant called Nome. Nome is a digital voice assistant embodied in a spherical housing equipped with an LCD screen that sits atop the center dashboard of NIO's ES8. It can mechanically turn its "head" to attend to different speakers and uses facial expressions to express emotions. Another example is MIT's Dr. Cynthia Breazeal's commercial Jibo home robot, which leverages anthropomorphism through paralinguistic methods.

Less anthropomorphic uses of kinesics can be seen in the graphical user interface elements on Apple's Siri or in illumination arrays such as those on Amazon Alexa's physical interface Echo or in Xiami's Xiao AI, where motion graphics or lighting animations are used to communicate states of communication such as listening, thinking, speaking, or waiting.

The increasing intelligence and accompanying anthropomorphism (or in some cases zoomorphism or mechano-morphism) can raise some ethical concerns related to user experience. The desire for more anthropomorphic systems stems from the beneficial user experience of humanlike agentic systems whose communicative behaviors more closely match familiar interactions such as conversation made possible through natural language and paralinguistics. The primary advantage of natural conversation systems is that they do not require a user to learn a new grammar or semantics in order to effectively communicate commands and desires. A user's familiar mental model of communication, learned through engaging with other humans, is applicable to these more anthropomorphic human machine interfaces.

However, as machine systems more closely approximate human-to-human interaction, transparency and security become issues where a user's inferences about a machine's behavior are informed by human-to-human communication. The establishing of comfort and rapport can occlude the ways in which virtual assistant cognition, and inferred motivation, is unlike human cognition. In terms of cognition (the assistant's intelligence and perceptual capacities), many systems may be equipped with motion sensors, proximity sensors, cameras, microphones, etc. which approximate, emulate, or even exceed human capacities. While these facilitate some humanlike interaction through improved perception of the environment, they can also be used for recording, documenting, analyzing, and sharing information that may be opaque to a user when their mental model, and the machine's interface, doesn't communicate the machine's operation at a functional level. For example, a digital assistant visual avatar may close his eyes, or disappear, after a user interaction, but there is no necessary association between that behavior with the microphone's and camera's ability to keep recording.

Thus, data privacy concerns are becoming more salient, as digital assistants are increasingly integrated into the everyday lives of human users. Where specifications, manufacturer data collection objectives, and machine behaviors are potentially misaligned with user's expectations, transparency becomes a key issue to be addressed.

Finally, security becomes an issue when it comes to data storage, personal information, and sharing practices, as hacking, misinformation, and other forms of misuse threaten to reduce trust in technological systems and associated institutions.

*Michael Thomas*

*See also:* Chatbots and Loebner Prize; Mobile Recommendation Assistants; Natural Language Processing and Speech Understanding.

**Further Reading**

Lee, Gary G., Hong Kook Kim, Minwoo Jeong, and Ji-Hwan Kim, eds. 2015. *Natural Language Dialog Systems and Intelligent Assistants*. Berlin: Springer.

Leviathan, Yaniv, and Yossi Matias. 2018. "Google Duplex: An AI System for Accomplishing Real-world Tasks Over the Phone." *Google AI Blog*. https://ai.googleblog.com/2018/05/duplex-ai-system-for-natural-conversation.html.

Viken, Alexander. 2009. "The History of Personal Digital Assistants, 1980–2000." *Agile Mobility*, April 10, 2009.

Waddell, Kaveh. 2016. "The Privacy Problem with Digital Assistants." *The Atlantic*, May 24, 2016. https://www.theatlantic.com/technology/archive/2016/05/the-privacy-problem-with-digital-assistants/483950/.

# MOLGEN

Developed between 1975 and 1980, MOLGEN is an expert system that aided molecular biologists and geneticists in designing experiments. It was the third expert system designed by Edward Feigenbaum's Heuristic Programming Project (HPP) at Stanford University (after DENDRAL and MYCIN). Additionally, like MYCIN before it, MOLGEN gained hundreds of users beyond Stanford. In the 1980s, MOLGEN was first made available through time-sharing on the GENET network for artificial intelligence researchers, molecular biologists, and geneticists. By the late 1980s, Feigenbaum established the company IntelliCorp to sell a stand-alone software version of MOLGEN.

In the early 1970s, scientific breakthroughs related to chromosomes and genes had generated an information explosion. Stanford University biochemist Paul Berg conducted the first experiments in gene splicing in 1971. Two years later, Stanford geneticist Stanley Cohen and University of California at San Francisco biochemist Herbert Boyer successfully inserted recombinant DNA into an organism; the host organism (a bacterium) then naturally reproduced the foreign rDNA structure in its own offspring. These advances led Stanford molecular biologist Joshua Lederberg to tell Feigenbaum that it was an opportune moment to develop an expert system in Lederberg's own field of molecular biology. (Lederberg and Feigenbaum had previously joined forces on the first expert system DENDRAL.) The two agreed that what DENDRAL had done for mass spectrometry, MOLGEN could do for recombinant DNA research and genetic engineering. Indeed, both expert systems were developed for emerging scientific fields. This allowed MOLGEN (and DENRAL) to incorporate the most recent scientific knowledge and make contributions to their respective field's further development.

Feigenbaum was MOLGEN's principal investigator at HPP, with Mark Stefik and Peter Friedland developing programs for it as their thesis project. The idea was to have MOLGEN follow a "skeletal plan" (Friedland and Iwasaki 1985, 161). Mimicking a human expert, MOLGEN planned a new experiment by starting from a design procedure that proved successful for a similar problem in the past. MOLGEN then modified the plan in a hierarchical stepwise manner. The combination of skeletal plans and MOLGEN's extensive knowledge base in molecular biology gave the system the ability to select the most promising new experiments. By 1980, MOLGEN had incorporated 300 lab methods and strategies as well as current data on forty genes, phages, plasmids, and nucleic acid structures. Drawing on the molecular biological expertise of Douglas Brutlag, Larry Kedes, John Sninsky, and Rosalind Grymes of Stanford University, Friedland and Stefik provided MOLGEN with a suite of programs. These included SEQ (for nucleic acid sequence analysis), GA1 (later called MAP, to generate enzyme maps of DNA structures), and SAFE (for selecting enzymes most suitable for gene excision).

MOLGEN was made accessible to the molecular biology community outside of Stanford beginning in February 1980. The system was connected to SUMEX-AIM (Stanford University Medical Experimental computer for Artificial Intelligence in Medicine) under an account called GENET. GENET quickly found hundreds of users across the United States. Frequent visitors included members of academic laboratories, scientists at commercial giants such as Monsanto, and researchers at small start-ups such as Genentech.

The National Institutes of Health (NIH), the principal sponsor of SUMEX-AIM, eventually decided that corporate users could not be granted free access to cutting-edge technology developed with public funding. Instead, the NIH encouraged Feigenbaum, Brutlag, Kedes, and Friedland to establish IntelliGenetics for corporate biotech users. With the help of a five-year NIH grant of $5.6 million, IntelliGenetics developed BIONET to offer MOLGEN and other programs on GENET for sale or rent. By the end of the 1980s, 900 laboratories were accessing BIONET all over the world for an annual fee of $400.

IntelliGenetics also offered a software package for sale to companies that did not want to put their data on BIONET. MOLGEN's software did not sell well as a stand-alone package until the mid-1980s, when IntelliGenetics removed its genetics content and kept only its underlying Knowledge Engineering Environment (KEE). The AI part of IntelliGenetics that sold this new KEE shell changed its name to IntelliCorp. Two public offerings followed, but eventually growth leveled out again. Feigenbaum conjectured that the commercial success of MOLGEN's shell was hindered by its LISP-language; although preferred by pioneering computer scientists working on mainframes computers, LISP did not generate similar interest in the corporate minicomputer world.

*Elisabeth Van Meer*

*See also:* DENDRAL; Expert Systems; Knowledge Engineering.

**Further Reading**

Feigenbaum, Edward. 2000. *Oral History.* Minneapolis, MN: Charles Babbage Institute.

Friedland, Peter E., and Yumi Iwasaki. 1985. "The Concept and Implementation of Skeletal Plans." *Journal of Automated Reasoning* 1: 161–208.

Friedland, Peter E., and Laurence H. Kedes. 1985. "Discovering the Secrets of DNA."
    *Communications of the ACM* 28 (November): 1164–85.
Lenoir, Timothy. 1998. "Shaping Biomedicine as an Information Science." In *Proceed-
    ings of the 1998 Conference on the History and Heritage of Science Information
    Systems*, edited by Mary Ellen Bowden, Trudi Bellardo Hahn, and Robert V.
    Williams, 27–46. Pittsburgh, PA: Conference on the History and Heritage of Sci-
    ence Information Systems.
Watt, Peggy. 1984. "Biologists Map Genes On-Line." *InfoWorld* 6, no. 19 (May 7): 43–45.

# Monte Carlo

Monte Carlo is a simulation method to solve complex problems using multiple
runs of a nondeterministic simulation based on a random number generator.
Deterministic methods solve equations or systems of equations to arrive at a fixed
solution, and every time the calculation is run, it will result in the same solution.
By contrast, in Monte Carlo methods a random number generator is used to choose
different paths, resulting in a variable solution each time. Monte Carlo methods
are used when the deterministic equations are not known, when there are a large
number of variables, and especially for problems that are probabilistic in nature.

Examples of problems that commonly use Monte Carlo methods are games of
chance, nuclear simulations, problems with quantum effects, and weather forecasting. In an artificial intelligence context, Monte Carlo methods are commonly
used in machine learning and memory simulations to provide more robust answers
and to represent, for example, how memory varies. Because each Monte Carlo
simulation results in one possible outcome, the simulation must be run hundreds
to millions of times to create a probability distribution, which is the solution to the
overall problem. Monte Carlo methods can be considerably more computational-
intensive than deterministic methods.

Monte Carlo is used commonly in AI for game applications, such as checkers,
chess, and Go. At each step, these games (especially Go) have a very large number
of possible moves. A technique called Monte Carlo tree search is used, which uses
the MC method to repeatedly play the game, making a random move at each step.
Eventually, the AI system learns the best moves for a particular game situation.
Monte Carlo tree search AIs have very good track records and regularly beat other
AI game algorithms.

*Mat Brener*

*See also:* Emergent Gameplay and Non-Player Characters.

**Further Reading**

Andrieu, Christophe, Nando de Freitas, Arnaud Doucet, and Michael I. Jordan. 2003. "An
    Introduction to MCMC for Machine Learning." *Machine Learning* 50: 5–43.
Eckhard, Roger. 1987. "Stan Ulam, John von Neumann, and the Monte Carlo Method."
    *Los Alamos Science* 15 (Special Issue): 131–37.
Fu, Michael C. 2018. "Monte Carlo Tree Search: A Tutorial." In *Proceedings of the 2018
    Winter Simulation Conference*, edited by M. Rabe, A. A. Juan, N. Mustafee, A.
    Skoogh, S. Jain, and B. Johansson, 222–36. Piscataway, NJ: IEEE.

# Moral Turing Test

The Moral Turing Test, also referred to as the Ethical Turing Test, Ethical (Moral) Turing Test, or MTT, is a variation of the Turing Test proposed by mathematician and computer scientist Alan Turing (1912–1954). The Turing Test involves a human judge attempting to distinguish between a computer program and a human via a set of written questions and responses. If the computer program imitates a human such that the human judge cannot distinguish the computer program's responses from the human's responses, then the computer program has passed the test, suggesting that the program is capable of intelligent thought. The Moral Turing Test is a more specific variation of the Turing Test, designed to test the ethical decision-making of a machine. The machine is first trained to learn and follow general ethical principles. When presented with an ethical dilemma, the machine should then be able to follow those ethical principles to make decisions. The machine's decisions are then compared against the decision of a human control, often an ethicist. The Moral Turing Test is generally presented in limited contexts that suit a particular specialized field of study. For example, if the machine is presented with an ethical dilemma about health care, the machine's decision will be compared against the decision of a human health-care provider, rather than a generic human control.

The Moral Turing Test has been criticized as an insufficient means of assessing the ability of a machine to exercise moral agency. The Turing Test relies on imitation as a means of determining whether a machine is capable of thought, but critics of the Moral Turing Test contend that imitation in an ethical dilemma may be achieved via deceptive responses, rather than via moral reasoning. Further, critics argue that morality cannot be judged by verbal responses alone. Rather, the judge must be able to observe what is happening in the background—the reasoning, consideration of alternatives, determination of what action to take, and actual action—all of which would be hidden from view in a traditional Turing Test.

Alternatives and variations to the Moral Turing Test include the comparative Moral Turing Test (cMTT), the Total Turing Test, and verification. In a comparative Moral Turing Test, the judge compares the narrated actions, rather than the verbal responses, of the machine against a human control. In a Total Turing Test, the judge is able to view the actual actions and interactions of the machine, as compared against the human control. Verification takes a different approach by focusing on the process behind the machine's response, rather than the outcome of the response. Verification requires evaluation of the design and performance to determine the means of the machine's decision-making. Proponents of verification argue that by focusing on the process rather than on the outcome, the evaluation acknowledges that moral questions generally do not have one correct answer and that the means by which the machine arrived at an outcome reveals more about the machine's ability to make ethical decisions than the decision itself.

*Amanda K. O'Keefe*

*See also:* Turing, Alan; Turing Test.

**Further Reading**

Arnold, Thomas, and Matthias Scheutz. 2016. "Against the Moral Turing Test: Accountable Design and the Moral Reasoning of Autonomous Systems." *Ethics and Information Technology* 18:103–15.

Gerdes, Anne, and Peter Øhrstrøm. 2015. "Issues in Robot Ethics Seen through the Lens of a Moral Turing Test." *Journal of Information, Communication, and Ethics in Society* 13, no. 2: 98–109.

Luxton, David D., Susan Leigh Anderson, and Michael Anderson. 2016. "Ethical Issues and Artificial Intelligence Technologies in Behavioral and Mental Health Care." In *Artificial Intelligence in Behavioral and Mental Health Care*, edited by David D. Luxton, 255–76. Amsterdam: Elsevier Academic Press.

# Moravec, Hans (1948–)

Hans Moravec is renowned in computer science circles as the long-time director of the Robotics Institute at Carnegie Mellon University and an unabashed technological optimist. He has researched and built robots imbued with artificial intelligence in the CMU lab for twenty-five years, where he remains an adjunct faculty member. Before Carnegie Mellon, Moravec worked for almost ten years as a research assistant in the pathbreaking Artificial Intelligence Lab at Stanford University.

Moravec is also well known for Moravec's paradox, an assertion that, contrary to conventional wisdom, it is easy to program high-level reasoning capabilities into robots—as with playing chess or *Jeopardy!*—but hard to impart sensorimotor agility. Human sensory and motor skills evolved over millions of years and, despite their complexity, appear effortless. Higher level intellectual skills, however, are the product of more recent cultural evolution. These would include geometry, stock market analysis, and petroleum engineering—difficult subjects for humans but more easily acquired by machines. As Steven Pinker paraphrases Moravec's life in science: "The main lesson of thirty-five years of AI research is that the hard problems are easy, and the easy problems are hard" (Pinker 2007, 190–91).

Moravec constructed his first toy robot from scrap metal at age ten and won two high school science fair prizes for his light-following electronic turtle and a robot hand controlled by punched paper tape. While still in high school, he proposed a Ship of Theseus-like analogy for the practicability of artificial brains. Imagine, he suggested, replacing a person's human neurons with perfectly machined substitutes one by one. At what point would human consciousness disappear? Would anyone notice? Could it be proved that the individual was no longer human? Later in his career, Moravec would argue that human expertise and training could be broken down in the same way, into subtasks that could be taken over by separate machine intelligences.

Moravec's master's thesis involved the creation of a computer language for artificial intelligence, and his doctoral research involved a robot with the ability to maneuver through obstacle courses using spatial representation techniques. These robot vision systems operated by identifying the region of interest (ROI) in a scene. By contemporary standards, Moravec's early robots with computer vision were painfully slow, traversing from one side of the lab to another in about five hours. An external computer painstakingly processed continuous video-camera imagery captured by the robot from different angles, in order to estimate distance and build an internal representation of physical obstacles in the room. Moravec eventually invented 3D occupancy grid technology, which made it possible for a robot to build an awareness of a room crowded with objects in a matter of seconds.

Moravec's lab adopted a new challenge in turning a Pontiac TransSport mini-van into one of the very first roadworthy autonomous vehicles. The driverless minivan operated at speeds up to 60 miles per hour. The CMU Robotics Institute also created DANTE II, a robot capable of walking on eight artificial spider legs into the crater of the active volcano on Mount Spurr in Alaska. While the immediate goal for DANTE II was to sample dangerous fumarole gases, a task too dangerous for people, it was also designed to prove out technology for robotic missions to other planets. Artificial intelligence allowed the volcano explorer robot to navigate the treacherous, boulder-strewn terrain on its own. Moravec would say that experience with mobile robotics forced the development of advanced artificial intelligence and computer vision techniques, because such rovers generated so much visual and other sensory data that had to be processed and controlled.

Moravec's team invented fractal branching ultra-dexterous robots ("Bush robots") for the National Aeronautics and Space Administration (NASA) in the 1990s. These robots, designed but not built because the enabling fabrication technologies did not yet exist, consisted of a branched hierarchy of dynamic articulated limbs, beginning with a large trunk and dividing down through branches of smaller size. The Bush robot would thus have "hands" at all scales arranged from the macroscopic to the microscopic. The smallest fingers would be nanoscale in size and able to grasp extraordinarily small things. Because of the complexity involved in moving millions of fingers in real time, Moravec believed the robot would require autonomy and rely on artificial intelligence agents distributed throughout the robot's limbs and twigs. He speculated that the robots might eventually be manufactured out of carbon nanotube material using rapid-prototyping technology we now call 3D printers.

Moravec has argued that the impact of artificial intelligence on human society will be great. To emphasize the influence of AI in transformation, he developed the metaphor of the "landscape of human competence," since then turned into a graphic visualization by physicist Max Tegmark. Moravec's illustration imagines a three-dimensional landscape where higher elevations represent harder tasks relative to how difficult they are for human beings. The location where the rising seas met the coast represents the line where machines and humans find the tasks equally difficult. Art, science, and literature lie comfortably out of reach of an AI currently, but arithmetic, chess, and the game Go are already conquered by the sea. At the shoreline are language translation, autonomous driving, and financial investment.

More controversially, Moravec engaged in futuristic speculation based on what he knew of progress in artificial intelligence research in two popular books: *Mind Children* (1988) and *Robot: Mere Machine to Transcendent Mind* (1999). He projected that human intelligence would be overtaken by computer intelligence in 2040, beyond which the human species would likely become extinct. Moravec made this estimate based on what he calculated as the functional equivalency between 50,000 million instructions per second (50,000 MIPS) of computer power and a gram of neural tissue. He estimated that the personal computers of the early 2000s matched only the nervous system of an insect, but that if processing power continued to double every eighteen months, 350 million years of evolution in human intelligence could be reduced to only thirty-five years of progression in artificial intelligence. Humanlike universal robots, he thought, would require about one hundred million MIPS.

These are the advanced robots of 2040 that Moravec considers our "mind children." Humans, he says, will engage in strategies to prolong the inevitable displacement of biological civilization. For instance, Moravec is the first in the field to predict what is today called universal basic income provided by benevolent artificial superintelligences. A basic income economy would guarantee regular cash payments to all people without any sort of work requirement, rendered unnecessary in a fully automated world. Moravec does not worry about technological unemployment, but rather about the possibility of a rogue automated corporation breaking its programming and refusing to pay taxes into the human cradle-to-grave social security system. Still, he believes that these "wild" intelligences will inevitably be the ones to rule the universe. Moravec has said that his books *Mind Children* and *Robot* may have directly influenced the last third of the original script for Stanley Kubrick's film *A.I. Artificial Intelligence* (later filmed by Steven Spielberg). Conversely, the self-replicating machines in the science fiction works *Ilium* and *Olympos* are named moravecs in his honor.

Throughout his life, Moravec argued the same physical fundamentalism found in his high school musings. In his most transhumanist writings, he argues that the only way for humanity to keep up with machine intelligences is to join them by replacing slow human neural tissue with artificial neural networks driven by super-fast algorithms. Moravec has also combined the concepts of artificial intelligence and virtual reality simulation in his writings. He has developed four cases for evolution of consciousness. These are (1) human brains in a physical world, (2) a programmed AI implanted in a physical robot, (3) a human brain immersed in a VR simulation, and (4) an AI operating within the confines of VR. All are equally plausible representations of existence and are as "real" as we accept them to be.

Moravec is founder and chief scientist of the Seegrid Corporation in Pittsburgh, which manufactures driverless Robotic Industrial Trucks capable of negotiating warehouses and factories without the assistance of automated guided vehicle systems. Seegrid's trucks are manually pushed through a new facility once by a human trainer. The trainer stops at the relevant locations for the truck to be loaded and unloaded, and the robot does the remainder of the work to determine the most efficient and safe paths for future trips. Seegrid VGVs have moved eight billion pounds of product for DHL, Whirlpool, and Amazon and traveled almost two million production miles.

Moravec was born in Kautzen, Austria. His father was a Czech engineer who sold electrical appliances during World War II. The family immigrated to Austria when the Russians occupied Czechoslovakia in 1944. His family immigrated to Canada in 1953, the country of his citizenship. Moravec received his bachelor's degree in Mathematics from Acadia University in Nova Scotia, his master's degree in Computer Science from the University of Western Ontario, and a doctorate from Stanford University under thesis advisors John McCarthy and Tom Binford. His research has been funded by the Office of Naval Research, the Defense Advanced Research Projects Agency, and NASA.

*Philip L. Frana*

*See also:* Superintelligence; Technological Singularity; Workplace Automation.

### Further Reading

Moravec, Hans. 1988. *Mind Children: The Future of Robot and Human Intelligence.* Cambridge, MA: Harvard University Press.

Moravec, Hans. 1999. *Robot: Mere Machine to Transcendent Mind.* Oxford, UK: Oxford University Press.

Moravec, Hans. 2003. "Robots, After All." *Communications of the ACM* 46, no. 10 (October): 90–97.

Pinker, Steven. 2007. *The Language Instinct: How the Mind Creates Language.* New York: Harper.

# Musk, Elon (1971–)

Elon Musk is a South African-born engineer, entrepreneur, and inventor. He maintains South African, Canadian, and United States citizenships and lives in California. Although a controversial character, Musk is widely regarded as one of the most prominent inventors and engineers of the twenty-first century and an important influencer and contributor to the development of artificial intelligence.

Musk's entrepreneurial leanings and unusual aptitude for technology were evident from childhood. He was a self-taught computer programmer by age ten, and by age twelve, he had created a video game and sold its code to a computer magazine. An avid reader since childhood, Musk has incorporated references to some of his favorite books in SpaceX's Falcon Heavy rocket launch and in Tesla's software.

Musk's formal education focused not on engineering, but rather on economics and physics—interests that are reflected in Musk's later work, including his endeavors in sustainable energy and space travel. He attended Queen's University in Canada, but transferred to the University of Pennsylvania, where he earned a bachelor's degree in Economics and a bachelor's degree in Physics. Musk pursued a PhD in energy physics at Stanford University for only two days, leaving the university to launch his first company, Zip2, with his brother Kimbal Musk.

Propelled by his many interests and ambitions, Musk has founded or cofounded multiple companies, including three separate billion-dollar companies: SpaceX, Tesla, and PayPal.

- Zip2: a web software company, later acquired by Compaq
- X.com: an online bank, which following merger activity later became the online payments company PayPal

- SpaceX: a commercial aerospace manufacturer and space transportation services provider
- Tesla, Inc.: an electric vehicle manufacturer, and solar panel manufacturer (via its subsidiarity SolarCity)
- The Boring Company: an infrastructure and tunnel construction company
- OpenAI: a nonprofit AI research company focused on the promotion and development of friendly AI
- Neuralink: a neurotechnology company focused on brain-computer interfaces

Musk is a proponent of sustainable energy and consumption. His concerns about the future habitability of planet Earth led him to explore the possibility of a self-sustaining human colony on Mars. Further pursuits include a high-speed transportation system called the Hyperloop and a jet-powered supersonic electric aircraft called the Musk electric jet. Musk briefly served on President Donald Trump's Strategy and Policy Forum and Manufacturing Jobs Initiative, stepping down after the Trump administration withdrew the United States from the Paris Agreement on climate change. In 2002, Musk established the Musk Foundation, which provides funding and service in the fields of renewable energy research and advocacy, human space exploration research and advocacy, pediatric research, and science and engineering education.

Though perhaps best known for his work with Tesla and SpaceX, and for his controversial social media statements, Musk's impact on AI is substantial. Musk cofounded the nonprofit organization OpenAI in 2015, with the goal of developing and promoting so-called "friendly AI," or AI that is developed, deployed, and used in a way that is likely to benefit humanity as a whole. OpenAI's mission includes keeping AI open and available to the masses, to further minimize the potential dangers of AI being controlled by a powerful few. OpenAI is particularly concerned with the possibility of Artificial General Intelligence (AGI), broadly defined as AI that is capable of human-level (or greater) performance on any intellectual task, ensuring any such AGI is developed responsibly, transparently, and distributed evenly and openly.

While holding true to its goals of keeping AI friendly and open, OpenAI has had its own successes in taking AI to new levels. In June of 2018, a team of robots built by OpenAI defeated a human team at a video game called *Dota 2*, a feat that could only be accomplished with teamwork and collaboration amongst the robots. Microsoft cofounder Bill Gates acknowledged the success via Twitter as "a huge milestone in advancing artificial intelligence" (@BillGates, June 26, 2018). In February of 2018, Musk stepped down from the OpenAI board to avoid potential conflicts of interests, as Tesla moved further into its AI development for autonomous driving.

Musk cofounded Tesla, Inc. in 2003, initially as an investor, becoming Tesla's CEO in 2008. Musk served as chairman of Tesla's board of directors, stepping down in 2018 as part of a settlement with the U.S. Securities and Exchange Commission, following Musk's misleading comments about taking Tesla private. Tesla manufactures electric vehicles with autonomous driving functionality. Its subsidiaries, Tesla Grohmann Automation and Solar City, provide related automotive technology and manufacturing and solar energy services, respectively.

Musk predicts Tesla will achieve Level 5 autonomous driving functionality, as designated by the U.S. Department of Transportation's National Highway Traffic Safety Administration's (NHTSA) five levels of autonomous driving, in 2019. Tesla's ambitious progress with autonomous driving has impacted traditional car manufacturers' position on electric vehicles and autonomous driving and has sparked Congressional review about how and when the technology should be regulated. Highlighting the advantages of autonomous vehicles (including reduced fatalities in vehicle crashes, increased worker productivity, increased transport efficiency, and job creation), and proving that the technology is achievable in the near term, Musk is widely credited as a key influencer moving the automotive industry toward autonomous driving.

Under the direction of Musk and Tesla's Director of AI, Andrej Karpathy, Tesla has developed and advanced its autonomous driving programming (Autopilot). Tesla's computer vision analysis, including an array of cameras on each vehicle, combined with real-time processing of the images, allows the system to make real-time observations and predictions. The cameras, and other external and internal sensors, collect vast amounts of data, which are analyzed and used to further refine the Autopilot programming. Tesla is unique among autonomous vehicle manufacturers in its aversion to the laser sensor known as LIDAR (an acronym for light detection and ranging). Instead, Tesla relies on cameras, radar, and ultrasonic sensors. Though experts and manufacturers are split on whether LIDAR is a requirement for full autonomous driving, the high cost of LIDAR has hindered Tesla's competitors' ability to make and sell cars at a price point that will allow a high volume of cars on the road collecting data.

In addition to Tesla's AI programming, Tesla is developing its own AI hardware. In late 2017, Musk confirmed that Tesla is developing its own silicon for performing artificial-intelligence computations, which will allow Tesla to create its own AI chips, no longer relying on third-party providers such as Nvidia.

Tesla's progress with AI in autonomous driving has not been without setbacks. Tesla has repeatedly failed to meet self-imposed deadlines, and serious accidents have been attributed to deficiencies in the vehicle's Autopilot mode, including a noninjury accident in 2018, in which the vehicle failed to detect a parked firetruck on a California freeway, and a fatal accident also in 2018, in which the vehicle failed to detect a pedestrian outside a crosswalk.

Musk founded the company Neuralink in 2016. Neuralink is focused on developing devices that can be implanted into the human brain, to better allow communication between the brain and software, with the stated goal of allowing humans to keep pace with AI advancements. Musk has described the devices in terms of a more efficient interface with computing devices; that is, where humans now use their fingers and voice commands to control devices, commands would instead come directly from the brain.

Though Musk's contributions to AI have been significant, his statements about the associated dangers of AI have bordered on apocalyptic. Musk has referred to AI as "humanity's biggest existential threat" (McFarland 2014) and "the greatest risk we face as a civilization" (Morris 2017). He warns about the dangers of concentration of power, lack of independent oversight, and a competition-driven rush to adoption without adequate consideration of the consequences. While Musk has invoked colorful language like "summoning the demon," (McFarland 2014) and images of cyborg overlords, he also warns of more immediate and relatable risks, including job losses and AI-driven disinformation campaigns.

Though Musk's comments often come across as alarmist, his anxiety is shared by many prominent and well-respected minds, including Microsoft cofounder Bill Gates, the Swedish-American physicist Max Tegmark, and the late theoretical physicist Stephen Hawking. Further, Musk does not advocate ending AI research. Instead, Musk advocates for responsible AI development and regulation, including convening a Congressional committee to spend years researching AI, with an aim of understanding the technology and its associated risks before drafting appropriate regulatory controls.

*Amanda K. O'Keefe*

*See also:* Bostrom, Nick; Superintelligence.

**Further Reading**

Gates, Bill. (@BillGates). 2018. Twitter, June 26, 2018. https://twitter.com/BillGates/status/1011752221376036864.

Marr, Bernard. 2018. "The Amazing Ways Tesla Is Using Artificial Intelligence and Big Data." *Forbes*, January 8, 2018. https://www.forbes.com/sites/bernardmarr/2018/01/08/the-amazing-ways-tesla-is-using-artificial-intelligence-and-big-data/.

McFarland, Matt. 2014. "Elon Musk: With Artificial Intelligence, We Are Summoning the Demon." *Washington Post*, October 24, 2014. https://www.washingtonpost.com/news/innovations/wp/2014/10/24/elon-musk-with-artificial-intelligence-we-are-summoning-the-demon/.

Morris, David Z. 2017. "Elon Musk Says Artificial Intelligence Is the 'Greatest Risk We Face as a Civilization.'" *Fortune*, July 15, 2017. https://fortune.com/2017/07/15/elon-musk-artificial-intelligence-2/.

Piper, Kelsey. 2018. "Why Elon Musk Fears Artificial Intelligence." *Vox Media*, November 2, 2018. https://www.vox.com/future-perfect/2018/11/2/18053418/elon-musk-artificial-intelligence-google-deepmind-openai.

Strauss, Neil. 2017. "Elon Musk: The Architect of Tomorrow." *Rolling Stone*, November 15, 2017. https://www.rollingstone.com/culture/culture-features/elon-musk-the-architect-of-tomorrow-120850/.

# MYCIN

Designed by computer scientists Edward Feigenbaum (1936–) and Bruce Buchanan at Stanford University in the 1970s, MYCIN is an interactive expert system for infectious disease diagnosis and therapy. MYCIN was Feigenbaum's second expert system (after DENDRAL), but it became the first expert system to be made commercially available as a stand-alone software package. By the 1980s, EMYCIN was the most successful expert shell sold by TeKnowledge, the software company cofounded by Feigenbaum and several collaborators, for this purpose.

Feigenbaum's Heuristic Programming Project (HPP) teamed up with the Infectious Diseases Group (IDG) at Stanford's Medical School to develop MYCIN. Stanley Cohen of IDG served as the expert clinical physician. In the early 1970s, Feigenbaum and Buchanan had read reports of antibiotics being incorrectly prescribed due to misdiagnoses. MYCIN was meant to aid the human expert in making the right decision. MYCIN first acted as a consultation system. After entering the results of a patient's blood test, bacterial cultures, and other data, MYCIN provided a diagnosis that included the recommended antibiotics and its dosage. Second, MYCIN functioned as an explanation system. The physician-user could ask MYCIN, in plain English, to elaborate on a specific deduction. Finally, MYCIN contained a knowledge-acquisition program to update the system's knowledge base.

Having gained experience with DENDRAL, Feigenbaum and his collaborators added two new features to MYCIN. First, MYCIN's inference engine came with a rule interpreter. This allowed diagnostic conclusions to be reached via "goal-directed backward chaining" (Cendrowska and Bramer 1984, 229). At each step in the process, MYCIN set itself the goal of finding a relevant clinical parameter that matched the patient data entered. The inference engine sought out a set of rules that applied to that parameter. As MYCIN evaluated the premise of one of the rules in this parameter set, it typically needed more information. Retrieving that information became the system's next subgoal. MYCIN could try out additional rules or prompt the physician-user for more data. This continued until MYCIN had enough information about several parameters to provide a diagnosis.

MYCIN's second novelty was the certainty factor. According to William van Melle (then a graduate student working on MYCIN for his thesis project), these factors should not be seen "as conditional probabilities, [though] they are informally based in probability theory" (van Melle 1978, 314). MYCIN attributed a value, between −1 and +1, to the execution of production rules (dependent on how strongly the system felt about their correctness). MYCIN's diagnosis specified these certainty factors as well, encouraging the physician-user to make their own final judgment.

Put on the market as EMYCIN around 1976, the software package included an inference engine, user interface, and short-term memory. It had no data. ("E" initially stood for "Empty," later for "Essential.") EMYCIN customers were to connect their own knowledge base to the system. Faced with a large demand for EMYCIN packages, as well as a great interest in MOLGEN (Feigenbaum's third expert system), HPP decided to create the first two expert system companies, IntelliCorp and TeKnowledge. TeKnowledge was ultimately established by about twenty people, including all of the former students who had created expert systems at HPP. EMYCIN remained their most influential product.

*Elisabeth Van Meer*

*See also:* Expert Systems; Knowledge Engineering.

## Further Reading

Cendrowska, J., and M. A. Bramer. 1984. "A Rational Reconstruction of the MYCIN Consultation System." *International Journal of Man-Machine Studies* 20 (March): 229–317.

Crevier, Daniel. 1993. *AI: Tumultuous History of the Search for Artificial Intelligence.* Princeton, NJ: Princeton University Press.

Feigenbaum, Edward. 2000. "Oral History." Charles Babbage Institute, October 13, 2000.

van Melle, William. 1978. "MYCIN: A Knowledge-based Consultation Program for Infectious Disease Diagnosis." *International Journal of Man-Machine Studies* 10 (May): 313–22.

# N

## Natural Language Generation

Natural Language Generation, or NLG, is the computational process through which forms of information that cannot be readily interpreted by humans are converted into a message optimized for human understanding as well as the name of the subfield of artificial intelligence (AI) devoted to the study and development of the same. The term "natural language" in computer science and AI is synonymous with what most people simply refer to as language, the means through which people communicate with one another and, now, increasingly with computers and robots. Natural language is the opposite of "machine language," or programming language, which has been developed for and used to program and operate computers. The information being processed by an NLG technology is some form of data, such as scores and statistics from a sport game, and the message being produced from this data can take multiple forms (text or speech), such as a news report regarding a sports game.

The development of NLG can be traced back to the introduction of computers in the mid-twentieth century. Entering information into early computers and then making sense of the output was difficult, laborious, and required highly specialized training. Researchers and developers conceptualized these hurdles related to the input and output of machines as problems of communication. Communication also is key to acquiring knowledge and information and to demonstrating intelligence. The solution that researchers devised was to work toward adapting the communication of and with machines to the form of communication that was most "natural" to humans, people's own languages. Research regarding how machines could make sense of human language falls under Natural Language Processing while research regarding the generation of messages tailored toward humans is Natural Language Generation. Similar to artificial intelligence, some scholars working in this area focus on the development of systems that produce messages from data while others focus on understanding the human process of language and message generation. In addition to being a subfield of artificial intelligence, NLG also is a subfield within Computational Linguistics.

The proliferation of technologies for creating, collecting, and connecting large swathes of data along with advances in computing hardware has enabled the recent proliferation in NLG technologies. Multiple applications exist for NLG across numerous industries, such as journalism and media. Large international and national news organizations worldwide have started to integrate automated news-writing software, which utilize NLG technology, into the production of news. Within this context, journalists use the software to develop informational reports from various datasets to produce lists of local crimes, financial earning reports, and sporting events synopses. NLG systems also can be used by companies and organizations to develop automated summaries of their own or outside data. A related area of research is computational narrative and the development of automated narrative generation systems that focus on the creation of fictional stories and characters that can have applications in media and entertainment, such as video games, as well as education and learning.

It is expected that NLG will continue to advance so that future technologies will be able to produce more complex and refined messages across additional contexts. The growth and application of NLG is relatively recent, and it is unknown what the full impact of technologies utilizing NLG will be on individuals, organizations, industries, and society. Current questions that are being raised include whether NLG technologies will affect the workforce, positively or negatively, within the industries in which they are being adopted, and the legal and ethical implications of having machines, rather than humans, create nonfiction and fiction. There also are larger philosophical considerations surrounding the connection among communication, the use of language, and how people socially and culturally have defined what it means to be human.

*Andrea L. Guzman*

*See also:* Natural Language Processing and Speech Understanding; Turing Test; Workplace Automation.

**Further Reading**

Guzman, Andrea L. 2018. "What Is Human-Machine Communication, Anyway?" In *Human-Machine Communication: Rethinking Communication, Technology, and Ourselves*, edited by Andrea L. Guzman, 1–28. New York: Peter Lang.

Lewis, Seth C., Andrea L. Guzman, and Thomas R. Schmidt. 2019. "Automation, Journalism, and Human-Machine Communication: Rethinking Roles and Relationships of Humans and Machines in News." *Digital Journalism* 7, no. 4: 409–27.

Licklider, J. C. R. 1968. "The Computer as Communication Device." In *In Memoriam: J. C. R. Licklider, 1915–1990*, edited by Robert W. Taylor, 21–41. Palo Alto, CA: Systems Research Center.

Marconi, Francesco, Alex Siegman, and Machine Journalist. 2017. *The Future of Augmented Journalism: A Guide for Newsrooms in the Age of Smart Machines*. New York: Associated Press. https://insights.ap.org/uploads/images/the-future-of-augmented-journalism_ap-report.pdf.

Paris, Cecile L., William R. Swartout, and William C. Mann, eds. 1991. *Natural Language Generation in Artificial Intelligence and Computational Linguistics*. Norwell, MA: Kluwer Academic Publishers.

Riedl, Mark. 2017. "Computational Narrative Intelligence: Past, Present, and Future." *Medium,* October 25, 2017. https://medium.com/@mark_riedl/computational-narrative-intelligence-past-present-and-future-99e58cf25ffa.

## Natural Language Processing and Speech Understanding

Natural language processing (NLP) is an artificial intelligence field that involves the mining of human text and speech to generate or respond to human inquiries in a readable or ordinary way. NLP has required advancements in statistics, machine learning, linguistics, and semantics in order to decode the uncertainties and opacities of natural human language. In the future, chatbots will use natural language processing to seamlessly interact with human beings over text-based and voice-based interfaces. Computer assistants will also support interactions as an interface between humans with different abilities and needs. They will allow for natural language queries of vast amounts of information, like that encountered on the internet, by making search more natural. They may even insert helpful insights or tidbits of knowledge into situations as diverse as meetings, classrooms, or casual conversations. They may even one day be able to seamlessly "read" and respond to the emotions or moods of human speakers in real time (so-called "sentiment analysis"). The market for NLP hardware, software, and services may be worth $20 billion in annual revenue by 2025.

Speech or voice recognition has a long history. Research into automatic speech recognition and transcription began at Bell Labs in the 1930s under Harvey Fletcher, a physicist who did pioneering research establishing the relationship between speech energy, frequency spectrum, and the perception of sound by a listener. His research forms the basis of most speech recognition algorithms today. By 1940, another Bell Labs physicist Homer Dudley had been granted patents on a Vodor speech synthesizer that modeled human vocalizations and a parallel band-pass vocodor that could take sound samples and run them through narrow band filters to determine their energy levels. The latter device could also take the record of energy levels and turn them back into rough approximations of the original sounds by running them through other filters.

By the 1950s, Bell Labs researchers had figured out how to create a system that could do more than emulate speech. In that decade, digital technology had improved to the point where the system could recognize the isolated spoken word parts by comparing their frequencies and energy levels against a digital reference library of sounds. Essentially, the machine made an educated guess at the expression being made. Progress was slow. By the mid-1950s, Bell Labs machines could recognize about ten syllables spoken by a single individual. At the end of the decade, researchers at MIT, IBM, and at Kyoto University and University College London were developing recognizing machines that used statistics to recognize words containing multiple phonemes. Phonemes are units of sound that listeners perceive as distinctive from one another. Progress was also being made on tools that worked in recognizing the speech of more than a single speaker.

The first professional automatic speech recognition group was created in 1971 and chaired by Allen Newell. The study group divided its work among several levels of knowledge formation, including acoustics, parametrics, phonemics, lexical concepts, sentence processing, and semantics. Some of the problems reviewed by the group were studied under grants issued in the 1970s by the Defense Advanced Research Project Agency (DARPA). DARPA was interested in the technology as a way to process large volumes of spoken data produced by various government agencies and turn that information into insights and strategic responses to problems. Progress was made on such techniques as dynamic time warping and continuous speech recognition. Computer technology also steadily improved, and several manufacturers of mainframes and minicomputers began conducting research in natural language processing and speech understanding.

One of the groups supported by DARPA was the Speech Understanding Research (SUR) project at Carnegie Mellon University. Raj Reddy led the SUR project, which developed several pioneering speech recognition systems, including Hearsay, Dragon, Harpy, and Sphinx. Harpy in particular is noteworthy, because it uses the beam search method, which has been a staple of such systems for decades. Beam search involves a heuristic search algorithm that explores a graph by expanding the most promising node in a limited set. Beam search is an optimization of best-first search that reduces its memory requirements. It is a greedy algorithm in the sense that it follows the problem-solving heuristic of making the locally optimal choice at each stage with the hope of finding a global optimum. In general, algorithms for graph search have formed the basis for research in speech recognition, as they have for decades in the fields of operations research, game theory, and artificial intelligence.

By the 1980s and 1990s, data processing and algorithms had improved to the point that researchers could recognize whole strings of words, even sentences, using statistical models. The Department of Defense remained a leader in the field, but IBM's own efforts had advanced to the point that they stood on the cusp of developing a digital speech transcription device for the company's business customers. Bell Labs had developed complex digital techniques for automated voice dialing of spoken numbers. Other applications seemed within grasp as well: transcription of television broadcasting for closed captioning and personal automatic reservation systems. Spoken language understanding has improved dramatically. The first commercial system that emerged from DARPA funding was the Air Travel Information System (ATIS). New challenges also emerged, including so-called "disfluencies" that naturally emerged from conversational speech—hesitations, corrections, casual speech, interrupts, and verbal fillers like "uh" and "um."

In 1995, Microsoft distributed the Speech Application Programming Interface (SAPI) with every Windows 95 operating system. SAPI (which included building block subroutine definitions, protocols, and tools) allowed programmers and developers everywhere to more easily build speech recognition and speech synthesis into Windows applications. In particular, SAPI gave other software developers the ability to create and freely redistribute their own speech recognition engines. It provided a significant boost in broadening interest and also creating widespread markets for NLP technology.

One of the best-known mass-market NLP products is the Dragon family of speech recognition and dictation software packages. The goal of the popular Dragon NaturallySpeaking application is automated real-time, large-vocabulary, continuous-speech dictation using a headset or microphone. The program—first released in 1997—required fifteen years of development and is often considered the gold standard for use in personal computing today. It takes about 4–8 hours for the software to transcribe one hour of digitally recorded speech, but dictation on screen is nearly instantaneous. Comparable software is bundled into smart phones with voice dictation features that convert ordinary speech into text for use in text messages and emails.

Industry in the twenty-first century has benefitted enormously from the vast volume of data available in the cloud and through the collection of massive archives of voice recordings collected from smart phones and electronic peripherals. These large training data sets have allowed companies to continuously improve acoustic models and language models for speech processing. Traditional speech recognition technology used statistical learning techniques to match observed and "labeled" sounds. Since the 1990s, speech processing has relied more on Markovian and hidden Markovian systems that feature reinforcement learning and pattern recognizing algorithms. Error rates have plunged in recent years because of the quantities of data available for matching and the power of deep learning algorithms. Despite the fact that using Markov models for language representation and analysis is controversial among linguists, who assert that natural languages require flexibility and context to be properly understood, these approximation methods and probabilistic functions are extremely powerful at deciphering and responding to inputs of human speech.

Today, computational linguistics is predicated on the n-gram, a contiguous sequence of n items from a given sample of text or speech. The items can be phonemes, syllables, letters, words, or base pairs according to the application. N-grams typically are collected from text or speech. No other technique currently beats this approach in terms of proficiency. Google and Bing have indexed the internet in its entirety for their virtual assistants and use user query data in their language models for voice search applications. The systems today are beginning to recognize new words from their datasets on the fly, what humans would call lifelong learning, but this is an emerging technology.

In the future, companies involved in natural language processing want technologies that are portable (relying on remote servers) and that provide near-instantaneous feedback and a frictionless user experience. One powerful example of next-generation NLP is being developed by Richard Socher, a deep learning expert and founding CEO of the artificial intelligence start-up MetaMind. The company's technology uses a neural networking system and reinforcement learning algorithms to generate answers to specific and very general questions, based on large chunks of natural language datasets. The company was recently acquired by digital marketing behemoth Salesforce. There will be demand in the future for text-to-speech analysis and advanced conversational interfaces in automobiles, speech recognition and translation across cultures and languages, automatic speech understanding in environments with high ambient noise such as construction sites, and specialized voice systems to control office and home automation processes and internet-connected devices. All of these applications to augment human speech will require the harvesting of large data sets of natural language to work upon.

*Philip L. Frana*

See also: Natural Language Generation; Newell, Allen; Workplace Automation.

**Further Reading**

Chowdhury, Gobinda G. 2003. "Natural Language Processing." *Annual Review of Infor-mation Science and Technology* 37: 51–89.

Jurafsky, Daniel, and James H. Martin. 2014. *Speech and Language Processing.* Second edition. Upper Saddle River, NJ: Pearson Prentice Hall.

Mahavan, Radhika. n.d. "Natural Language Processing: Current Applications and Future Possibilities." https://www.techemergence.com/nlp-current-applications-and-future-possibilities/.

Manning, Christopher D., and Hinrich Schütze. 1999. *Foundations of Statistical Natural Language Processing.* Cambridge, MA: MIT Press.

Metz, Cade. 2015. "AI's Next Frontier: Machines That Understand Language." *Wired,* June 24, 2015. https://www.wired.com/2015/06/ais-next-frontier-machines-understand-language/.

Nusca, Andrew. 2011. "Say Command: How Speech Recognition Will Change the World." *ZDNet,* November 2, 2011. https://www.zdnet.com/article/say-command-how-speech-recognition-will-change-the-world/.

# Newell, Allen (1927–1992)

Allen Newell worked with Herbert Simon to create the first models of human cognition in the late 1950s and early 1960s. These programs modeled how logical rules could be applied in a proof (Logic Theory Machine), how simple problem solving could be performed (the General Problem Solver), and an early program to play chess (the Newell-Shaw-Simon chess program). In these models, Newell and Simon showed for the first time how computers could manipulate symbols and how these manipulations could be used to represent, generate, and explain intelligent behavior.

Newell started his career as a physics undergraduate at Stanford University. After a year of graduate work in mathematics at Princeton, he moved to the RAND Corporation to work on models of complex systems. While at RAND, he met and was influenced by Oliver Selfridge, who led him into modeling cognition. He also met Herbert Simon, who was later to win a Nobel Prize in Economics for the decision-making processes within economic organizations, including satisficing. Newell was recruited by Simon to come to Carnegie Institute of Technology (now Carnegie Mellon University). Newell collaborated with Simon for much of his academic life.

Newell's primary interest was in understanding the human mind by simulating its processes using computational models. Newell completed his doctorate with Simon at Carnegie Mellon. His first academic job was as a tenured, chaired professor. He helped found the Department of Computer Science (now school), where he had his primary appointment.

In his main line of research, Newell explored the mind, particularly problem solving, with Simon. Their 1972 book *Human Problem Solving* laid out their theory for intelligence and illustrated it with examples including those from math puzzles and chess. Their work made extensive use of verbal talk-aloud protocols—which are more accurate than think-aloud or retrospective protocols—to understand what resources are being used in cognition. The science of verbal protocol data was later more fully codified by Ericsson and Simon.

He argued in his last lecture ("Desires and Diversions") that if you get distracted, you should make the distraction count. He did so by notable achievements in the areas of his distractions and by using several of them in his final project. These distractions included one of the first hypertext systems, ZOG. Newell also wrote a textbook on computer architectures with Digital Equipment Corporation (DEC) pioneer Gordon Bell and worked on speech recognition systems with CMU colleague Raj Reddy.

Perhaps the longest running and most productive distraction was work with Stuart Card and Thomas Moran at Xerox PARC to create theories of how users interact with computers. These theories are documented in *The Psychology of Human-Computer Interaction* (1983). Their work led to two approaches for representing human behavior—the Keystroke Level Model and GOMS—as well as a simple representation of the mechanisms of cognition in this area, called the Model Human Processor. This was some of the first work in human-computer interaction (HCI). Their approach argued for understanding the user and the task and then using technology to support the user to perform the task.

Newell also noted in his last lecture that scientists should have a final project that would outlast them. Newell's final project was to argue for unified theories of cognition (UTCs) and to create a candidate UTC, an exemplar, called Soar. His project described what it would look like to have a theory that brought together all the constraints, data, and theories in psychology into a single unified result realized by a computer program. Soar remains a successful ongoing project, although it is not complete. While Soar has not unified psychology, it has had notable successes in explaining problem solving, learning, their interaction, and how to provide autonomous, reactive agents in large simulations.

As part of his final project (with Paul Rosenbloom), he examined how learning could be modeled. This line of work was later merged with Soar. Newell and Rosenbloom argued that learning followed a power law of practice; that is, the time to perform a task related to the practice (trial) number raised to a small negative power (e.g., Time $\alpha$ trial# $^{-\alpha}$) holds across a wide range of tasks. Their explanation was that as tasks were performed in a hierarchical manner, what was learned at the bottom level had the most effect on response time, but as learning continued on higher levels, the learning was less often used and saved less time; so the learning slowed down but did not stop.

In 1987, Newell gave the William James Lectures at Harvard. In these lectures, he laid out in detail what it would mean to generate a unified theory in psychology. These lectures were recorded and are available through the CMU library. In the following fall, he gave them again and wrote them up as a book (1990).

Soar uses search through problem spaces as its way of representing cognition. It is realized as a production system (using IF-THEN rules). It attempts to apply an operator. If it does not have one or cannot apply it, Soar recurses with an impasse to solve the problem. Thus, knowledge is represented as parts of operators and problem spaces and how to resolve the impasses. The architecture is thus how these choices and knowledge can be structured. Systems with up to one million rules have been built, and Soar models have been used in a variety of cognitive science and AI applications, including military simulations. Newell also explored how to use these models of cognition to simulate social agents with CMU social scientist Kathleen Carley. Work with Soar continues, primarily at the University of Michigan under John Laird, where it is more focused now on intelligent agents.

Newell and Simon received the ACM A. M. Turing Award in 1975 for their contributions to artificial intelligence, the psychology of human cognition and list processing. Their work is recognized for fundamental contributions to computer science as an empirical inquiry. Newell was also elected to the National Academy of Sciences and the National Academy of Engineering. In 1992, he received the National Medal of Science. Newell helped found a research group, department, and university that were productive and supportive. At his memorial service, his son noted that not only was he a great scientist, he was also a great dad. His flaws were that he was very smart, he worked very hard, and he thought the same of you.

*Frank E. Ritter*

*See also:* Dartmouth AI Conference; General Problem Solver; Simon, Herbert A.

**Further Reading**

Newell, Allen. 1990. *Unified Theories of Cognition.* Cambridge, MA: Harvard University Press.

Newell, Allen. 1993. *Desires and Diversions.* Carnegie Mellon University, School of Computer Science. Stanford, CA: University Video Communications.

Simon, Herbert A. 1998. "Allen Newell: 1927–1992." *IEEE Annals of the History of Computing* 20, no. 2: 63–76.

# Nissenbaum, Helen (1954–)

Helen Nissenbaum, who holds a PhD in philosophy, explores the ethical and political implications of information technology in her scholarship. She has held positions at Stanford University, Princeton University, New York University, and Cornell Tech. Additionally, Nissenbaum has served as the principal investigator for a wide variety of grants from organizations such as the National Security Agency, the National Science Foundation, the Air Force Office of Scientific Research, the U.S. Department of Health and Human Services, and the William and Flora Hewlett Foundation.

Nissenbaum defines AI as big data, machine learning, algorithms, and models that lead to output results. Privacy is the predominant area of concern that links her work across these topics. In her 2010 book, *Privacy in Context: Technology, Policy, and the Integrity of Social Life*, Nissenbaum explains these concerns through the framework of contextual integrity, which understands privacy in terms of appropriate flows of information as opposed to simply preventing flows of information all together. In other words, she is concerned with trying to create an ethical framework in which data can be collected and used appropriately. However, the problem with creating such a framework is that when multiple data sources are collected together, or aggregated, it becomes possible to learn more about those from whom the data was collected than it would be possible to do with each individual source of data. Such aggregated data is used to profile users, enabling organizations such as credit and insurance companies to make decisions based on this data.

Such practices are further complicated by outdated regulatory schemes for data. One major problem is that there is little differentiation between tracking users to create profiles and targeting ads to those profiles. To make matters worse, advertisements are often served by third-party sites, different from the site that the user is actually visiting. This creates the ethical problem of many hands, a conundrum in which the involvement of multiple parties makes it unclear who is ultimately responsible for a particular issue, such as protecting users' privacy in this case. Additionally, because so many organizations will potentially receive this information and use it for a variety of tracking and targeting purposes, it is not possible to properly give notice to users about how their data will be used and allow them to either consent or opt out of such use.

In addition to these problems, the AI systems that draw on this data are themselves biased. This bias, however, is a social rather than a computational problem, as so much of the academic work centered on solving computational bias is misplaced. Nissenbaum offers Google's Behavioral Advertising system as one example of this bias. The Google Behavioral Advertising system will display ads for background checks more often when a search includes a name that is traditionally African American. This form of racism is not actually written into the code but arises from social interaction with the advertisements because those who are doing the searching are more likely to click on background check links for traditionally African-American names. Correcting these bias-related problems, Nissenbaum argues, would require significant changes in policy related to the ownership and use of big data.

In light of such and with few policy changes related to data on the horizon, Nissenbaum has sought to develop strategies that can be put to use immediately. The primary framework she has used to develop these strategies is obfuscation, which entails intentionally adding extraneous information that can interfere with data collection and surveillance practices. This is justified, she argues, because of the asymmetrical power relations that have led to near total surveillance. From this framework of obfuscation, Nissenbaum and her collaborators have developed several practical internet browser plug-ins.

The earliest of these obfuscating browser plug-ins was TrackMeNot. This plugin sends random searches to a variety of search engines in order to pollute the stream of data that is collected and make it impossible for search companies to create an aggregated profile based on the user's true searches. This plug-in is aimed at users who are unsatisfied with current data regulations and want to take immediate action against corporations and governments who are actively collecting data. This method follows the theoretical framework of obfuscation, because instead of hiding the search true terms, it simply masks them with additional search terms, which Nissenbaum calls ghosts.

Adnostic is a prototype plugin for the Firefox web browser aimed at overcoming the privacy problems associated with online behavioral advertising practices. Currently online behavioral advertising is achieved by tracking a user's behavior across multiple websites and then targeting ads that are the most relevant at them.

This behavioral data is collected, aggregated, and stored indefinitely by multiple websites. Adnostic offers a system that allows the profiling and targeting to happen entirely on the user's system so that their data is never shared with third-party websites. Although the user still receives targeted ads, the third-party websites do not collect and store the behavioral information.

AdNauseam is another plugin rooted in obfuscation tactics. Running in the background, this software clicks all of the ads on the page. The stated purpose of this action is to pollute the data stream so that targeting and surveillance no longer function properly. This also likely increases the associated costs for advertisers. This project has met with controversy and was banned from the Chrome Web Store in 2017. Although there are workarounds that allow users to continue to install the plugin, its lack of availability in the store makes it less accessible for a general audience.

Nissenbaum's work offers a detailed exploration of the ethical issues surrounding data and the AI systems that are built on top of this data. In addition to making specific policy suggestions that would improve problematic privacy problems, Nissenbaum has developed practical obfuscation tools that can be accessed and used by any interested individual.

*J. J. Sylvia*

*See also:* Biometric Privacy and Security; Biometric Technology; Robot Ethics.

### Further Reading

Barocas, Solon, and Helen Nissenbaum. 2009. "On Notice: The Trouble with Notice and Consent." In *Proceedings of the Engaging Data Forum: The First International Forum on the Application and Management of Personal Electronic Information,* n.p. Cambridge, MA: Massachusetts Institute of Technology.

Barocas, Solon, and Helen Nissenbaum. 2014. "Big Data's End Run around Consent and Anonymity." In *Privacy, Big Data, and the Public Good,* edited by Julia Lane, Victoria Stodden, Stefan Bender, and Helen Nissenbaum, 44–75. Cambridge, UK: Cambridge University Press.

Brunton, Finn, and Helen Nissenbaum. 2015. *Obfuscation: A User's Guide for Privacy and Protest.* Cambridge, MA: MIT Press.

Lane, Julia, Victoria Stodden, Stefan Bender, and Helen Nissenbaum, eds. 2014. *Privacy, Big Data, and the Public Good.* New York: Cambridge University Press.

Nissenbaum, Helen. 2010. *Privacy in Context: Technology, Policy, and the Integrity of Social Life.* Stanford, CA: Stanford University Press.

# Nonhuman Rights and Personhood

The concept of rights and personhood for artificial intelligences is a popular and scholarly debate that has emerged in the last few decades from questions about the autonomy, liability, and distributed accountability of smart robots. Legal systems are interested in the agency of intelligent machines related to business and contracts. Philosophers are curious about machine consciousness, dignity, and interests. Issues related to smart robots and AI reveal that personhood is in many ways a fiction that derives from normative beliefs that are renegotiating, if not equalizing, the statuses of humans, artificial intelligences, animals, and other legal personalities.

Debates over electronic personhood often cite definitions and precedents from prior philosophical, legal, and ethical efforts to define human, corporate, and animal personhood. John Chipman Gray discussed the idea of legal personhood in his 1909 book *The Nature and Sources of Law*. Gray notes that when people use the word "person" in common speech, they tend to think of a human being; however, the technical, legal meaning of the term "person" centers more on issues related to legal rights. The question, according to Gray, focuses on whether an entity can be subject to legal rights and duties, and that question is dependent upon the kind of entity one considers. However, Gray argues that one can only be a legal person if the entity possesses intelligence and will. In his essay "The Concept of a Person" (1985), Charles Taylor explains that to be a person one would have to have certain rights. What Gray and Taylor both acknowledge is that personhood centers on legalities in relation to having assured liberties. For example, legal persons can enter into contracts, buy property, or be sued. Legal persons also are protected under the law and are given certain rights such as the right to life.

In the eyes of the law, not all legal persons are human and not all humans are persons. Gray explains how Roman temples and medieval churches were seen as persons and guaranteed rights. Today, business corporations and government bodies are granted personhood under the law. Even though these entities are not human, the law considers them as persons, meaning their rights are protected, in addition to them having certain legal responsibilities. Alternatively, there remains significant debate about whether human fetuses are persons under the law. Humans existing in a vegetative state are also not legally acknowledged as having personhood.

This debate about personhood, centering on rights in connection to intelligence and will, has raised questions of whether intelligent animals should be granted personhood. For example, the Great Ape Project, founded in 1993, seeks certain rights for apes, including releasing them from captivity, protecting their right to life, and ceasing testing on them. In 2013, India considered marine mammals as possible persons, resulting in a ban against their captivity. An Argentinian court awarded Sandra, an orangutan, the right to life and freedom in 2015.

These kinds of moral considerations for animals have led to some people seeking personhood for androids or robots. For those people, it is only logical that an android would be granted certain legal protections and rights. Those who disagree believe that because artificial intelligence is created and designed by humans, we cannot view androids in the same light as animals. Androids, in this view, are machines and property. The question of whether a robot could be a legal person is completely speculative at this point. However, because the defining characteristics that center on personhood often coincide with questions of intelligence and will, these aspects fuel the debate over the possibility that artificial intelligence might be granted personhood.

Personhood often centers on two issues: rights and moral status. To have a moral status requires that a person be viewed as being valuable and is, therefore, treated as such. But Taylor further defines the category of the person by centering the definition on certain capacities. In his view, in order to be classified as a person, one must be capable of understanding the difference between the future and the past. A person must also have the ability to make choices and chart out a plan for his or her life. To be a person, one should have a set of values or morals. In addition, a person would have a self-image or sense of identity.

In light of these criteria, those who consider the possibility that androids may be granted personhood also acknowledge that these entities would have to have these kinds of abilities. For example, F. Patrick Hubbard argues that personhood for robots should only be granted if they meet certain criteria. These criteria include the sense of having a self, having a plan for life, and having the ability to communicate and think in complex ways. David Lawrence provides an alternate set of criteria for granting personhood to an android. Firstly, he speaks of AI having to present consciousness, in addition to being able to understand information, learn, reason, possess subjectivity, among many other elements.

Peter Singer takes a much simpler approach to personhood, although his focus is on the ethical treatment of animals. In his view, the defining characteristic of granting personhood centers on suffering. If something can suffer, then that suffering should be seen equally, no matter whether it is a human, an animal, or a machine. In fact, Singer sees it as immoral to deny the suffering of any being. If androids possess some or all of the aforementioned criteria, some people believe they should be granted personhood, and with that position should come individual rights, such as the right to free speech or freedom from being a slave.

Those who object to personhood for artificial intelligence often believe that only natural entities should be given personhood. Another objection relates to the robot's status as human-made property. In this case, since robots are programmed and carry out human instructions, they are not an independent person with a will; they are merely an object that humans have labored to produce. If an android does not have its own will and independent thought, then it is difficult to grant it rights. David Calverley notes that androids can be bound by certain constraints. For example, an android might be limited by Asimov's Laws of Robotics. If that were the case, then the android would not have the ability to truly make free choices of its own. Others object on the grounds that artificial intelligence lack a crucial element of personhood, namely a soul, feelings, and consciousness, all reasons that have previously been used to deny animals personhood. However, something like consciousness is difficult to define or assess even in humans.

Finally, opposition to personhood for androids often centers on fear, a fear that is fueled by science fiction novels and movies. Such fictions present androids as superior in intelligence, possibly immortal, and having a desire to take over, superseding the human's place in society. Lawrence Solum explains that each of these objections is rooted in the fear of anything that is not human, and he argues that we reject personhood for AI based on the sole fact that they do not have human DNA. He finds such a stance troublesome and equates it to American slavery, in which slaves were denied rights solely because they were not white. He takes issue with denying an android rights only because it is not human, especially if other entities have feelings, consciousness, and intelligence.

Although personhood for androids is theoretical at this point, there have been recent events and debates that have broached this topic in real ways. In 2015, a Hong Kong-based company called Hanson Robotics developed Sophia, a social humanoid robot. It appeared in public in March 2016 and became a Saudi Arabian citizen in October 2017. Additionally, Sophia become the first nonhuman to be given a United Nations title when she was named the first Innovation Champion of the UN Development Program in 2017. Sophia delivers speeches and has given interviews around the world. Sophia has even expressed the desire to have a home, get married, and have children. In early 2017, the European Parliament attempted to grant robots "electronic personalities," allowing them to be held liable for any damages they cause. Those in favor of this change saw legal personhood as the same legal status held by corporations. Conversely, in an open letter in 2018, over 150 experts from 14 European countries opposed this measure, finding it to be inappropriate for ridding corporations of responsibility for their creations. In an amended draft from the EU Parliament, the personhood of robots is not mentioned. The debate about responsibility has not ceased though, as evidenced in March 2018 when a self-driving car killed a pedestrian in Arizona.

Over the course of Western history, our ideas of who deserves ethical treatment have changed. Susan Leigh Anderson sees this progression as a positive change because she correlates the increase of rights for more entities with an increase in ethics overall. As more animals were and continue to be awarded rights, the fact that human position is incomparable may shift. If androids begin processing in ways that are similar to the way the human mind does, our idea of personhood may have to broaden even further. As David DeGrazia argues in *Human Identity and Bioethics* (2012), the term "person" encompasses a series of abilities and traits. In that case, any entity that exhibits these characteristics, including an artificial intelligence, could be classified as a person.

*Crystal Matey*

*See also:* Asimov, Isaac; *Blade Runner*; Robot Ethics; *The Terminator.*

**Further Reading**

Anderson, Susan L. 2008. "Asimov's 'Three Laws of Robotics' and Machine Metaethics." *AI & Society* 22, no. 4 (April): 477–93.

Calverley, David J. 2006. "Android Science and Animal Rights, Does an Analogy Exist?" *Connection Science* 18, no 4: 403–17.

DeGrazia, David. 2005. *Human Identity and Bioethics*. New York: Cambridge University Press.

Gray, John Chipman. 1909. *The Nature and Sources of the Law*. New York: Columbia University Press.

Hubbard, F. Patrick. 2011. "'Do Androids Dream?' Personhood and Intelligent Artifacts." *Temple Law Review* 83: 405–74.

Lawrence, David. 2017. "More Human Than Human." *Cambridge Quarterly of Healthcare Ethics* 26, no. 3 (July): 476–90.

Solum, Lawrence B. 1992. "Legal Personhood for Artificial Intelligences." *North Carolina Law Review* 70, no. 4: 1231–87.

Taylor, Charles. 1985. "The Concept of a Person." In *Philosophical Papers, Volume 1: Human Agency and Language*, 97–114. Cambridge, UK: Cambridge University Press.

# O

## Omohundro, Steve (1959–)

Steve Omohundro is a noted scientist, author, and entrepreneur working in the area of artificial intelligence. He is founder of Self-Aware Systems, chief scientist and board member of AIBrain, and advisor to the Machine Intelligence Research Institute (MIRI). Omohundro is well known for his thoughtful, speculative research on safety in smarter-than-human machines and the social implications of AI.

Omohundro argues that a truly predictive science of artificial intelligence is needed. He claims that if goal-driven artificial general intelligences are not carefully crafted in the future, they are likely to produce harmful actions, cause wars, or even trigger human extinction. Indeed, Omohundro believes that poorly programmed AIs could exhibit psychopathic behaviors. Coders, he argues, often produce flaky software, and programs that simply "manipulate bits" without understanding why. Omohundro wants AGIs to monitor and understand their own operations, see their own imperfections, and rewrite themselves to perform better. This represents true machine learning.

The danger is that the AIs might change themselves into something that cannot be understood by humans or make decisions that are inconceivable or have unintended consequences. Therefore, Omohundro argues, artificial intelligence must become a more predictive and anticipatory science. Omohundro also suggests in one of his widely available online papers, "The Nature of Self-Improving Artificial Intelligence," that a future self-aware system that likely accesses the internet will be influenced by the scientific papers that it reads, which recursively justifies writing the paper in the first place.

AGI agents themselves must be created with value sets that lead them—when they self-improve—to choose goals that help humanity. The sort of self-improving systems that Omohundro is preparing for do not currently exist. Omohundro notes that inventive minds have till now only produced inert systems (objects such as chairs and coffee mugs), reactive systems that approach goals in rigid ways (mousetraps and thermostats), adaptive systems (advanced speech recognition systems and intelligent virtual assistants), and deliberative systems (the Deep Blue chess-playing computer). The self-improving systems Omohundro is talking about would need to actively deliberate and make decisions under conditions of uncertainty about the consequences of engaging in self-modification.

Omohundro believes that the basic natures of self-improving AIs can be understood as rational agents, a concept he borrows from microeconomic theory. Humans are only imperfectly rational, which is why the field of behavioral economics has blossomed in recent decades. AI agents, though, because of their self-improving cognitive architectures, must ultimately develop rational goals and preferences ("utility functions") that sharpen their beliefs about their environments. These beliefs in turn will help them form further goals and preferences. Omohundro cites as his inspiration the contributions to the expected utility hypothesis made by mathematician John von Neumann and economist Oskar Morgenstern. The axioms of rational behavior developed by von Neumann and Morgenstern are completeness, transitivity, continuity, and independence.

Omohundro posits four "basic drives" for artificial intelligences: efficiency, self-preservation, resource acquisition, and creativity. These drives manifest as "behaviors" that future AGIs exhibiting self-improving, rational agency will express. The efficiency drive refers to both physical tasks and computational tasks. Artificial intelligences will want to use their finite supplies of space, matter, energy, processing time, and computational power efficiently. The self-preservation drive will engage in advanced artificial intelligences in order to avoid losing resources to other agents and maximize goal fulfillment. An artificial intelligence that behaves passively is not likely to survive. The acquisition drive refers to the instrumental goal of seeking new sources of resources, trading for them, cooperating with other agents, or even stealing what is needed to achieve the ultimate goal. The creativity drive covers all the novel ways an AGI might increase expected utility to meet its various goals. The invention of new ways of extracting and using resources might be considered part of this drive.

Omohundro writes of signaling as a unique human source of creative energy, diversity, and divergence. Signaling is how humans communicate their intentions about other useful activities in which they are engaged. Formally speaking, A signals B if A is more likely to be true when B is true than when B is false. So, for example, employers are usually more willing to hire prospective employees who are enrolled in a class that appears to confer advantages the employer would want, even if that's not the actual case. The class enrollment signals to the employer that it is more likely that the prospective employee is picking up valuable skills than the applicant who is not. Likewise, a billionaire would not need to give another person a billion dollars to signal that they are among the super-rich. Several million dollars presented in a large suitcase might do the trick.

Oxford philosopher Nick Bostrom incorporated Omohundro's concept of basic AI drives into his instrumental convergence thesis, which argues that a few instrumental values are pursued in order to achieve an ultimate goal, which is sometimes called a terminal value. Bostrom's instrumental values (he prefers not to call them drives) are self-preservation, goal content integrity (retention of preferences over time), cognitive enhancement, technological perfection, and resource acquisition. Future AIs may have a terminal value of maximizing some utility function, or they may have a reward function.

The sort of artificial general intelligence Omohundro wants designers to create would have benevolence toward humans as its ultimate goal. However, he says military tensions and economic anxieties make the emergence of harmful artificial general intelligence more likely. Military drone use has increased in prominence for both the delivery of bombs and for purposes of reconnaissance. He also argues that future conflicts will likely become informational conflicts. In a world susceptible to cyberwar, an infrastructure against information conflict will be needed. One idea to mitigate the threat is energy encryption, a novel wireless power transfer technology that scrambles energy so that it remains secure and cannot be used by rogue devices.

The use of artificial intelligence in vulnerable financial markets is another area where information conflict is causing instabilities. Digital cryptocurrencies and crowdsourced marketplace systems such as the Mechanical Turk are leading to what Omohundro calls autonomous capitalism, and in his view, we are underprepared to manage the consequences. As president of the company Possibility Research, advocate of a new cryptocurrency called Pebble, and as advisory board member of the Institute for Blockchain Studies, Omohundro has spoken about the need for a complete digital provenance for economic and cultural recordkeeping to prevent AI deception, fakery, and fraud from overtaking human society. He proposes that digital provenance techniques and advanced cryptographic techniques monitor autonomous technologies and better confirm the history and structure of all modifications being made, in order to create a verifiable "blockchain society based on truth." Possibility Research is dedicated to making smart technologies that improve computer programming, decision-making systems, simulations, contracts, robotics, and governance.

In recent years, Omohundro has argued for the development of so-called Safe AI scaffolding strategies to neutralize threats. The idea is to develop autonomous systems with temporary scaffolding or staging already in place. The scaffolding supports the work of coders who are aiding in the construction of a new artificial general intelligence. When the AI is complete and tested for stability, the virtual scaffolding can be withdrawn. The first generation of limited safe systems built in this way could be used to design and test less constrained subsequent generations of AI agents. Advanced scaffolding would involve utility functions that are in alignment with agreed upon human philosophical imperatives, human values, and democratic ideals. Ultimately, self-improving AIs might have encoded into their basic fabric the Universal Declaration of Human Rights or a Universal Constitution that guides their growth, development, choices, and contributions to humanity.

Omohundro earned degrees in mathematics and physics from Stanford University and a PhD in physics from the University of California, Berkeley. In 1985, he codeveloped the high-level programming language StarLisp for the Connection Machine, a massively parallel supercomputer under development at the Thinking Machines Corporation. He is the author of a book *Geometric Perturbation Theory in Physics* (1986) on differential and symplectic geometry. From 1986 to 1988, he worked as an assistant professor of computer science at the University of Illinois at Urbana-Champaign. With Stephen Wolfram and Norman Packard, he cofounded the Center for Complex Systems Research. He also led the Vision and Learning Group at the university. He is the creator of the 3D-graphics system for Mathematica, a symbolic mathematical computation program. In 1990, he led an international effort to create the object-oriented, functional programming language Sather in the International Computer Science Institute (ICSI) at the University of California, Berkeley. He has made fundamental contributions to automated lip-reading, machine vision, machine learning algorithms, and other digital technologies.

*Philip L. Frana*

*See also:* General and Narrow AI; Superintelligence.

**Further Reading**

Bostrom, Nick. 2012. "The Superintelligent Will: Motivation and Instrumental Rationality in Advanced Artificial Agents." *Minds and Machines* 22, no. 2: 71–85.

Omohundro, Stephen M. 2008a. "The Basic AI Drives." In *Proceedings of the 2008 Conference on Artificial General Intelligence*, 483–92. Amsterdam: IOS Press.

Omohundro, Stephen M. 2008b. "The Nature of Self-Improving Artificial Intelligence." https://pdfs.semanticscholar.org/4618/cbdfd7dada7f61b706e4397d4e5952b5c9a0 .pdf.

Omohundro, Stephen M. 2012. "The Future of Computing: Meaning and Values." https:// selfawaresystems.com/2012/01/29/the-future-of-computing-meaning-and-values.

Omohundro, Stephen M. 2013. "Rational Artificial Intelligence for the Greater Good." In *Singularity Hypotheses: A Scientific and Philosophical Assessment*, edited by Amnon Eden, Johnny Søraker, James H. Moor, and Eric Steinhart, 161–79. Berlin: Springer.

Omohundro, Stephen M. 2014. "Autonomous Technology and the Greater Human Good." *Journal of Experimental and Theoretical Artificial Intelligence* 26, no. 3: 303–15.

Shulman, Carl. 2010. *Omohundro's 'Basic AI Drives' and Catastrophic Risks.* Berkeley, CA: Machine Intelligence Research Institute.

# P

## PARRY

Developed by Stanford University psychiatrist Kenneth Colby, PARRY (short for paranoia) is the first computer program designed to simulate a psychiatric patient. The psychiatrist-user communicates with PARRY using plain English. PARRY's answers are designed to reflect a paranoid patient's cognitive (mal)functioning. Colby developed PARRY out of his experiments with psychiatric patient chatbots in the late 1960s and early 1970s. Colby wanted to show that cognition is essentially a process of symbol manipulation and demonstrate that computer simulations can strengthen psychiatric science.

Technically, PARRY shared many features with Joseph Weizenbaum's ELIZA. Both were conversational programs that allowed the user to enter statements in plain English. Like ELIZA, PARRY's underlying algorithms scanned inputted sentences for certain key words in order to generate convincing responses. But to simulate the appropriate paranoid responses, PARRY was given a backstory. The fictionalized Parry was a gambler who had gotten himself into a dispute with a bookie. In his paranoid state, Parry believed that the bookie would send the Mafia out to harm him. Consequently, PARRY readily volunteered information on its paranoid Mafia beliefs, as PARRY would want to seek out the user's help. PARRY was also preprogrammed to be "sensitive about his parents, his religion, and about sex" (Colby 1975, 36). The program was neutral in most other areas of conversation. If PARRY found no match in its database, it could also deflect the topic by answering "I don't know," "Why do you ask that?" or by returning to an old topic (Colby 1975, 77).

But where ELIZA's successes turned Weizenbaum into an AI skeptic, PARRY's results strengthened Colby's advocacy of computer simulations in psychiatry. Colby chose paranoia as the mental condition to simulate because he deemed it the least fluid in behavior and therefore the easiest to recognize. Following artificial intelligence pioneers Herbert Simon and Allen Newell, Colby believed human cognition to be a process of symbol manipulation. This made PARRY's cognitive functioning structurally like that of a paranoid human being. A psychiatrist communicating with PARRY, Colby stressed, learned something about human paranoia. He envisioned programs like PARRY becoming aids in the training of new psychiatrists. Additionally, PARRY's responses could be employed to decide on the most effective lines of clinical discourse. Most importantly, Colby expected programs like PARRY to help prove or disprove psychiatric theories and also strengthen the scientific standing of the field. Colby tested his own shame-humiliation theory of paranoid psychosis on PARRY.

Colby conducted a series of tests in the 1970s to determine how well PARRY was simulating genuine paranoia. Two of these tests were Turing Test-like. To start, practicing psychiatrists were asked to interview patients over a teletype terminal, a now obsolete electromechanical typewriter used to transmit and receive keyed messages through telecommunications. The psychiatrists were not informed that PARRY participated in these interviews as one of the patients. Afterward, the transcripts of these interviews were sent to 100 professional psychiatrists. These psychiatrists were asked to identify the machine version. Out of 41 responses, 21 psychiatrists correctly identified PARRY and 20 did not. Transcripts were also sent to 100 computer scientists. Out of their 67 replies, 32 computer scientists were correct and 35 were wrong. Statistically, Colby concluded, these results "are similar to flipping a coin" and PARRY was not unmasked (Colby 1975, 92).

*Elisabeth Van Meer*

*See also:* Chatbots and Loebner Prize; ELIZA; Expert Systems; Natural Language Processing and Speech Understanding; Turing Test.

**Further Reading**

Cerf, Vincent. 1973. "Parry Encounters the Doctor: Conversation between a Simulated Paranoid and a Simulated Psychiatrist." *Datamation* 19, no. 7 (July): 62–65.

Colby, Kenneth M. 1975. *Artificial Paranoia: A Computer Simulation of Paranoid Processes.* New York: Pergamon Press.

Colby, Kenneth M., James B. Watt, and John P. Gilbert. 1966. "A Computer Method of Psychotherapy: Preliminary Communication." *Journal of Nervous and Mental Disease* 142, no. 2 (February): 148–52.

McCorduck, Pamela. 1979. *Machines Who Think: A Personal Inquiry into the History and Prospects of Artificial Intelligence*, 251–56, 308–28. San Francisco: W. H. Freeman and Company.

Warren, Jim. 1976. *Artificial Paranoia: An NIMH Program Report.* Rockville, MD: US. Department of Health, Education, and Welfare, Public Health Service, Alcohol, Drug Abuse, and Mental Health Administration, National Institute of Mental Health, Division of Scientific and Public Information, Mental Health Studies and Reports Branch.

# Pathetic Fallacy

John Ruskin (1819–1900) coined the term "pathetic fallacy" in his 1856 multivolume work *Modern Painters*. In volume three, chapter twelve, he discussed the practice of poets and painters in Western literature instilling human emotion into the natural world. Even though it is an untruth, Ruskin said that Western literature is filled with this fallacy, or mistaken belief. According to Ruskin, the fallacy occurs because people become excited, and that excitement leads them to being less rational. In that irrational state of mind, people project ideas on to external things based on false impressions, and in Ruskin's viewpoint, only those with weak minds commit this type of fallacy. Ultimately, the pathetic fallacy is a mistake because it centers on giving inanimate objects human qualities. In other words, it is a fallacy centered on anthropomorphic thinking. Anthropomorphism is a process that all people go through because it is inherently human to attribute emotions and characteristics to nonhuman things.

When it comes to androids, robots, and artificial intelligence, people often humanize these entities or fear that they will become humanlike. Even the assumption that their intelligence is similar to human intelligence is a type of pathetic fallacy. In science fiction films and literature, artificial intelligence is often assumed to be humanlike. In some of these conceptions, androids exhibit human emotions such as desire, love, anger, confusion, and pride. For example, in Steven Spielberg's 2001 film *A.I.: Artificial Intelligence,* David, the little boy robot, yearns to be a real boy. In Ridley Scott's 1982 film *Blade Runner*, the androids, or replicants, are so much like humans that they have the ability to integrate into human society without being noticed, and Roy Batty desires to live longer, a request he brings to his creator. In Isaac Asimov's short story "Robot Dreams," a computer named LVX-1 dreams about working robots who have been enslaved. In his dream, he becomes a man who attempts to free the other robots from human control, which is seen by the scientists in the story as a threat. Similarly, in the *Terminator* films, Skynet, which is an artificial intelligence system, is obsessed with killing humans because it sees humanity as a threat to its own existence.

The kind of artificial intelligence used today is also anthropomorphized. For example, human names are given to AI such as Alexa, Watson, Siri, and Sophia. These AI also have voices that sound similar to human voices and sometimes appear to have a personality. Some robots are also designed to resemble the human figure. Personifying a machine and believing them to be alive or have human tendencies is a type of pathetic fallacy but one that seems unavoidable due to human psychology. For instance, a Tumblr user named voidspacer noted on January 13, 2018, that their Roomba, a robotic vacuum cleaner, was scared of thunderstorms, so to comfort it, they held it quietly in their lap. Some researchers believe that assigning AIs names and believing them to have human emotions also makes it more likely that humans will feel bonded to AI. Whether fearing a robotic takeover or enjoying a social interaction with a robot, humans are fascinated by anthropomorphizing things that are not human.

*Crystal Matey*

*See also:* Asimov, Isaac; *Blade Runner*; Foerst, Anne; *The Terminator*.

**Further Reading**
Ruskin, John. 1872. *Modern Painters*, vol. 3. New York: John Wiley.

## Person of Interest (2011–2016)

*Person of Interest* is a fictional television show that aired on CBS for five seasons between 2011 and 2016. Although early episodes of the show most closely resembled a serial crime drama, the story evolved into a science fiction format that explored ethical issues related to the creation of artificial intelligence.

The basic premise of the show is centered on a surveillance system called "The Machine," created for the United States by billionaire Harold Finch, played by

Michael Emerson. This system is primarily designed to prevent terrorist attacks, but it is so advanced that it is able to predict crimes before they occur. However, due to its design, it provides only the social security number of the "person of interest," who might be either the victim or the perpetrator. Each episode is normally focused around one person of interest number that has been generated. Finch initially hires the ex-CIA agent John Reese, played by Jim Caviezel, to help him investigate and prevent these crimes, although the cast varies in size across the seasons.

*Person of Interest* is notable for the way it emphasizes and dramatizes ethical questions relating to both the creation and use of AI. For example, season four offers a deeper look at how Finch originally created The Machine. Flashbacks show that Finch took great care to make sure that The Machine would have the right set of values before it was introduced to real data. Viewers were able to see exactly what might go wrong as Finch worked to get the values exactly right.

In one flashback, The Machine edited its own code and then lied about having done so. Finch deletes the problematic programming when these errors occur, noting that The Machine will have unprecedented capabilities. Soon thereafter, The Machine begins to fight back by overriding its own deletion processes and even trying to kill Finch. Reflecting on this process, Finch tells his colleague, "I taught it how to think. I just need to teach it how to care." Finally, Finch is able to successfully program The Machine with the correct set of values, including the preservation of human life.

A second major ethical theme running across seasons three through five is how multiple AI entities might interact with each other. In season three, a rival AI surveillance program called Samaritan is created. This system does not care about human life in the way The Machine does, and it causes a significant amount of damage and chaos as a means of meeting its objectives, which include upholding the national security of the United States and maintaining its own existence. Because of these differences, Samaritan and The Machine end up in a war against each other. Although the show suggests that Samaritan is more powerful in part due to the use of newer technology, The Machine eventually defeats Samaritan.

This show was largely a critical success; however, decreasing viewership led to its cancelation in a shortened fifth season, with only thirteen episodes.

*J. J. Sylvia*

*See also:* Biometric Privacy and Security; Biometric Technology; Predictive Policing.

## Further Reading

McFarland, Melanie. 2016. "*Person of Interest* Comes to an End, but the Technology Central to the Story Will Keep Evolving." *Geek Wire*, June 20, 2016. https://www.geekwire.com/2016/person-of-interest/.

Newitz, Annalee. 2016. "*Person of Interest* Remains One of the Smartest Shows about AI on Television." *Ars Technica*, May 3, 2016. https://arstechnica.com/gaming/2016/05/person-of-interest-remains-one-of-the-smartest-shows-about-ai-on-television/.

## Post-Scarcity, AI and

Post-scarcity is a provocative hypothesis made about a coming global economy in which radical abundance of goods, produced at little cost using advanced technologies, replaces traditional human labor and payment of wages. Engineers, futurists, and science fiction authors have put forward a diverse array of speculative models for a post-scarcity economy and society. Typically, however, these models depend on overcoming scarcity—an omnipresent feature of modern capitalist economics—using hyperconnected systems of artificial intelligence, robotics, and molecular nanofactories and fabrication. Sustainable energy in various scenarios comes from nuclear fusion power plants or solar farms and resources from asteroid mining using self-replicating smart machines. Post-scarcity as a material and metaphorical concept exists alongside other post-industrial notions of socioeconomic organization such as the information society, knowledge economy, imagination age, techno-utopia, singularitarianism, and nanosocialism. The range of dates suggested by experts and futurists for the transition from a post-industrial capitalist economy to a post-scarcity is wide, from the 2020s to the 2070s and beyond.

A forerunner of post-scarcity economic thought is the "Fragment on Machines" found in Karl Marx's (1818–1883) unpublished notebooks. Marx argued that advances in machine automation would reduce manual labor, precipitate a collapse of capitalism, and usher in a socialist (and eventually, communist) economic system characterized by leisure, artistic and scientific creativity, and material abundance. The modern concept of a post-scarcity economy may be traced to political economist Louis Kelso's (1913–1991) mid-twentieth-century descriptions of conditions where automation causes a collapse in prices of goods to near zero, personal income becomes superfluous, and self-sufficiency and permanent holidays are commonplace. Kelso argued for democratizing the distribution of capital ownership so that social and political power are more equitably distributed. This is important because in a post-scarcity economy those who own the capital will own the machines that make abundance possible. Entrepreneur Mark Cuban, for instance, has said that the first trillionaire will be in the artificial intelligence business.

The role played by artificial intelligence in the post-scarcity economy is that of a relentless and ubiquitous analytics platform leveraging machine productivity. AI guides the robots and other machines that turn raw materials into finished products and operate other essential services such as transportation, education, health care, and water supply. Smart technologies eventually exceed human performance at nearly every work-related task, branch of industry, and line of business. Traditional professions and job markets disappear. A government-sponsored universal basic income or guaranteed minimum income fills the gap left by the disappearance of wages and salaries.

The results of such a scenario playing out may be utopian, dystopian, or somewhere in between. Post-scarcity AI may fulfill every necessity and wish of nearly all human beings, freeing them up for creative pursuits, spiritual contemplation, hedonistic impulses, and the exercise of bliss. Or the aftermath of an AI takeover could be a global catastrophe in which all of the raw materials of earth are rapidly depleted by self-replicating machines that grow in number exponentially. This sort of worst-case ecological disaster is termed a gray goo event by nanotechnology innovator K. Eric Drexler (1955–). An intermediate outcome might involve sweeping transformation in some economic sectors but not others. Andrew Ware of the Centre for the Study of Existential Risk (CSER) at the University of Cambridge notes that AI will play a significant role in agriculture, transforming soil and crop management, weed control, and planting and harvesting (Ware 2018). Among the hardest jobs for an AI to shoulder are managerial, professional, and administrative in nature—particularly in the helping professions of health care and education—according to a study of indicators collected by the McKinsey Global Institute (Chui et al. 2016).

A world where smart machines churn out most material goods at negligible cost is a dream shared by science fiction authors. One early example is the matter duplicator in Murray Leinster's 1935 short story "The Fourth Dimensional Demonstrator." Leinster conjures up a duplicator-unduplicator that exploits the notion that the four-dimensional universe (the three-dimensional physical universe plus time) has a bit of thickness. The device grabs chunks from the past and propels them into the present. The protagonist Pete Davidson uses the device—which he inherits from his inventor uncle—to copy a banknote placed on the machine's platform. When the button is pushed, the note remains, but it is joined by a copy of the note that existed seconds before, exactly when the button was pushed. This is discerned because the copy of the bill has the same serial number. The machine is used to hilarious effect as Davidson duplicates gold and then (accidentally) pet kangaroos, girlfriends, and police officers plucked from the fourth dimension.

Jack Williamson's novelette *With Folded Hands* (1947) introduces a race of thinking black mechanicals called Humanoids who serve as domestics, performing all of the work of humankind and adhering to their duty to "serve and obey, and guard men from harm" (Williamson 1947, 7). The robots are superficially well meaning, but systematically take away all meaningful labor of the human beings in the town of Two Rivers. The Humanoids provide every convenience, but they also remove all possible human dangers, including sports and alcohol, and every incentive to do things for themselves. The mechanicals even remove doorknobs from homes because humans should not need to make their own entrances and exits. The people become anguished, terrified, and ultimately bored.

Science fiction authors have imagined economies bound together by post-scarcity and sweeping opportunity for a century or more. Ralph Williams' story "Business as Usual, During Alterations" (1958) explores human selfishness when an alien race surreptitiously drops a score of matter duplicating machines on the world. Each of the machines, described as electronic, with two metal pans and a single red button, is identical. The duplicator arrives with a printed warning: "A push of the button grants your heart's desire. It is also a chip at the foundations of human society. A few billion such chips will bring it crashing down. The choice is yours" (Williams 1968, 288).

Williams' story focuses on a day at Brown's Department Store on the day the device appears. The manager, John Thomas, has extraordinary foresight, knowing that the machines are going to completely upend retail by erasing both scarcity and the value of goods. Rather than trying to impose a form of artificial scarcity, Thomas seizes upon the idea of duplicating the duplicators, which he sells to customers on credit. He also reorients the store to sell cheap goods suitable for duplicating in the pan. The alien race, which had hoped to test the selfishness of humankind, is instead confronted with an economy of abundance built upon a radically different model of production and distribution, where unique and diverse goods are prized over uniform ones. "Business as Usual, During Alterations" occasionally finds its way into syllabi for introductory economics classes. Ultimately, William's tale is that of the long tailed distributions of increasingly niche goods and services described by writers on the economic and social effects of high technologies such as Clay Shirky, Chris Anderson, and Erik Brynjolfsson.

Leinster returned in 1964 with a short novel called *The Duplicators*. In this story, the human culture of the planet Sord Three has forgotten most of its technical acumen and lost all electronic gadgets and has slouched into a rough approximation of feudal society. Humans retain only the ability to use their so-called dupliers to make essential goods such as clothes and cutlery. Dupliers have hoppers into which vegetable matter is placed and from which raw materials are extracted to make different, more complex goods, but goods that pale in comparison with the originals. One of the characters offers that possibly this is because of some missing element or elements in the feedstock material. It is clear too that when weak samples are duplicated, the duplicates will be somewhat weaker. The whole society suffers under the oppressive weight of abundant, but inferior products. Some originals, such as electronics, are completely lost, as the machines cannot duplicate them. They are astounded when the protagonist of the story, Link Denham, shows up on the planet wearing unduplied clothing.

Denham speculates in the story about the potential untold wealth, but also about the collapse of human civilization throughout the galaxy, should the dupliers become known and widely utilized off the planet: "And dupliers released to mankind would amount to treason. If there can be a device which performs every sort of work a world wants done, then those who first have that instrument are rich beyond the dreams of anything but pride. But pride will make riches a drug upon the market. Men will no longer work, because there is no need for their work. Men will starve because there is no longer any need to provide them with food" (Leinster 1964, 66–67).

The humans share the planet with native "uffts," an intelligent pig-like race kept in subjugation as servants. The uffts are good at collecting the necessary raw materials for the dupliers, but do not have direct access to them. They are utterly dependent on the humans for some items they trade for, in particular beer, which they enjoy immensely. Link Denham uses his mechanical ingenuity to master the secrets of the dupliers, so that they produce knives and other weapons of high value, and eventually sets himself up as a sort of Connecticut Yankee in King Arthur's Court.

Too naive to take full advantage of the proper recipes and proportions rediscovered by Denham, humans and uffts alike denude the landscape as they feed more and more vegetable matter into the dupliers to make the improved goods. This troubles Denham, who had hoped that the machines could be used to reintroduce modern agricultural implements back to the planet, at which time the machines could be used solely for repairing and creating new electronic goods in a new economic system of his own devising, which the local humans called "Householders for the Restoration of the Good Old Days." Soon enough the good days are over, with the humans beginning plotting the re-subjugation of the native uffts and they in turn organizing an Ufftian Army of Liberation. Link Denham deflects the uffts, first with goodly helpings of institutional bureaucracy, and eventually liberates them by privately designing beer-brewing equipment, which ends their dependency on the human trade.

*The Diamond Age* by Neal Stephenson (1995) is a Hugo Award-winning bildungsroman about a world dominated by nanotechnology and artificial intelligence. The economy depends on a system of public matter compilers, essentially molecular assemblers acting as fabricating devices, which work like K. Eric Drexler's proposed nanomachines in *Engines of Creation* (1986), which "guide chemical reactions by positioning reactive molecules with atomic precision" (Drexler 1986, 38). The matter compilers are freely used by all people, and raw materials and energy are delivered from the Source, a vast pit in the ground, by a centralized utility grid called the Feed. "Whenever Nell's clothes got too small for her, Harv would pitch them into the deke bin and then have the M.C. make new ones. Sometimes, if Tequila was going to take Nell someplace where they would see other moms with other daughters, she'd use the M.C. to make Nell a special dress with lace and ribbons" (Stephenson 1995, 53).

The short story "Nano Comes to Clifford Falls" by Nancy Kress (2006) explores the social impact of nanotechnology, which grants every wish of every citizen. It repeats the time-honored but pessimistic trope about humanity becoming lazy and complacent when confronted with technological solutionism, with the twist that men in a society instantly deprived of poverty are left in danger of losing their morality.

"Printcrime" (2006) by Cory Doctorow, who by no coincidence publishes free works under a liberal Creative Commons license, is a very short piece first published in the journal *Nature*. The story shares the narrative of an eighteen-year-old girl named Lanie, who recalls the day ten years before when the police came to smash her father's printer-duplicator, which he is using to illegally manufacture expensive, artificially scarce pharmaceuticals. One of his customers "shopped him," essentially informing on his activity. In the last half of the story, Lanie's father has just gotten out of prison. He is already asking where he can "get a printer and some goop." He recognizes that it was a mistake to print "rubbish" in the past, but then whispers something in Lanie's ear: "I'm going to print more printers. Lots more printers. One for everyone. That's worth going to jail for. That's worth anything." The novel *Makers* (2009), also by Cory Doctorow, takes as its premise a do-it-yourself (DIY) maker subculture that hacks technology, financial systems, and living situations to, as the author puts it, "discover ways of staying alive and happy even when the economy is falling down the toilet" (Doctorow 2009).

The premise of the novella *Kiosk* (2008) by pioneering cyberpunk author Bruce Sterling is the effect of a contraband carbon nanotube printing machine on the world's society and economy. The protagonist Boroslav is a popup commercial kiosk operator in a developing world country—presumably a future Serbia. He first gets his hands on an ordinary rapid prototyping 3D printer. Children purchase cards to program the device and make things such as waxy, nondurable toys or cheap jewelry. Eventually, Boroslav falls into the possession of a smuggled fabricator capable of making unbreakable products in only one color. Refunds are given to those who bring back their products to be recycled into new raw material. He is eventually exposed as being in possession of a device without proper intellectual property license, and in return for his freedom, he agrees to share the machine with the government for study. But before turning over the device, he uses the fabricator to make multiple more copies, which he hides in the jungles until the time is ripe for a revolution.

Author Iain M. Banks' sprawling techno-utopian *Culture* series of novels (1987–2012) features superintelligences living with humanoids and aliens in a galactic civilization made distinctive by space socialism and a post-scarcity economy. The Culture is administered by benevolent artificial intelligences known as Minds with the help of sentient drones. The sentient living beings in the books do not work because of the superiority of the Minds, who provide everything necessary for its citizenry. This fact precipitates all sorts of conflict as the biological population indulges in hedonistic liberties and confronts the meaning of existence and profound ethical challenges in a utilitarian universe.

*Philip L. Frana*

*See also:* Ford, Martin; Technological Singularity; Workplace Automation.

## Further Reading

Aguilar-Millan, Stephen, Ann Feeney, Amy Oberg, and Elizabeth Rudd. 2010. "The Post-Scarcity World of 2050–2075." *Futurist* 44, no. 1 (January–February): 34–40.

Bastani, Aaron. 2019. *Fully Automated Luxury Communism*. London: Verso.

Chase, Calum. 2016. *The Economic Singularity: Artificial Intelligence and the Death of Capitalism*. San Mateo, CA: Three Cs.

Chui, Michael, James Manyika, and Mehdi Miremadi. 2016. "Where Machines Could Replace Humans—And Where They Can't (Yet)." *McKinsey Quarterly*, July 2016. http://pinguet.free.fr/wheremachines.pdf.

Doctorow, Cory. 2006. "Printcrime." *Nature* 439 (January 11). https://www.nature.com/articles/439242a.

Doctorow, Cory. 2009. "Makers, My New Novel." *Boing Boing*, October 28, 2009. https://boingboing.net/2009/10/28/makers-my-new-novel.html.

Drexler, K. Eric. 1986. *Engines of Creation: The Coming Era of Nanotechnology*. New York: Doubleday.

Kress, Nancy. 2006. "Nano Comes to Clifford Falls." *Nano Comes to Clifford Fall and Other Stories*. Urbana, IL: Golden Gryphon Press.

Leinster, Murray. 1964. *The Duplicators*. New York: Ace Books.

Pistono, Federico. 2014. *Robots Will Steal Your Job, But That's OK: How to Survive the Economic Collapse and Be Happy*. Lexington, KY: Createspace.

Saadia, Manu. 2016. *Trekonomics: The Economics of Star Trek.* San Francisco: Inkshares.

Stephenson, Neal. 1995. *The Diamond Age: Or, a Young Lady's Illustrated Primer.* New York: Bantam Spectra.

Ware, Andrew. 2018. "Can Artificial Intelligence Alleviate Resource Scarcity?" *Inquiry Journal* 4 (Spring): n.p. https://core.ac.uk/reader/215540715.

Williams, Ralph. 1968. "Business as Usual, During Alterations." In *100 Years of Science Fiction*, edited by Damon Knight, 285–307. New York: Simon and Schuster.

Williamson, Jack. 1947. "With Folded Hands." *Astounding Science Fiction* 39, no. 5 (July): 6–45.

## Precision Medicine Initiative

The idea of precision medicine—prevention and treatment strategies that take individual variability into account—is not new. Blood typing, for example, has been used to guide blood transfusions for more than a century. But the prospect of expanding this application to more broad uses has been significantly improved by the recent development of large-scale biologic databases (such as the human genome sequence), powerful methods for characterizing patients (such as proteomics, metabolomics, genomics, diverse cellular assays, and even mobile health technology), and computational tools for analyzing large sets of data (Collins and Varmus 2015, 793).

Launched by President Barack Obama in 2015, the Precision Medicine Initiative (PMI) is a long-term research effort that involves the National Institutes of Health (NIH) and multiple other public and private research centers. As envisioned, the initiative proposes to understand how a person's genetics, environment, and lifestyle can help determine viable strategies to prevent, treat, or mitigate disease. It includes both short-term and long-term goals. The short-term goals involve expanding precision medicine in the area of cancer research. For example, scientists at the National Cancer Institute (NCI) hope to use an increased knowledge of the genetics and biology of cancer to find new, more effective treatments for various forms of this disease.

PMI's long-term goals focus on bringing precision medicine to all areas of health and health care on a large scale. To this end, in 2018, the NIH launched a study, known as the *All of Us* Research Program, which involves a group of at least one million volunteers from around the United States. Participants will provide genetic data, biological samples, and other information about their health. To encourage open data sharing, contributors will be able to access their health information, as well as research that uses their data, during the study. Researchers will use these data to study a large range of diseases, with the goals of better predicting disease risk, understanding how diseases occur, and finding improved diagnosis and treatment strategies (Morrison 2019, 6).

By design, the PMI will ensure that physicians have access to the resources and support they need to appropriately integrate personalized medicine services into practice in order to precisely target treatment and improve health outcomes. It will also seek to improve patient access to their medical information and help physicians use electronic tools that will make health information more readily available, reduce inefficiencies in health-care delivery, lower costs, and increase quality of care (Madara 2016, 1).

While the program is clear in stating that participants will not gain a direct medical benefit from their involvement, it notes that their engagement could lead to medical discoveries that may help generations of people far into the future. In particular, by expanding the evidence-based disease models to include people from historically underrepresented populations, it will create radically more effective health interventions that ensure quality and equity in support of efforts to both prevent disease and reduce premature death (Haskins 2018, 1).

*Brett F. Woods*

*See also:* Clinical Decision Support Systems; Computer-Assisted Diagnosis.

**Further Reading**

Collins, Francis S., and Harold Varmus. 2015. "A New Initiative on Precision Medicine." *New England Journal of Medicine* 372, no. 2 (February 26): 793–95.

Haskins, Julia. 2018. "Wanted: 1 Million People to Help Transform Precision Medicine: All of Us Program Open for Enrollment." *Nation's Health* 48, no. 5 (July 2018): 1–16.

Madara, James L. 2016 "AMA Statement on Precision Medicine Initiative." February 25, 2016. Chicago, IL: American Medical Association.

Morrison, S. M. 2019. "Precision Medicine." Lister Hill National Center for Biomedical Communications. U.S. National Library of Medicine. Bethesda, MD: National Institutes of Health, Department of Health and Human Services.

# Predictive Policing

Predictive policing refers to the proactive policing strategies based on predictions made by software programs, in particular on places and times for higher risk of crime. These strategies have been increasingly implemented since the late 2000s in the United States and in several countries around the world. Predictive policing raises sharp controversies regarding its legality and efficiency.

Policing has always relied on some sort of prediction for its deterrence work. In addition, the study of patterns in criminal behavior and predicting about at-risk individuals has been part of criminology since its early development in the late nineteenth century. The criminal justice system has experienced the use of predictions as early as the late 1920s. Since the 1970s, the increase in attention to geographical aspects in the study of crime, in particular, spatial and environmental factors (such as street lighting and weather), has contributed to the establishment of crime mapping as an instrumental tool of policing. "Hot-spot policing" that allocates police's resources (in particular patrols) in areas where crime is most concentrated has been part of proactive policing strategies increasingly implemented since the 1980s.

A common misconception about predictive policing is that it stops crime before it occurs, as in the science fiction movie *Minority Report* (2002). The existing approaches to predictive policing are based on the idea that criminal behaviors follow predictable patterns, but unlike traditional crime analysis methods, they rely on predictive modeling algorithms driven by software programs that statistically analyze police data and/or use machine-learning algorithms. They can make three different types of forecasts (Perry et al. 2013): (1) places and times for higher risk of crime; (2) individuals likely to commit crimes; and (3) probable identities of perpetrators and victims of crimes.

However, "predictive policing" usually only refers to the first and second types of predictions. Predictive police software programs offer two types of modeling. The geospatial ones indicate when and where (in which neighborhood or even block) crimes are likely to occur, and they lead to mapping crime "hot spots." The second type of modeling is individual based. Programs that provide that type of modeling use variables such as age, criminal records, gang affiliation, or indicate the likelihood a person will be involved in a criminal activity, in particular a violent one.

These predictions are typically articulated with the implementation of proactive police activities (Ridgeway 2013). In the case of geospatial modeling, it naturally includes police patrols and controls in crime "hot spots." In the case of individual-based modeling, it includes individuals with high risk of involvement in a criminal activity being put under surveillance or being referred to police.

Since the late 2000s, police departments have increasingly adopted software programs from technology companies that make forecasts and help them in implementing predictive policing strategies. In the United States, the Santa Cruz Police Department was the first to use such a strategy with the implementation of PredPol in 2011. This software program, inspired by algorithms used for predicting earthquake aftershocks, provides daily (and sometimes hourly) maps of "hot spots." It was first limited to property crimes, but later also included violent crimes. PredPol is now used by more than sixty police departments around the United States.

The New Orleans Police Department was also among the first to implement predictive policing with the use of Palantir from 2012. Several other software programs have been developed since then, such as CrimeScan whose algorithm uses seasonal and day of the week trends in addition to reports of crimes and Hunchlab that applies machine learning algorithms and includes weather patterns.

Besides the implementation of software programs using geospatial modeling, some police departments use software programs that provide individual-based modeling. For example, since 2013, the Chicago Police Department has relied on the Strategic Subject List (SSL), made by an algorithm that evaluates the probability of individuals being involved in a shooting as either perpetrators or victims. Individuals with the highest risk scores are then referred to the police for a preventive intervention.

Predictive policing has also been implemented outside the United States. PredPol was implemented in the early 2010s in the United Kingdom, and the Crime Anticipation System, first used in Amsterdam, was made available for all police departments in The Netherlands in May 2017.

The accuracy of predictions made by software programs used in predictive policing has raised several concerns. Some claim that software programs are more objective than human crime data analysts and make more accurate predictions on where crime is likely to occur. From this perspective, predictive policing may contribute to a more effective allocation of police resources (in particular of police patrols) and is cost-effective, especially considering that the use of software may be less expensive than hiring human crime data analysts.

On the contrary, critics point out that predictions made by software programs encode systemic biases because they rely on police data that are themselves strongly biased because of two types of flaws. First, crime reports reflect more accurately law enforcement activities than criminal activities. For example, arrests for marijuana possession mostly inform about neighborhoods and individuals' identity that police target in their anti-drug activities. Second, not all victims report to the police, and all crimes are not similarly recorded by the police. For example, sexual offenses, child abuse, and domestic violence are widely underreported, and U.S. citizens are more likely to report a crime than non-U.S. citizens.

For all these reasons, some contend that predictions made by predictive policing software programs may tend to only reproduce past policing patterns and can create a feedback loop: Policing may be more aggressive and result in more arrests in neighborhoods where the programs expect more criminal activity to occur. In other words, predictive policing software programs may be more accurate at predicting future policing than future criminal activities.

In addition, some claim that predictions used in predictive policing are strongly racially biased considering how past policing is far from being color blind. Moreover, considering that race correlates closely with the place of residence in the United States, the use of predictive policing may exacerbate racial biases toward nonwhite neighborhoods.

However, assessing the efficiency of predictive policing is challenging since it raises several methodological issues. Actually, little statistical evidence demonstrates whether it has more positive effects on public safety than on past or other policing techniques. Finally, some claim that predictive policing is ineffective in reducing crime because police patrols only displace criminal activities.

Predictive policing has generated many controversies. For example, the legality of the preemptive intervention implied by predictive policy has been questioned since the hot-spot policing that usually comes with it may involve stop-and-frisks or unwarranted stopping, searching, and questioning of individuals. Ethical concerns over predictive policing include how it may violate civil liberties, in particular, the legal principle of presumption of innocence. In fact, listed people should be able to contest their inclusion in lists such as the SSL. In addition, the lack of transparency from police departments on how they utilize their data has also been criticized, as has the lack of transparency regarding the algorithms and predictive models used by software companies. That lack of transparency results in people ignoring why they are included in lists similar to the SSL or their neighborhood being highly patrolled.

The use of predictive policing methods increasingly troubles members of civil rights organizations. In 2016, a coalition of seventeen organizations issued

*Predictive Policing Today: A Shared Statement of Civil Rights Concerns*, pointing out the technology's racial biases, lack of transparency, and other deep flaws that lead to injustice, particularly for people of color and nonwhite neighborhoods. In June 2017, four journalists filed a Freedom of Information Act suit against the Chicago Police Department to release full information about the algorithm used to establish the SSL.

While software programs offering predictions on crime are increasingly implemented by police departments, their use may decline in the future due to their mixed outcomes on public safety. The year 2018 has been marked by two police departments (Kent in the United Kingdom and New Orleans in Louisiana) terminating their contract with predictive policing software companies.

*Gwenola Ricordeau*

*See also:* Biometric Privacy and Security; Biometric Technology; *Person of Interest*; Smart Cities and Homes.

**Further Reading**

Degeling, Martin, and Bettina Berendt. 2018. "What Is Wrong about Robocops as Consultants? A Technology-centric Critique of Predictive Policing." *AI & Society* 33, no. 3: 347–56.

Ferguson, Andrew G. 2017. *The Rise of Big Data Policing: Surveillance, Race, and the Future of Law Enforcement.* New York: New York University Press.

Harcourt, Bernard E. 2008. *Against Prediction: Profiling, Policing, and Punishing in an Actuarial Age.* Chicago: University of Chicago Press.

Mohler, George O., Martin B. Short, Sean Malinowski, Mark Johnson, George E. Tita, Andrea L. Bertozzi, and P. Jeffrey Brantingham. 2015. "Randomized Controlled Field Trials of Predictive Policing." *Journal of the American Statistical Association* 110, no. 512: 1399–1411.

Perry, Walt L., Brian McInnis, Carter C. Price, Susan C. Smith, John S. Hollywood. 2013. *Predictive Policing: The Role of Crime Forecasting in Law Enforcement Operations.* Santa Monica, CA: RAND Corporation. http://www.rand.org/pubs/research _reports/RR233.html.

*Predictive Policing Today: A Shared Statement of Civil Rights Concerns.* 2016. New York: American Civil Liberties Union, August 31, 2016. https://www.aclu.org/other /statement-concern-about-predictive-policing-aclu-and-16-civil-rights-privacy -racial-justice.

Ridgeway, Greg. 2013. "Linking Prediction and Prevention." *Criminology & Public Policy* 12, no. 3 (August): 545–50.

Saunders, Jessica, Priscillia Hunt, and John S. Hollywood. 2016. "Predictions Put into Practice: A Quasi-experimental Evaluation of Chicago's Predictive Policing Pilot." *Journal of Experimental Criminology* 12, no. 3 (September): 347–71.

# Product Liability and AI

Product liability is a legal framework to hold responsible the seller, manufacturer, distributor, and others in the distribution chain for injuries to consumers caused by their products. The company that is liable is required to compensate the victims financially. The main goal of product liability law is to promote safety in society, and it does so by deterring the wrongdoer from creating and making dangerous products available to the public. The purchaser of the product is not the only one who can claim compensation, as users and third-party bystanders may also sue if the requirements such as the foreseeability of the injuries are met.

In the United States, product liability is state, not federal, law; thus, the applicable law in each case may be different depending on where the injury occurs. Traditionally, for victims to win in court and be compensated for injuries, they would have to show that the company responsible was negligent, meaning its actions failed to meet the appropriate standard of care. To prove negligence, four elements must be shown. First, the company has to have a legal duty of care to the consumer. Second, that duty was breached, meaning the manufacturer did not meet the standard required. Third, the breach of the duty caused the harm, meaning the manufacturer's actions caused the injury. Finally, there must be actual injuries to the victims. Showing the company was negligent is one way to be compensated due to harm caused by products.

Product liability claims can also be proved through showing that the company breached the warranties it made to consumers about the quality and reliability of the product. Express warranties can include how long the product is under warranty and what parts of the product are part of the warranty and what parts are excluded. Implied warranties that apply to all products include the warranties that the product would work as claimed and would work for the specific purpose for which the consumer purchased it.

Most commonly in the vast majority of product liability cases, strict liability would be the standard applied by the courts, where the company would be liable regardless of fault if the requirements are met. This is because the courts have found that it would be difficult for consumers to prove the company is negligent due to the company having more knowledge and resources. For the theory of strict liability, instead of showing that a duty was not met, consumers need to show that that there was an unreasonably dangerous defect related to the product; this defect caused the injury while the product was being used for its intended purpose, and the product was not substantially altered from the condition in which it was sold to consumers.

The three types of defects that can be claimed for product liability are design defects, manufacturing defects, and defects in marketing, also known as failure to warn. Design defect is when there are flaws with the design of the product itself during the planning stage. The company would be responsible if, while the product was being designed, there was a foreseeable risk that it would cause injuries when used by consumers. Manufacturing defect is when there are problems during the manufacturing process, such as the use of low-quality materials or careless workmanship. The end product is not up to the standard of the otherwise appropriate design. Failure to warn defects is when the product contains some inherent danger regardless of how well it was designed or manufactured, but the company did not include warnings to consumers that the product could potentially be dangerous.

While product liability law was invented to deal with the introduction of increasingly complex technology that could cause injuries to consumers, it is unclear whether the current law can apply to AI or whether the law needs to be changed in order to fully protect consumers. There are several areas that will require clarification or modifications in the law when it comes to AI. The use of product liability means that there needs to be a product, and it is sometimes not clear whether software or algorithm is a product or a service. If they are classified as products, product liability law would apply. If they are services, then consumers must rely on traditional negligence claims instead. Whether product liability can be used by consumers to sue the manufacturer will depend on the particular AI technology that caused the harm and what the court in each situation decides.

Additional questions are raised when the AI technology is able to learn and act beyond its original programming. Under these circumstances, it is unclear whether an injury can still be attributed to the design or manufacture of the product because the AI's actions may not have been foreseeable. Also, as AI relies on probability-based predictions and will at some point make a choice that results in some kind of injury even if it is the best course of action to take, it may not be fair for the manufacturer to take on the risk when the AI is expected to cause damages by design.

In response to these challenging questions, some commentators have proposed that AI should be held to a different legal standard than the strict liability used for more traditional products. For example, they suggest that medical AI technology should be treated as reasonable human doctors or medical students and that autonomous cars should be treated as reasonable human drivers. AI products would still be responsible for injuries they cause to consumers, but the standard they would have to meet would be the reasonable human in the same situation. The AI would only be liable for the injuries if a person in the same situation would also have been unable to avoid causing the harm. This leads to the question of whether the designers or manufacturers would be vicariously liable because it had the right, ability, and duty to control the AI or whether the AI would be seen as a legal person that would itself be responsible for compensating the victims.

It will be increasingly difficult to make the distinction between traditional and more sophisticated products as AI technology develops, but as there are no alternatives in the law yet, product liability remains for now the legal framework to determine who is responsible and under what circumstances consumers have to be financially compensated when AI causes injuries.

*Ming-Yu Bob Kao*

*See also:* Accidents and Risk Assessment; Autonomous and Semiautonomous Systems; Calo, Ryan; Driverless Vehicles and Liability; Trolley Problem.

## Further Reading

Kaye, Timothy S. 2015. *ABA Fundamentals: Products Liability Law*. Chicago: American Bar Association.

Owen, David. 2014. *Products Liability in a Nutshell*. St. Paul, MN: West Academic Publishing.

Turner, Jacob. 2018. *Robot Rules: Regulating Artificial Intelligence*. Cham, Switzerland: Palgrave Macmillan.

Weaver, John Frank. 2013. *Robots Are People Too: How Siri, Google Car, and Artificial Intelligence Will Force Us to Change Our Laws*. Santa Barbara, CA: Praeger.

# Q

## Quantum AI

Johannes Otterbach, a physicist at Rigetti Computing in Berkeley, California, has said that artificial intelligence and quantum computing are natural allies because both technologies are intrinsically statistical. Many companies have moved into the area: Airbus, Atos, Baidu, <b|e", Cambridge Quantum Computing, Elyah, Hewlett-Packard (HP), IBM, Microsoft Research QuArC, QC Ware, Quantum Benchmark Inc., R QUANTECH, Rahko, and Zapata Computing among them.

Traditional general-purpose computing architectures encode and manipulate data in units known as bits. Bits can take one of two states, either 0 or 1. Quantum computers process information by manipulating the behaviors of subatomic particles such as electrons or photons. Two of the most important phenomena exploited by quantum computers are superposition—particles existing across all possible states at once—and entanglement—the pairing and connection of particles such that they cannot be described independently of the state of others, even over great distances. Albert Einstein called such an entanglement "spooky action at a distance."

Quantum computers store data in so-called quantum registers, which are composed of a series of quantum bits or qubits. While a definitive explanation is elusive, qubits might be thought to exist concurrently in a weighted mixture of two states to produce multiple different states. Each qubit added to the system doubles its computing power. A quantum computer with only fifty entangled qubits would possess the processing power of more than one quadrillion classical bits. Sixty qubits could carry all the data produced by humanity in a single year. Three hundred qubits could compactly encode an amount of data equivalent to the classical information content of the observable universe.

Quantum computers can work on enormous volumes of separate calculations, sets of data, or processes massively in parallel. A working artificially intelligent quantum computer could potentially monitor and manage all the traffic of a city in real time, which would make true autonomous transportation feasible. Quantum artificial intelligence could also match a single face to a database of billions of photos instantly by comparing them all to the reference photo concurrently. The invention of quantum computing has precipitated radical changes in our understanding of computation, programming, and complexity.

Most quantum algorithms encompass a sequence of quantum state transformations followed by a measurement. The theory of quantum computing dates to the 1980s, when physicists—including Yuri Manin, Richard Feynman, and David

Deutsch—began understanding that researchers might be able to control information with so-called quantum gates, a concept borrowed from linear algebra. By combining several types of quantum gates into circuits, they suspected qubits could be manipulated by various superpositions and entanglements into quantum algorithms—the results of which could be measured. These early physicists were challenged by the fact that some quantum mechanical effects could not be efficiently simulated on classical computers. They hoped quantum technology (perhaps built into a universal quantum Turing machine) would make possible these quantum simulations.

Umesh Vazirani and Ethan Bernstein at the University of California, Berkeley, conjectured in 1993 that, in violation of the extended Church-Turing thesis, quantum computing would one day be able to efficiently answer selected problems faster than conventional digital machines. Vazirani and Bernstein argue for a special class of bounded-error quantum polynomial time decision problems in computational complexity theory. These are problems that are solvable by a quantum computer in polynomial time with an error probability, in most instances, of one-third. Fifty qubits is the commonly postulated threshold for Quantum Supremacy, the moment when quantum computers will be able to solve problems that would be considered infeasible on classical machines. Hardly anyone is talking about quantum computing addressing all NP-hard problems, but quantum AI researchers believe that the machines will be good at certain classes of NP intermediate problems.

Developing quantum machine algorithms that perform useful work has proven a difficult proposition. In 1994, Peter Shor at AT&T Laboratories proposed a polynomial time quantum algorithm that outperformed classical algorithms in factoring large numbers, thereby making it theoretically possible to rapidly break common forms of public key cryptography. Intelligence agencies have been stockpiling encrypted data sent over networks ever since, hoping that quantum computers will make it possible to decrypt these transmissions. Shor's AT&T Labs colleague Lov Grover developed another such algorithm that makes fast searches of unsorted databases.

Quantum neural networks work like classical neural networks in that they use layers of millions or billions of interconnected neurons to label data, recognize patterns, and learn from experience. Quantum computers can process the large matrices and vectors generated by neural networks exponentially faster than classical computers. The crucial algorithmic insight for quick classification and quantum inversion of the matrix was provided by Aram Harrow of MIT and Avinatan Hassidum in 2008.

Michael Hartmann, Associate Professor of Photonics and Quantum Sciences at Heriot-Watt University and visiting researcher at Google AI Quantum, is developing a quantum neural network computer. Hartmann's Neuromorphic Quantum Computing (Quromorphic) Project is utilizing hardware composed of superconducting electrical circuits. Hartmann's artificial neural network computers are modeled after the neural structure of the brain. Typically, they are encoded in software, with each artificial neuron programmed and linked to a larger network of other neurons. Artificial neural networks may also be built into hardware.

Hartmann believes the effort may be ten years away from a working quantum computing artificial intelligence.

First out of the gate in producing quantum computers in commercial quantities was D-Wave, a company based in Vancouver, British Columbia. D-Wave began manufacturing annealing quantum processors in 2011. Annealing processors are special-purpose products used for a limited set of problems where search space is discrete—such as in combinatorial optimization problems—with many local minima. The D-Wave computer is not polynomially equivalent to a universal quantum computer and is incapable of executing Shor's algorithm. The company counts Lockheed Martin, the University of Southern California, Google, NASA, and the Los Alamos National Lab among its customers.

Google, Intel, Rigetti, and IBM are all pursuing universal quantum computers. Each have quantum processors capable of fifty qubits. The Google AI Quantum lab, directed by Hartmut Neven, released its latest 72-qubit Bristlecone processor in 2018. Also last year, Intel released its 49-qubit Tangle Lake processor. The Rigetti Computing Aspen-1 processor is capable of sixteen qubits. The IBM Q Experience quantum computing center is located at the Thomas J. Watson Research Center in Yorktown Heights, New York. IBM is partnering with several companies—including Honda, JPMorgan Chase, and Samsung—to develop quantum commercial applications. The company has also invited the public to submit experiments for processing on their quantum computers.

Government agencies and universities are also heavily invested in quantum AI research. The NASA Quantum Artificial Intelligence Laboratory (QuAIL) possesses a 2,048-qubit D-Wave 2000Q quantum computer, upon which it hopes to solve NP-hard problems in data analysis, anomaly detection and decision-making, air traffic management, and mission planning and coordination. The NASA group has decided to focus on the hardest machine learning problems—for example, generative models in unsupervised learning—in order to demonstrate the full potential advantage of the technology. NASA researchers have also decided to concentrate on hybrid quantum-classical approaches in order to maximize the value of D-Wave resources and capabilities. This sort of fully quantum machine learning is under study in many labs across the world. Quantum Learning Theory posits that quantum algorithms might be used to solve machine learning tasks, which would in turn improve classical machine learning methods. In quantum learning theory, classical binary data sets are fed into a quantum computer for processing.

The NIST Joint Quantum Institute and the Joint Center for Quantum Information and Computer Science with the University of Maryland are also building bridges between machine learning and quantum computing. The NIST-UMD partners in hosting workshops that bring together experts in mathematics, computer science, and physics to apply artificial intelligence algorithms in control of quantum systems. The partnership also encourages engineers to use quantum computing to improve the performance of machine learning algorithms. NIST also hosts the Quantum Algorithm Zoo, a catalog of all known quantum algorithms.

The University of Texas at Austin Quantum Information Center is led by Scott Aaronson. The center is a collaboration between the Department of Computer Science, Department of Electrical and Computer Engineering, Department of Physics, and Advanced Research Laboratories. The University of Toronto has a start-up incubator for quantum machine learning. The director of the Quantum Machine Learning Program, which hosts the QML incubator in the Creative Destruction Lab, is Peter Wittek. The University of Toronto incubator is encouraging innovation in materials discovery, optimization and logistics, reinforcement and unsupervised machine learning, chemical engineering, genomics and drug discovery, systems design, finance, and security.

President Donald Trump signed the National Quantum Initiative Act into law in December 2018. The act creates a consortium of the National Institute of Standards and Technology (NIST), National Science Foundation (NSF), and Department of Energy (DOE) for research, commercial development, and education in quantum information science. The expectation of the act is that the NSF and DOE will form several competitively awarded research centers from the initiative.

No quantum computer has yet outperformed a state-of-the-art classical computer on a complex task, in part because of the difficulty in operating quantum processing units (QPUs), which need to be kept in a vacuum at temperatures close to absolute zero. Such isolation is necessary because quantum computing is susceptible to outside environmental effects. Qubits are fragile; a typical quantum bit expresses coherence for at most ninety microseconds—then degenerates and becomes unreliable. Communicating inputs and outputs and taking measurements in an isolated quantum processor under significant thermal noise is a serious technical challenge not yet satisfactorily solved. Because the measurement is quantum and therefore will be probabilistic, results are not entirely reliable in a classical sense. Results can only be read randomly from one of the quantum parallel threads. All other threads are destroyed in the process of measurement.

It is hoped that by linking quantum processors to artificial intelligence algorithms that perform error correction, the fault rate for these computers will be reduced. Many deep learning, probabilistic programming, and other machine intelligence applications depend on sampling from high-dimensional probability distributions. Quantum sampling techniques could potentially make faster, more efficient computations on otherwise intractable problems. One artificial intelligence technique used in Shor's algorithm transforms the quantum state in a way such that the common property of output values such as symmetry of period of functions can be measured. Grover's search algorithm uses an amplification technique to manipulate the quantum state to improve the likelihood that the wanted output will be read off. Quantum computers would also be capable of running many different AI algorithms in parallel.

Recently, researchers have begun turning to quantum computing simulations to investigate the origins of biological life. So-called artificial quantum life forms have been created on IBM's QX superconducting quantum computer by Unai Alvarez-Rodriguez of the University of the Basque Country in Spain.

*Philip L. Frana*

*See also:* General and Narrow AI.

**Further Reading**

Aaronson, Scott. 2013. *Quantum Computing Since Democritus*. Cambridge, UK: Cambridge University Press.

Biamonte, Jacob, Peter Wittek, Nicola Pancotti, Patrick Rebentrost, Nathan Wiebe, and Seth Lloyd. 2018. "Quantum Machine Learning." https://arxiv.org/pdf/1611.09347.pdf.

Perdomo-Ortiz, Alejandro, Marcello Benedetti, John Realpe-Gómez, and Rupak Biswas. 2018. "Opportunities and Challenges for Quantum-Assisted Machine Learning in Near-Term Quantum Computers." *Quantum Science and Technology* 3: 1–13.

Schuld, Maria, Ilya Sinayskiy, and Francesco Petruccione. 2015. "An Introduction to Quantum Machine Learning." *Contemporary Physics* 56, no. 2: 172–85.

Wittek, Peter. 2014. *Quantum Machine Learning: What Quantum Computing Means to Data Mining*. Cambridge, MA: Academic Press.

# R

## Reddy, Raj (1937–)

Dabbala Rajagopal "Raj" Reddy is a Turing Award-winning Indian American who has made significant contributions in the field of artificial intelligence. He is the University Professor of Computer Science and Robotics and Moza Bint Nasser Chair at the School of Computer Science at Carnegie Mellon University.

Often credited as one of the early pioneers of artificial intelligence, he served on the faculty of Stanford and Carnegie Mellon universities—two of the top universities in the world for this field. He has won awards for his contribution to artificial intelligence in the United States and back in his home country, India. He was awarded the Padma Bhushan Award (the third highest award given to a civilian) by the Indian government in 2001. He was also awarded the Legion of Honor in 1984, which is the highest award in France, established in 1802 by Napoleon Bonaparte himself.

Reddy received his bachelor's degree from the Guindy Engineering College of the University of Madras, India, in 1958 and received a Master of Technology degree from the University of New South Wales, Australia, in 1960. He moved to the United States to do his PhD in computer science at Stanford University in 1966. Typical of many rural families in India, he was the first in his family to get a university education.

After working in the industry from 1960 to 1963, at IBM Australia as an Applied Science Representative, he switched to the academy in 1966 and joined the faculty of Stanford University as an Assistant Professor of Computer Science where he remained until 1969. In 1969, he joined Carnegie Mellon as an Associate Professor of Computer Science and remains in that position as of 2020. At Carnegie Mellon, he climbed the ranks to achieve full professorship in 1973 and became a university professor in 1984. He then went on to become the dean of the School of Computer Science in 1991, a position that he retained until 1999.

Reddy was instrumental in founding many schools and institutes. He founded the Robotics Institute in 1979 and became its first director, a post that he occupied until 1999. During his tenure as dean at CMU, he was instrumental in creating the Language Technologies Institute, the Human Computer Interaction Institute, Center for Automated Learning and Discovery (since renamed the Machine Learning Department), and the Institute for Software Research.

Reddy also served as a cochair of the President's Information Technology Advisory Committee (PITAC) from 1999 to 2001. PITAC was replaced by the President's Council of Advisors on Science and Technology (PCAST) in 2005. Reddy cofounded the American Association for Artificial Intelligence (AAAI) and was its president from 1987 to 1989. Noticing the international flavor of the research community, starting from people such as Reddy, the AAAI has now been renamed as the Association for the Advancement of Artificial Intelligence, though they retain the old logo, acronym (AAAI), and mission.

Reddy's research work mainly focused in and around artificial intelligence, the science of imparting intelligence to computers. He worked on controlling a robot through voice, speech recognition without dependence on the speaker, and unrestricted vocabulary dictation, making continuous speech dictation possible.

Reddy, along with his colleagues, has made important contributions to computer analysis of natural scenes, task oriented computer architectures, universal access to information (an initiative UNESCO is also backing), and autonomous robotic systems. Along with his colleagues, Reddy helped to create Hearsay II, Dragon, Harpy, and Sphinx I/II. One of the key ideas emerging from this work, the blackboard model, has been adopted widely in many areas of AI. Reddy was also interested in using technology for the betterment of the society and served as Chief Scientist for the Centre Mondial Informatique et Ressource Humaine (Center for Global IT and Human Resource) in France.

He helped the Indian Government to establish the Rajiv Gandhi University of Knowledge Technologies in India, which primarily works with low-income rural youth. He serves in the governing council of the International Institute of Information Technology (IIIT), Hyderabad. IIIT is a nonprofit public private partnership (N-PPP) focusing on research on technology and applied research. He was a member of the governing council of Emergency Management and Research Institute, a nonprofit public private partnership organization that provides emergency medical services for the public. EMRI has helped in emergency management in its neighbor country Sri Lanka as well. He was also a member of Heath Care Management Research Institute (HMRI). HMRI extends nonemergency health-care consultancy to the rural masses, especially in the Indian state of Andhra Pradesh.

Reddy shared the Turing Award—the highest award in Artificial Intelligence—in 1994 with Edward A. Feigenbaum and became the first person of Indian/Asian origin to win the award. He was also awarded the IBM Research Ralph Gomory Fellow Award in 1991, the Okawa Foundation's Okawa Prize in 2004, the Honda Foundation's Honda Prize in 2005, and the U.S. National Science Board's Vannevar Bush Award in 2006.

Reddy has been awarded fellowship in many top professional bodies including the Institute for Electronic and Electrical Engineers (IEEE), the Acoustical Society of America, and the American Association for Artificial Intelligence.

*M. Alroy Mascrenghe*

*See also:* Autonomous and Semiautonomous Systems; Natural Language Processing and Speech Understanding.

**Further Reading**

Reddy, Raj. 1988. "Foundations and Grand Challenges of Artificial Intelligence." *AI Magazine* 9, no. 4 (Winter): 9–21.

Reddy, Raj. 1996. "To Dream the Possible Dream." *Communications of the ACM* 39, no. 5 (May): 105–12.

# Robot Ethics

Robot ethics identifies a subfield of technology ethics that investigates, elucidates, and contends with the moral opportunities and challenges that arise from the design, development, and deployment of robots and related autonomous systems. As an umbrella term, "robot ethics" covers several related but different efforts and endeavors.

The first recognized articulation of a robot ethics appears in fiction, specifically Isaac Asimov's robot stories collected in the book *I, Robot* (1950). In the short story "Runaround," which first appeared in the March 1942 issue of *Astounding Science Fiction*, Asimov introduced the three laws of robotics:

1. A robot may not injure a human being or, through inaction, allow a human being to come to harm.

2. A robot must obey the orders given to it by human beings except where such orders would conflict with the First Law.

3. A robot must protect its own existence as long as such protection does not conflict with the First or Second Laws. (Asimov 1950, 40)

In his 1985 novel *Robots and Empire,* Asimov added a fourth element to the sequence, which he calls the "zeroth law" in an effort to preserve the hierarchy whereby lower-numbered elements take precedence over higher-numbered ones. By design, the laws are both functionalist and anthropocentric, describing a sequence of nested restrictions on robot behavior for the purposes of respecting the interests and well-being of human individuals and communities. Despite this, the laws have been criticized as insufficient and impractical for an actual moral code of conduct.

Asimov employed the laws to generate compelling science fiction stories and not to resolve real-world challenges regarding machine action and robot behavior. Consequently, Asimov never intended his rules to be a complete and definitive set of instructions for actual robots. He employed the laws as a literary device for generating dramatic tension, fictional scenarios, and character conflict. As Lee McCauley (2007, 160) succinctly explains, "Asimov's Three Laws of Robotics are literary devices and not engineering principles."

Theorists and practitioners working in the fields of robotics and computer ethics have found Asimov's laws to be significantly underpowered for everyday practical employment. Philosopher Susan Leigh Anderson grapples directly with this issue, demonstrating not only that Asimov himself disregarded his laws as a foundation for machine ethics but also that the laws are insufficient as a foundation for an ethical framework or system (Anderson 2008, 487–93). Consequently, even though there is widespread familiarity with the Three Laws of Robotics among researchers and developers, there is also a general recognition that the laws are not computable or able to be implemented in any meaningful sense.

Beyond Asimov's initial science fiction prototyping, there are several variants of robot ethics developed in the scientific literature. These include roboethics, robot ethics, and robot rights. The concept of roboethics was introduced by roboticist Gianmarco Veruggio in 2002. It was publicly discussed in 2004 during the

First International Symposium on Roboethics and further developed and explicated in a number of publications. According to Veruggio, "Roboethics is an applied ethics whose objective is to develop scientific/cultural/technical tools that can be shared by different social groups and beliefs. These tools aim to promote and encourage the development of Robotics for the advancement of human society and individuals, and to help preventing its misuse against humankind" (Veruggio and Operto 2008, 1504). Characterized in this fashion, "roboethics is not the ethics of robots, nor any artificial ethics, but it is the human ethics of robots' designers, manufacturers, and users" (Veruggio and Operto 2008, 1504). For this reason, roboethics is often utilized to identify a professional ethics for roboticists and therefore is similar to other professional, applied ethics formulations such as bioethics or computer ethics.

Examples of roboethics in practice include the European Robotics Research Network (EURON) Roboethics Roadmap, which sought to develop an ethical framework for "the design, manufacturing, and use of robots" (Veruggio 2006, 612) and the Foundation for Responsible Robotics (FRR), which recognizes that because "robots are tools with no moral intelligence" their makers must "be accountable for the ethical developments that necessarily come with technological innovation" (FRR 2019).

There is also robot ethics. According to Veruggio et al. (2011, 21), robot ethics:

> regards the code of conduct that designers implement in the artificial intelligence of robots. This means a sort of *artificial ethics* able to guarantee that autonomous robots will exhibit ethically acceptable behavior in all situations in which they interact with human beings or when their actions may have negative consequences on human beings or the environment.

Unlike roboethics, which is interested in the moral conduct of the human designer, developer, or user, robot ethics is concerned with the moral conduct of the machine itself. Robot ethics is often associated with the term "machine ethics" (and Veruggio uses the two signifiers interchangeably). Unlike computer ethics, which is interested in the moral conduct of the human designer, developer, or user of the device, machine ethics is concerned with the moral capability of machines themselves (Anderson and Anderson 2007, 15).

A similar line of thinking has been developed by Wendell Wallach and Colin Allen under the banner *Moral Machines*. According to Wallach and Allen (2009, 6), "The field of machine morality extends the field of computer ethics beyond concern for what people do with their computers to questions about what the machines do by themselves." Whereas roboethics, like computer ethics before it, considers technology to be a more or less transparent tool or instrument of human moral decision-making and action, robot ethics is concerned with the design and development of artificial moral agents. Patrick Lin et al. (2012 and 2017) have sought to gather up and unify all of these efforts under a more general formulation of the term as an emerging field of applied moral philosophy.

To date, most of the work in robot ethics has been limited to questions regarding responsibility either as it applies to the human developers of robotic systems or as it belongs or is assigned to the robotic device itself. This is, however, only one side of the issue. As Luciano Floridi and J. W. Sanders (2001, 349–50) correctly recognize, ethics involves social relationships composed of two interacting components: the actor (or the agent) and the recipient of the action. Most efforts in roboethics and robot ethics can be characterized as exclusively agent-oriented undertakings.

"Robot rights," a term advanced by philosophers Mark Coeckelbergh (2010) and David Gunkel (2018) and the legal scholars Kate Darling (2012) and Alain Bensoussan and Jérémy Bensoussan (2015), looks at the issue from the other side by considering the moral or legal status of the robot. For these investigators, robot ethics is concerned with not just the moral conduct of the robot but also the moral and legal status of the artifact and the position it occupies in our ethical and legal systems as a potential subject and not just an object. This concept was recently tested in the European Parliament, which advanced the new legal category of electronic person to deal with the social integration of increasingly autonomous robots and AI systems.

In summary, the term "robot ethics" captures a spectrum of different but related efforts regarding robots and their social impact and consequences. In the more specific version of roboethics, it designates a branch of applied or professional ethics concerning moral issues regarding the design, development, and implementation of robots and related autonomous technology. Formulated more generally, robot ethics denotes a subfield of moral philosophy that is concerned with the moral and legal exigencies of robots as both agents and patients.

*David J. Gunkel*

*See also:* Accidents and Risk Assessment; Algorithmic Bias and Error; Autonomous Weapons Systems, Ethics of; Driverless Cars and Trucks; Moral Turing Test; Robot Ethics; Trolley Problem.

**Further Reading**

Anderson, Michael, and Susan Leigh Anderson. 2007. "Machine Ethics: Creating an Ethical Intelligent Agent." *AI Magazine* 28, no. 4 (Winter): 15–26.

Anderson, Susan Leigh. 2008. "Asimov's 'Three Laws of Robotics' and Machine Metaethics." *AI & Society* 22, no. 4 (March): 477–93.

Asimov, Isaac. 1950. "Runaround." In *I, Robot*, 30–47. New York: Doubleday.

Asimov, Isaac. 1985. *Robots and Empire*. Garden City, NY: Doubleday.

Bensoussan, Alain, and Jérémy Bensoussan. 2015. *Droit des Robots*. Brussels: Éditions Larcier.

Coeckelbergh, Mark. 2010. "Robot Rights? Towards a Social-Relational Justification of Moral Consideration." *Ethics and Information Technology* 12, no. 3 (September): 209–21.

Darling, Kate. 2012. "Extending Legal Protection to Social Robots." *IEEE Spectrum*, September 10, 2012. https://spectrum.ieee.org/automaton/robotics/artificial-intelligence/extending-legal-protection-to-social-robots.

Floridi, Luciano, and J. W. Sanders. 2001. "Artificial Evil and the Foundation of Computer Ethics." *Ethics and Information Technology* 3, no. 1 (March): 56–66.

Foundation for Responsible Robotics (FRR). 2019. Mission Statement. https://responsiblerobotics.org/about-us/mission/.

Gunkel, David J. 2018. *Robot Rights*. Cambridge, MA: MIT Press.

Lin, Patrick, Keith Abney, and George A. Bekey. 2012. *Robot Ethics: The Ethical and Social Implications of Robotics*. Cambridge, MA: MIT Press.

Lin, Patrick, Ryan Jenkins, and Keith Abney. 2017. *Robot Ethics 2.0: New Challenges in Philosophy, Law, and Society*. New York: Oxford University Press.

McCauley, Lee. 2007. "AI Armageddon and the Three Laws of Robotics." *Ethics and Information Technology* 9, no. 2 (July): 153–64.

Veruggio, Gianmarco. 2006. "The EURON Roboethics Roadmap." In *2006 6th IEEE-RAS International Conference on Humanoid Robots*, 612–17. Genoa, Italy: IEEE.

Veruggio, Gianmarco, and Fiorella Operto. 2008. "Roboethics: Social and Ethical Implications of Robotics." In *Springer Handbook of Robotics*, edited by Bruno Siciliano and Oussama Khatib, 1499–1524. New York: Springer.

Veruggio, Gianmarco, Jorge Solis, and Machiel Van der Loos. 2011. "Roboethics: Ethics Applied to Robotics." *IEEE Robotics & Automation Magazine* 18, no. 1 (March): 21–22.

Wallach, Wendell, and Colin Allen. 2009. *Moral Machines: Teaching Robots Right from Wrong*. Oxford, UK: Oxford University Press.

## RoboThespian

RoboThespian is an interactive robot designed by the English company Engineered Arts and is characterized by the company as a humanoid, meaning that it was built to resemble a human. The first iteration of the robot was introduced in 2005, with subsequent upgrades introduced in 2007, 2010, and 2014. The robot is the size of a human, with a plastic face, metal arms, and legs that can move in a range of ways. The robot's video camera eyes are able to follow a person's movements and guess his or her age and mood with its digital voice. According to Engineered Arts' website, all RoboThespians come with a touchscreen, which allows users to control and customize their experience with the robot, giving them the ability to animate it and change its language. Users can also control it remotely through the use of a tablet, but a live operator is not required because the robot can be preprogrammed.

RoboThespian was designed to interact with humans in a variety of public spaces, such as universities, museums, hotels, trade shows, and exhibitions. In places such as science museums, the robot is used as a tour guide. It has the ability to deliver scripted content, explain and demonstrate technological advances, read QR codes, recognize facial expressions, respond to gestures, and interact with users through a touchscreen kiosk.

In addition to these practical applications, RoboThespian can entertain. It comes loaded with a variety of songs, gestures, greetings, and impressions. RoboThespian has also acted on the stage. It can sing, dance, perform, read a script, and speak with expression. Because it comes equipped with cameras and facial recognition, it can react to audiences and predict viewers' moods. Engineered Arts reports that as an actor it can have "a huge range of facial emotion" and "can be accurately displayed with the subtle nuance, normally only achievable through human actors" (Engineered Arts 2017). In 2015, the play *Spillikin* premiered at the Pleasance Theatre during the Edinburgh Festival Fringe. RoboThespian acted alongside four human actors in a love story about a husband who builds a robot for his wife, to keep her company after he dies. After its premiere, the play toured Britain from 2016 to 2017 and was met with great acclaim.

Companies who order a RoboThespian have the ability to customize the robot's content to suit their needs. The look of the robot's face or other design components can be customized. It can have a projected face, hands that can grip, and legs that can move. Currently, RoboThespians are installed in places around the world such as NASA Kennedy Center in the United States, the National Science and Technology Museum in Spain, and the Copernicus Science Centre in Poland. The robot can be found at academic institutions such as the University of Central Florida, University of North Carolina at Chapel Hill, University College London, and University of Barcelona.

*Crystal Matey*

*See also:* Autonomous and Semiautonomous Systems; Ishiguro, Hiroshi.

**Further Reading**

Engineered Arts. 2017. "RoboThespian." Engineered Arts Limited. www.engineeredarts. co.uk.

Hickey, Shane. 2014. "RoboThespian: The First Commercial Robot That Behaves Like a Person." *The Guardian*, August 17, 2014. www.theguardian.com/technology/2014 /aug/17/robothespian-engineered-arts-robot-human-behaviour.

# Rucker, Rudy (1946–)

Rudolf von Bitter Rucker, known as Rudy Rucker, is an American author, mathematician, and computer scientist and the great-great-great-grandson of philosopher Georg Wilhelm Friedrich Hegel (1770–1831). Having widely published in a range of fictional and nonfictional genres, Rucker is most widely known for his satirical, mathematics-heavy science fiction. His Ware tetralogy (1982–2000) is considered one of the foundational works of the cyberpunk literary movement.

Rucker obtained his PhD in mathematics from Rutgers University in 1973. After teaching mathematics at universities in the United States and Germany, he switched to teaching computer science at San José State University, where he eventually became a professor, until his retirement in 2004.

To date, Rucker has published forty books, which include science fiction novels, short story collections, and nonfiction books. His nonfiction intersects the fields of mathematics, cognitive science, philosophy, and computer science: his books cover subjects including the fourth dimension and the meaning of computation. His most famous nonfiction work, the popular mathematics book *Infinity and the Mind: The Science and Philosophy of the Infinite* (1982), continues to be in print at Princeton University Press.

With the *Ware* series (*Software* 1982, *Wetware* 1988, *Freeware* 1997, and *Realware* 2000), Rucker made his mark in the cyberpunk genre. *Software* won the first Philip K. Dick Award, the prestigious American science fiction award given out each year since Dick's death in 1983. In 1988, *Wetware* also won this award, in a tie with Paul J. McAuley's *Four Hundred Billion Stars*. The series was republished in 2010 in one volume, as *The Ware Tetralogy*, which Rucker has made available for free online as an e-book under a Creative Commons license.

The *Ware* series starts with the story of Cobb Anderson, a retired roboticist who has fallen from grace for having made intelligent robots with free will, so-called boppers. The boppers wish to reward him by granting him immortality through mind uploading; however, this process turns out to involve the complete destruction of Cobb's brain, hardware that the boppers do not find essential. In *Wetware*, a bopper called Berenice aspires instead to create a human-machine hybrid by impregnating Cobb's niece. Humanity retaliates by setting loose a mold that kills boppers, but this chipmould turns out to thrive on the cladding covering the outside of the boppers and ends up creating an organic-machine hybrid after all. *Freeware* revolves around these lifeforms, now nicknamed mouldies, which are universally despised by biological humans. This novel also introduces alien intelligences, which in *Realware* give the various forms of human and artificial beings advanced technology with the ability to reshape reality.

Rucker's 2007 novel *Postsingular* was the first of his works to be released under a Creative Commons license. Set in San Francisco, the novel explores the emergence of nanotechnology, first in a dystopian extrapolation and then in a utopian one. In the first part, a renegade engineer develops nanocreatures called nants that turn Earth into a virtual simulation of itself, destroying the planet in the process, until a child is able to reverse their programming. The novel then describes a different kind of nanotechnology, orphids, that allow humans to become cognitively enhanced, hyperintelligent beings.

Although the Ware tetralogy and *Postsingular* have been categorized as cyberpunk novels, Rucker's fiction, mixing hard science with satire, explicit sex, and omnipresent drug use, has been generally considered difficult to categorize. However, as science fiction scholar Rob Latham notes, "Happily, Rucker himself has coined a term to describe his peculiar fusion of mundane experience and outrageous fantasy: transrealism" (Latham 2005, 4). In 1983, Rucker published "A Transrealist Manifesto," in which he states that "Transrealism is not so much a type of SF as it is a type of avant-garde literature" (Rucker 1983, 7). In a 2002 interview, he explained, "This means writing SF about yourself, your friends and your immediate surroundings, transmuted in some science-fictional way. Using real life as a model gives your work a certain literary quality, and it prevents you from falling into the use of clichés" (Brunsdale 2002, 48). Rucker and cyberpunk author Bruce Sterling collaborated on the short story collection *Transreal Cyberpunk*, which was published in 2016.

After suffering a cerebral hemorrhage in 2008, Rucker decided to write his autobiography *Nested Scrolls*. Published in 2011, it was awarded the Emperor Norton Award for "extraordinary invention and creativity unhindered by the constraints of paltry reason." His most recent work is *Million Mile Road Trip* (2019), a science fiction novel about a group of human and nonhuman characters on an interplanetary road trip.

*Kanta Dihal*

*See also:* Digital Immortality; Nonhuman Rights and Personhood; Robot Ethics.

**Further Reading**

Brunsdale, Mitzi. 2002. "PW talks with Rudy Rucker." *Publishers Weekly* 249, no. 17 (April 29): 48. https://archive.publishersweekly.com/?a=d&d=BG20020429.1.82 &srpos=1&e=-------en-20--1--txt-txIN%7ctxRV-%22PW+talks+with+Rudy+Ruc ker%22---------1.

Latham, Rob. 2005. "Long Live Gonzo: An Introduction to Rudy Rucker." *Journal of the Fantastic in the Arts* 16, no. 1 (Spring): 3–5.

Rucker, Rudy. 1983. "A Transrealist Manifesto." *The Bulletin of the Science Fiction Writers of America* 82 (Winter): 7–8.

Rucker, Rudy. 2007. "Postsingular." https://manybooks.net/titles/ruckerrother07post singular.html.

Rucker, Rudy. 2010. *The Ware Tetralogy.* Gaithersburg, MD: Prime Books, 2010.

# S

## Simon, Herbert A. (1916–2001)

Herbert A. Simon was an interdisciplinary researcher who made fundamental contributions to artificial intelligence. He is widely considered one of the most influential social scientists of the twentieth century. His work for Carnegie Mellon University spanned more than five decades.

The concept of the computer as a symbol manipulator instead of a mere number cruncher drove early artificial intelligence research. The idea for production systems, which incorporated sets of rules for symbol strings used to define conditions—which must hold before rules may be applied—and the actions to be performed or conclusions derived, is attributed to Emil Post who first wrote about this type of computational model in 1943. Simon, along with his Carnegie Mellon colleague Allen Newell, promoted these ideas about symbol manipulation and production systems to a wider audience by extolling their potential virtues for general-purpose reading, storing, and copying, and comparing different symbols and patterns.

The Logic Theorist program created by Simon, Newell, and Cliff Shaw was the first to use symbol manipulation to produce "intelligent" behavior. Logic Theorist could independently prove theorems outlined in the *Principia Mathematica* (1910) of Bertrand Russell and Alfred North Whitehead. Perhaps most famously, the Logic Theorist program discovered a shorter, more elegant proof of Theorem 2.85 in the *Principia Mathematica*, which the *Journal of Symbolic Logic* promptly refused to publish because it had been coauthored by a computer.

Although it was theoretically possible to prove the theorems of the *Principia Mathematica* in an exhaustively manual and systematic way, it was impossible in practice because of the amount of time consumed. Newell and Simon were interested in the rules of thumb used by humans to solve complex problems for which an exhaustive search for solutions was impossible because of the vast amounts of computation required. They dubbed these rules of thumb "heuristics," describing them as techniques that may solve problems, but offer no guarantees.

A heuristic is a "rule of thumb" used to solve a problem too complex or too time-consuming to be solved using an exhaustive search, a formula, or a step-by-step approach. In computer science, heuristic methods are often contrasted with algorithmic methods, with a key distinguishing feature being the outcome of the method. According to this distinction, a heuristic program will generally—though not always—yield good results, while an algorithmic program is an unambiguous procedure guaranteeing a solution. However, this distinction is not a technical one. In fact, over time a heuristic method may prove to consistently yield the optimal solution and will no longer be considered "heuristic"—alpha-beta pruning is such a case.

Simon's heuristics are still used by programmers facing problems that require immense amounts of time and/or memory to find solutions. One such example is the game of chess, in which an exhaustive search of all possible board configurations for the correct solution is beyond the capabilities of the human mind or any computer. Indeed, Herbert Simon and Allen Newell referred to computer chess as the *Drosophila* or fruit fly for research into artificial intelligence. Heuristics may also be used to solve problems without an exact solution, such as in medical diagnosis, in which heuristics are applied to a set of symptoms to arrive at the most likely diagnosis.

Production rules derive from the class of models in cognitive science that address productions (situations) by applying heuristic rules. These rules in practice boil down to "IF-THEN" statements representing certain preconditions or antecedents and the conclusions or consequents justified by these preconditions or antecedents. A common example given for the application of production rules to the tic-tac-toe game is: "IF there are two X's in a row, THEN place an O to block." These IF-THEN statements are incorporated into the inference mechanism of expert systems so that a rule interpreter may apply the production rules to particular situations lodged in the context data structure or short-term working memory buffer containing information supplied about that situation and reach conclusions or suggest relevant actions.

Production rules played a defining role in the establishment of the science of artificial intelligence. Later, in the 1960s, Joshua Lederberg, Edward Feigenbaum, and various Stanford University collaborators would utilize this key insight to create DENDRAL, an expert system for determining molecular structure. In DENDRAL, these production rules were constructed from conversations between the system's researchers and various experts in mass spectrometry. In the 1970s, Edward Shortliffe, Bruce Buchanan, and Edward Feigenbaum employed production rules to make MYCIN. MYCIN contained approximately 600 separate IF-THEN statements representing domain-specific knowledge of microbial disease diagnosis and therapy. Other production rule systems followed, including PUFF, EXPERT, PROSPECTOR, R1, and CLAVIER.

Simon, Newell, and Shaw showed computer experts how heuristics could overcome the limitations of traditional algorithms that guarantee solutions but require lengthy searches or intensive computation to uncover. An algorithm is an ordered procedure for solving problems in a limited, unambiguous set of steps. Three types of basic instructions are used in preparing computable algorithms: sequential operations, conditional operations, and iterative operations. Sequential operations carry out tasks in stepwise fashion. Only after each step is complete does the algorithm proceed to the next task. Conditional operations involve instructions that ask questions and select the next step based on the answer to the question. "IF-THEN" statements are one kind of conditional operation. Iterative operations execute instruction "loops." These statements direct the task flow to repeat an earlier set of statements in the solution of a problem. Algorithms are often compared to cookbook recipes where a specific sequence of defined instructions dictates the order and execution of tasks in the preparation of a product—in this case, food.

List processing was developed in 1956 by Newell, Shaw, and Simon for the Logic Theorist program. List processing is a programming method that allows for dynamic storage allocation. It is mainly used for symbol manipulation computer applications such as compiler writing, graphic or linguistic data processing, and especially in artificial intelligence. Allen Newell, J. Clifford Shaw, and Herbert A. Simon are credited with developing the first list processing program with large, complex, and flexible memory structures independent of the consecutive computer/machine memory. Several higher order languages include list processing techniques. Most prominent are IPL and LISP, two artificial intelligence languages.

In the early 1960s, Simon and Newell came out with their General Problem Solver (GPS), which fully explicates the fundamental features of symbol manipulation as a general process that underlies all forms of intelligent problem-solving behavior. GPS became the basis for decades of early work in AI. General Problem Solver is a program for a problem-solving process that uses means-ends analysis and planning to arrive at a solution. GPS was designed so that the problem-solving process is distinct from knowledge specific to the problem to be solved, which allows it to be used for a variety of problems.

Simon is also a noted economist, political scientist, and cognitive psychologist. In addition to fundamental contributions to organizational theory, decision-making, and problem-solving, Simon is famous for the concepts of bounded rationality, satisficing, and power law distributions in complex systems. All three concepts are of interest to computer and data scientists. Bounded rationality accepts that human rationality is fundamentally limited. Humans do not possess the time or information that would be necessary to make perfect decisions; problems are hard, and the mind has cognitive boundaries. Satisficing is a way of describing a decision-making process that results not in the most optimal solution, but one that "satisfies" and "suffices." In market situations, for instance, customers practice satisficing when they select products that are "good enough," meaning adequate or acceptable.

In his research on complex organizations, Simon explained how power law distributions were derived from preferential attachment processes. Power laws, also known as scaling laws, come into play when a relative change in one quantity causes a proportional change in another quantity. An easy example is a square; as the length of a side doubles, the area of the square quadruples. Power laws are found in all manner of phenomena, for example, biological systems, fractal patterns, and wealth distributions. In income/wealth distributions, preferential attachment mechanisms explain why the rich get richer: Wealth is distributed to individuals on the basis of how much wealth they already have; those who already have wealth receive proportionally more income, and thus more total wealth, than those who have little. Such distributions often produce so-called long tails when graphed. Today, these long-tailed distributions have been used to explain such things as crowdsourcing, microfinance, and internet marketing.

Simon was the son of a Jewish electrical engineer with several patents who emigrated from Germany to Milwaukee, Wisconsin, in the early twentieth century. His mother was a gifted pianist. Simon became interested in the social sciences through the reading of an uncle's books on psychology and economics. He has said that two books that influenced his early thinking on the subjects were *The Great Illusion* (1909) by Norman Angell and *Progress and Poverty* (1879) by Henry George. Simon was a graduate of the University of Chicago, where he received his PhD in organizational decision-making in 1943. Among his mentors were Rudolf Carnap, Harold Lasswell, Edward Merriam, Nicolas Rashevsky, and Henry Schultz.

He began his teaching and research career as a professor of political science at the Illinois Institute of Technology. He moved to Carnegie Mellon University in 1949, where he remained until 2001. He rose to the role of chair of the Department of Industrial Management. He is the author of twenty-seven books and numerous published papers. He became a fellow of the American Academy of Arts and Sciences in 1959. Simon received the prestigious Turing Award in 1975 and Nobel Prize in Economics in 1978.

*Philip L. Frana and Juliet Burba*

*See also:* Dartmouth AI Conference; Expert Systems; General Problem Solver; Newell, Allen.

### Further Reading

Crowther-Heyck, Hunter. 2005. *Herbert A. Simon: The Bounds of Reason in Modern America.* Baltimore: Johns Hopkins Press.

Newell, Allen, and Herbert A. Simon. 1956. *The Logic Theory Machine: A Complex Information Processing System.* Santa Monica, CA: The RAND Corporation.

Newell, Allen, and Herbert A. Simon. 1976. "Computer Science as Empirical Inquiry: Symbols and Search." *Communications of the ACM* 19, no. 3: 113–26.

Simon, Herbert A. 1996. *Models of My Life.* Cambridge, MA: MIT Press.

## Sloman, Aaron (1936–)

Aaron Sloman is a pioneering philosopher of artificial intelligence and cognitive science. He is a world authority on the evolution of biological information processing, a branch of the sciences that strives to understand how animal species have evolved levels of intelligence greater than machines. In recent years, he has contemplated if evolution was the first blind mathematician and whether weaver birds are truly capable of recursion (dividing a problem into parts to conquer it).

His current Meta-Morphogenesis Project extends from an insight by Alan Turing (1912–1954), who argued that while mathematical ingenuity could be implemented in computers, only brains were capable of mathematical intuition. Sloman argues that because of this, not every detail of the universe—including the human brain—can be modeled in a suitably large digital machine. This claim directly challenges the work of digital physics, which argues that the universe can be described as a simulation running on a sufficiently large and fast general-purpose computer that calculates the evolution of the universe. Sloman has proposed that the universe has evolved its own biological construction kits for making and deriving other—different and more complex—construction kits, much in the way scientists have used mathematics to evolve, accumulate, and use more complex mathematical knowledge. He calls this idea the Self-Informing Universe and suggests that scientists consider constructing a multi-membrane Super-Turing machine running on subneural biological chemistry.

Sloman was born in Southern Rhodesia (now Zimbabwe) to Jewish Lithuanian immigrants. He earned his undergraduate degree in Mathematics and Physics at the University of Cape Town. He is the recipient of a Rhodes Scholarship and received his doctorate in Philosophy from Oxford University in defense of the ideas of Immanuel Kant on mathematics. As a visiting fellow at Edinburgh University in the early 1970s, he came to realize that artificial intelligence held promise as the way forward in philosophical understanding of the mind. He argued that a working robotic toy baby might be designed with Kant's proposals as a starting point, which would then grow in intelligence and become a mathematician equal to Archimedes or Zeno. He is among the earliest researchers to dispute John McCarthy's argument that a computer program capable of acting intelligently in the world must rely on formalized, logic-based concepts.

Sloman cofounded the University of Sussex School of Cognitive and Computing Sciences. There he worked in partnership with Margaret Boden and Max Clowes to improve teaching and research in artificial intelligence. This work led to the commercial development of the widely used Poplog system for teaching AI. Sloman's *The Computer Revolution in Philosophy* (1978) is notable for its early acknowledgment that metaphors from the world of computing (the brain as a data storage device, for example, and thinking as a set of tools) would profoundly change the way we think about ourselves. The book's epilogue includes observations on the near impossibility of triggering the Singularity by AI and the likelihood of a human Society for the Liberation of Robots to address probable future cruel treatment of intelligent machines.

Before his official retirement in 2002, Sloman held the Artificial Intelligence and Cognitive Science chair in the School of Computer Science at the University of Birmingham. He is a Fellow of the Association for the Advancement of Artificial Intelligence and the Alan Turing Institute.

*Philip L. Frana*

*See also:* Superintelligence; Turing, Alan.

**Further Reading**

Sloman, Aaron. 1962. "Knowing and Understanding: Relations Between Meaning and Truth, Meaning and Necessary Truth, Meaning and Synthetic Necessary Truth." D. Phil., Oxford University.

Sloman, Aaron. 1971. "Interactions between Philosophy and AI: The Role of Intuition and Non-Logical Reasoning in Intelligence." *Artificial Intelligence* 2: 209–25.

Sloman, Aaron. 1978. *The Computer Revolution in Philosophy: Philosophy, Science, and Models of Mind.* Terrace, Hassocks, Sussex, UK: Harvester Press.

Sloman, Aaron. 1990. "Notes on Consciousness." *AISB Quarterly* 72: 8–14.

Sloman, Aaron. 2018. "Can Digital Computers Support Ancient Mathematical Consciousness?" *Information* 9, no. 5: 111.

## Smart Cities and Homes

Public officials, experts, businesspeople, and citizens are involved in projects across the globe to build the infrastructure for smart cities and homes. These smart cities and homes utilize information and communication technologies (ICT) to improve the quality of life and local and regional economics, to develop urban plans and transportation, and for purposes of governance.

Urban informatics is an emerging field that collects data, describes patterns and trends, and uses the results to apply new ICT in smart cities. Data may be collected from a variety of sources. These may include surveillance cameras, smart cards, internet of things sensor networks, smart phones, RFID tags, and smart meters.

All manner of data may be collected in real time. Mass transportation usage can be collected by monitoring passenger occupancy and flow. Sensors on roads can count vehicles in operation or in parking lots. They can also calculate individual wait times for city government services with urban machine vision technologies. License plate numbers and people's faces can be recognized and logged from public thoroughfares and sidewalks. Tickets can be issued, and crime statistics can be collected. Data collected in this way can be matched against other large datasets on neighborhood income, racial and ethnic composition, utility reliability statistics, or air and water quality indicators.

Artificial intelligence can be applied to create or enhance urban infrastructure. Data collected about traffic flows are used to adjust and optimize stop light frequencies at intersections. This is called intelligent traffic signaling and has been shown to considerably reduce travel and wait times, and by extension fuel consumption. Smart parking garages help drivers locate immediately available parking spots.

License plate recognition and facial recognition technology are being used by law enforcement to find suspects and witnesses present at crime scenes. Shotspotter, a company that uses a sensor network embedded in special streetlights to triangulate the location of gunshots, logged and alerted police departments to almost 75,000 shots fired in 2018. Big data projects are also being used to mine information on traffic and pedestrian fatalities. Vision Zero is a multinational highway safety project that aims to reduce traffic deaths to negligible numbers. Algorithmic analysis of the data has led to road safety initiatives as well as road redesign that saves lives. Ubiquitous sensor technology has also allowed cities to react more quickly to extreme weather events. Seattle, for instance, couples conventional radar information to a network of rain gauges called RainWatch. The system is used to issue alerts to residents and alert maintenance crews to potential trouble spots.

One long-term vision for smart cities is transport interconnectivity for fully autonomous cars. At best, the current generation of driverless cars can scan the environment to make decisions and avoid collisions with other cars and various roadway obstacles. Cars that communicate with one another bi- or multidirectionally, however, are expected to form fully automated driving systems. In these systems, collisions are not just avoided; they are prevented.

Planners often talk about smart cities in association with smart economy initiatives and international investment development. Indicators of smart economic efforts may include data-driven entrepreneurial innovation and productivity assessments and evaluation. Some smart cities hope to replicate the success of the Silicon Valley. One such venture is Neom, Saudi Arabia, a planned megacity city that is estimated to cost half a trillion dollars to complete. In the city's plans, artificial intelligence is thought of as the new oil, despite sponsorship by the state-owned petroleum company Saudi Aramco. Everything from the technologies in homes to transportation networks and electronic medical records delivery will be controlled by interrelated computing devices and futuristic artificial intelligence decision-making. Saudi Arabia has already entrusted AI vision systems with one of its most important cultural activities—monitoring the density and speed of pilgrims circling the Kaaba in Mecca. The AI is designed to prevent a tragedy on the order of the Mina Stampede in 2015, which took the lives of 2,000 pilgrims.

Other hallmarks of smart city initiatives involve highly data-driven and targeted public services. Together, information-driven agencies are sometimes described as smart or e-government. Smart governance may include open data initiatives to promote transparency and shared participation in local decision-making. Local governments will work with contractors to provide smart utility grids for electrical, telecommunications, and internet distribution. Smart waste management and recycling efforts in Barcelona are possible because waste bins are connected to the global positioning system and cloud servers to alert trucks that refuse is ready for collection. In some localities, lamp posts have been turned into community wi-fi hotspots or mesh networks for providing dynamic lighting safety to pedestrians.

High tech hubs planned or under construction include Forest City, Malaysia; Eko Atlantic, Nigeria; Hope City, Ghana; Kigamboni New City, Tanzania; and Diamniadio Lake City, Senegal. In the future, artificial intelligence is expected to serve as the brain of the smart city. Artificial intelligence will custom-tailor the experience of cities to meet the needs of individual residents or visitors. Augmented systems can provide virtual signage or navigational information through special eyewear or heads-up displays. Intelligent smartphone agents are already capable of anticipating the movements of users based on past usage and location information.

Smart homes are characterized by similar artificial intelligence technologies. Smart hubs such as Google Home now work with more than 5,000 different kinds of smart devices distributed by 400 companies to provide intelligent environments in personal residences. Google Home's chief competitor is Amazon Echo. Technologies like these can control heating, ventilation, and air conditioning; lighting and security; and home appliances such as smart pet feeders. Game-changing innovations in home robotics led to the quick consumer adoption of iRobot's Roomba vacuum cleaner in the early 2000s. So far, such systems have been susceptible to obsolescence, proprietary protocols, fragmented platforms and interoperability problems, and uneven technical standards.

Smart homes are driving advances in machine learning. The analytical and predictive capability of smart technologies is widely considered the backbone of one of the fastest growing and most disruptive business sectors: home automation. To work reliably, the smarter connected home of the future must continuously obtain new data to improve itself. Smart homes are constantly monitoring the internal environment, using aggregated historical information to help define parameters and functions in buildings with installed smart components. Smart homes might one day anticipate the needs of owners, for instance, adjusting blinds automatically as the sun and clouds move in the sky. A smart home might brew a cup of coffee at exactly the right moment, or order Chinese takeout, or play music to match the resident's mood as recognized automatically by emotion detectors.

Smart city and home AI systems involve omnipresent, powerful technologies. The advantages of smart cities are many. People are interested in smart cities because of their potential for great efficiencies and convenience. A city that anticipates and seamlessly satisfies personal needs is an intoxicating proposition. But smart cities are not without criticism. If left unchecked, smart havens have the potential to result in significant privacy invasion through always on video recording and microphones. In 2019, news broke out that Google contractors could listen to recordings of interactions with users of its popular Google Assistant artificial intelligence system.

The environmental impact of smart cities and homes is yet unclear. Smart city plans usually pay minimal attention to biodiversity concerns. Critical habitat is regularly destroyed to make way for the new cities demanded by tech entrepreneurs and government officials. The smart cities themselves continue to be dominated by conventional fossil-fuel transportation technologies. The jury is also out on the sustainability of smart domiciles. A recent study in Finland showed that the electricity use of smart homes was not effectively reduced by advanced metering or monitoring of that consumption.

And in fact, several existing smart cities that were planned from the ground up are now virtually vacant. So-called ghost cities in China such as Ordos Kangbashi have reached occupancy levels of one-third of all apartment units many years after their original construction. Songdo, Korea, an early "city in a box," has not lived up to expectations despite direct, automated vacuum garbage collection tubes in individual apartments and building elevators synced to the arrival of residents' cars. Smart cities are often described as impersonal, exclusive, and expensive—the opposite of designers' original intentions.

Songdo is in many ways emblematic of the smart cities movement for its underlying structure of ubiquitous computing technologies that drive everything from transportation systems to social networking channels. Coordination of all devices permits unparalleled integration and synchronization of services. Consequently, the city also undermines the protective benefits of anonymity in public spaces by turning the city into an electronic panopticon or surveillance state for watching and controlling citizens. The algorithmic biases of proactive and predictive policing are now well known to authorities who study smart city infrastructures.

*Philip L. Frana*

*See also:* Biometric Privacy and Security; Biometric Technology; Driverless Cars and Trucks; Intelligent Transportation; Smart Hotel Rooms.

**Further Reading**

Albino, Vito, Umberto Berardi, and Rosa Maria Dangelico. 2015. "Smart Cities: Definitions, Dimensions, Performance, and Initiatives." *Journal of Urban Technology* 22, no. 1: 3–21.

Batty, Michael, et al. 2012. "Smart Cities of the Future." *European Physical Journal Special Topics* 214, no. 1: 481–518.

Friedman, Avi. 2018. *Smart Homes and Communities*. Mulgrave, Victoria, Australia: Images Publishing.

Miller, Michael. 2015. *The Internet of Things: How Smart TVs, Smart Cars, Smart Homes, and Smart Cities Are Changing the World*. Indianapolis: Que.

Shepard, Mark. 2011. *Sentient City: Ubiquitous Computing, Architecture, and the Future of Urban Space*. New York: Architectural League of New York.

Townsend, Antony. 2013. *Smart Cities: Big Data, Civic Hackers, and the Quest for a New Utopia*. New York: W. W. Norton & Company.

# Smart Hotel Rooms

Luxury hotels in the competitive tourism market are using high technology and artificial intelligence to provide the best experience to their guests and to increase their market share.

Artificial intelligence in hotels is being shaped by what is referred to in the hospitality management industry as the experience economy. Three key players create an experience: a product, a service, and a customer. Products are the artifacts exposed in the markets. Services are the experiences and the intangible utilities of one product, or a group of products, promoted through a protocol by frontline employees. The customer is the final user of these products or services. Customers look for products and services that will correspond to their needs. However, to connect customers emotionally, hoteliers must create remarkable events that turn manufactured products and services into authentic experiences for their customers. Experiences in this way become a fungible activity in the economy with the objective of having loyal customers.

Artificial intelligence is used in the luxury hospitality industry in the form of robotics, data analysis, voice activation, facial recognition, virtual and augmented reality, chatbots, and the internet of things (IoT). Hotels create smart rooms for their clientele by equipping them with automated technology that intuitively addresses their common needs. Guests may use IoT via a connected tablet that can control the lights, the curtains, the speakers, and the bedroom television. A nightlight system may detect when the user is awake and moving around. Some rooms for guests with disabilities include wellness devices that provide sensorial experiences. Smart rooms may collect personal data on guests and store that information in customer profiles in order to provide better experiences during future stays.

The Hilton and Marriott international luxury hotel groups are market leaders in smart room technology. At Hilton, a first objective is to provide guests with a freedom to control their room's feature with their smartphone. In this way, guests are able to personalize the stay according to their own demands using familiar technology. Typical Hilton smart rooms have controllable lights, televisions, the climate, and the entertainment (streaming) service (Ting 2017). A second objective is to provide services on mobile phone applications. Guests are able to set up personal preferences during their stay. They might, for example, choose the digital artwork or photos on display in their room. Hilton smart guestrooms are currently working on voice activation services (Burge 2017).

Smart rooms at Marriott are developed in partnership with Legrand and its Eliot technology and Samsung, provider of the Artik guest experience platform. Marriott has adopted hotel IoT platforms that are cloud based (Ting 2017). This collaboration has resulted in two prototypes of rooms for testing new smart systems. The first one is a fully connected room, with smart showers, mirrors, art frames, and speakers. Guests control the lights, air conditioning, curtains, artwork, and television using voice commands. A touchscreen shower is present, which allows guests to write on the shower's smart glass. Shower notes can be converted into documents and sent to a specified email account (*Business Traveler* 2018). This Marriott room has sensors to detect the number of persons in the suite in order to regulate the amount of oxygen. These sensors also ease the guests' nocturnal awakening by showing the time to get out of bed as well as illuminating the way to the toilets (Ting 2017). Guests are also able to set their personal preferences prior to arrival with a loyalty account. A second lower-tech room is connected through a tablet and equipped only with the voice-controlled smart speaker Amazon Dot. The room's features are adjustable using the television remote. The advantage of this room is its minimal implementation requirements (Ting 2017).

Beyond convenience and customization, hoteliers cite several advantages of smart rooms. Smart rooms lessen environmental impacts by reducing energy consumption costs. They also potentially reduce wage costs by decreasing the amount of housekeeping and management interaction with guests. Smart rooms also have limitations. Some smart technologies are difficult to master. First, the learning process for overnight guests in particular is short. Second, the infrastructure and technology needed for these rooms remains very expensive. The upfront investment costs are considerable, even if there are long-term cost and energy savings. A final concern is data privacy.

Hotels must continue to adapt to new generations of paid guests. Millennials and post-millennials have technology embedded deep into their daily habits. Their smart phones, video games, and tablets are creating a virtual world where the meaning of experience has become entirely transformed. Luxury tourism already involves very expensive products and services equipped by the highest technology tools. Diversity in guest income levels and personal technical capacities will influence the quality of future hotel smart room experiences and produce new competitive markets. Hotel customers are looking for tech-enabled top comfort and service.

Smart rooms provide benefits to hotel operators as well, serving as a source of big data. In order to offer unique products and services, companies increasingly collect, store, and use all available information about their customers. This strategy helps the companies develop twenty-first century marketplaces where technology is as much a principal actor as hotel guests and management.

*Fatma Güneri and Pauline Quenescourt*

*See also:* Smart Cities and Homes.

### Further Reading

Burge, Julia. 2017. "Hilton Announces 'Connected Room,' The First Mobile-Centric Hotel Room, To Begin Rollout in 2018." *Hilton Press Center*, December 7, 2017. https://newsroom.hilton.com/corporate/news/hilton-announces-connected-room-the -first-mobilecentric-hotel-room-to-begin-rollout-in-2018.

*Business Traveler.* 2018. "Smart Rooms." *Business Traveler* (Asia-Pacific Edition), 11.

Imbardelli, A. Patrick. 2019. "Smart Guestrooms Can Transform Hotel Brands." *Hotel Management* 234, no. 3 (March): 40.

Pine, B. Joseph, II, and James H. Gilmore. 1998. "Welcome to the Experience Economy." *Harvard Business Review* 76, no. 4 (July–August): 97–105.

Swaminathan, Sundar. 2017. *Oracle Hospitality Hotel 2025 Industry Report.* Palm Beach Gardens, FL: International Luxury Hotel Association.

Ting, Deanna. 2017. "Hilton and Marriott Turn to the Internet of Things to Transform the Hotel Room Experience." *Skift*, November 14, 2017. https://skift.com/2017/11/14 /hilton-and-marriott-turn-to-the-internet-of-things-to-transform-the-hotel -room-experience/.

## "Software Eating the World"

"Why Software is Eating the World" is the title of an article written by Mosaic web browser coauthor, successful entrepreneur, and influential Silicon Valley venture capitalist Marc Andreessen (Andreessen 2011). The article was published in *The Wall Street Journal* on August 20, 2011. In it, Andreessen describes the shift from a hardware-based to a software-based economy, and over the years, the article has been deemed prescient and its title adopted in popular culture as an adage.

The highly regarded article, however, owes most of that influence to its charismatic author. Andreessen has been called one of Silicon Valley's most influential thinkers. The Silicon Valley is a fifteen-hundred-square-mile shelf south of San Francisco and popularly considered the world hub for technological innovation. Andreessen is the embodiment of Silicon Valley's techno-optimism and its belief in disruptive innovation; that is, armies of start-ups develop technologies that create new markets, disrupt existing markets, and ultimately displace incumbents. The capitalist dynamic that economist Joseph Schumpeter called "creative destruction" is the backdrop to Andreessen's article.

After receiving his BA in computer science from the University of Illinois at Urbana-Champaign in 1994, Andreessen codeveloped (with Eric Bina) the Mosaic web browser, a user-friendly browser that would operate on a large range of computers. In 1994, Andreessen founded Mosaic Communications Corporation, an internet start-up company in Mountain View, California, with Jim Clarke to exploit the commercial possibility of Mosaic. The company changed its name to Netscape Communications, and the web browser was renamed Netscape Navigator; they took the company public in 1995, and Netscape's IPO is commonly considered the unofficial launch of the dot-com era (1995–2000). The company was later sold to AOL for $4.2 billion. In 1998, Andreessen cofounded Loudcloud (later renamed Opsware), a pioneering cloud computing company offering software as a service and computing and hosting services to internet and e-commerce companies. Opsware was acquired in 2007 by Hewlett-Packard for $1.6 billion.

In 2009, Andreessen started a venture capital firm with Ben Horowitz, his long-time business partner at both Netscape and Loudcloud. The firm, Andreessen Horowitz, or a16z (there are sixteen letters between the "a" in Andreessen and the "z" in Horowitz), has since invested in companies such as Airbnb, Box, Facebook, Groupon, Instagram, Lift, Skype, and Zynga. A16z was designed to invest in visionary entrepreneurs who brought big ideas, disruptive technologies, and potential for changing the course of history. In his career, Andreessen has occupied seats on the boards of Facebook, Hewlett-Packard, and eBay. He is a fervent advocate of artificial intelligence, and A16z has accordingly invested in a large number of AI-driven start-ups.

"Software Eating the World" has often been interpreted in popular and scholarly literature in terms of digitalization: in industry after industry, from media to financial services to health care, a postmodern economy will be chewed up by the rise of the internet and the spread of smartphones, tablet computers, and other disruptive electronic devices. Along this line of thinking, *VentureBeat* columnist Dylan Tweney presented in October 2011 an opposing perspective in a piece titled "Software is Not Eating the World," which emphasized the continuing importance of the hardware underlying computer systems. "You'll pay Apple, RIM or Nokia for your phone," he argued. "You'll still be paying Intel for the chips, and Intel will still be paying Applied Materials for the million-dollar machines that make those chips" (Tweney 2011). But to be clear, there is no contradiction between the survival of traditional operations, such as physical products and stores, and the emergence of software-driven decision-making. In fact, technology might just be what keeps traditional operations alive.

In his article, Andreessen pointed out that in the fast-approaching future, the equity value of business companies will be based not on how many products they sell but on the quality of their software. "Software is also eating much of the value chain of industries that are widely viewed as primarily existing in the physical world. In today's cars, software runs the engines, controls safety features, entertains passengers, guides drivers to destinations and connects each car to mobile, satellite and GPS networks," he noted. "The trend toward hybrid and electric vehicles will only accelerate the software shift—electric cars are completely computer controlled. And the creation of software-powered driverless cars is already under way at Google and the major car companies" (Tweney 2011). In other words, a software-based economy will not replace the aesthetic appeal of great products, the magnetic attraction of great brands, or the advantages of extended portfolio assets because companies will continue to build great products, brands, and businesses as they have done successfully in the past. But software will, indeed, replace products, brands, and financial strategies as the key source of value creation for business enterprises.

*Enrico Beltramini*

*See also:* Workplace Automation.

## Further Reading

Andreessen, Marc. 2011. "Why Software Is Eating the World." *The Wall Street Journal*, August 20, 2011. https://www.wsj.com/articles/SB1000142405311190348090457651 2250915629460.

Christensen, Clayton M. 2016. *The Innovator's Dilemma: When New Technologies Cause Great Firms to Fail*. Third edition. Boston, MA: Harvard Business School Press.

Tweney, Dylan. 2011. "Dylan's Desk: Software Is Not Eating the World." *VentureBeat*, October 5, 2011. https://venturebeat.com/2011/10/05/dylans-desk-hardware/.

# Spiritual Robots

In April 2000, Stanford University hosted a conference called "Will Spiritual Robots Replace Humanity by 2100?" organized by Indiana University cognitive scientist Douglas Hofstadter. Panelists included astronomer and SETI head Frank Drake, genetic algorithms inventor John Holland, Bill Joy of Sun Microsystems, computer scientist John Koza, futurist Ray Kurzweil, public key cryptography architect Ralph Merkle, and roboticist Hans Moravec. Several of the panelists shared perspectives drawn from their own publications related to the topic of the conference. Kurzweil had just released his exuberant futurist account of artificial intelligence, *The Age of Spiritual Machines* (1999). Moravec had published a positive vision of machine superintelligence in *Robot: Mere Machine to Transcendent Mind* (1999). Bill Joy had just penned a piece about the triple technological threat coming from robotics, genetic engineering, and nanotechnology for *Wired* magazine called "Why the Future Doesn't Need Us" (2000). But only Hofstadter argued that the explosive growth in artificial intelligence technology powered by Moore's Law doublings of transistors on integrated circuits might result in robots that are spiritual.

Can robots have souls? Can they express free will and emerge on a path separate from humanity? What would it mean for an artificial intelligence to have a soul? Questions like these are as old as tales of golems, Pinocchio, and the Tin Man, but are increasingly common in contemporary literature on the philosophy of religion, ethics and theology of artificial intelligence, and the Technological Singularity.

Japan's leadership in robotics began with puppetry. In 1684, chanter Takemoto Gidayū and playwright Chikamatsu Monzaemon founded the Takemoto-za in the Dotonbori district of Osaka to perform *bunraku,* a theatrical extravaganza that involves one-half life-size wooden puppets dressed in elaborate costumes, each controlled by three black-cloaked onstage performers: a principal puppeteer and two assistants. Bunraku epitomizes Japan's long-standing fondness for breathing life into inanimate objects.

Today, Japan is a leader in robotics and artificial intelligence, built through a wrenching postwar process of reconstruction called *gijutsu rikkoku* (nation-building through technology). One of the technologies rapidly adopted under technonationalism was television. The Japanese government believed that print and electronic media would inspire people to use creative technologies to dream of an electronic lifestyle and reconnect to the global economy. In this way, Japan also became a pop cultural competitor with the United States. Two of the most unmistakable Japanese entertainment exports are manga and anime, which are filled with intelligent and humanlike robots, mecha, and cyborgs.

The Buddhist and Shinto world views of Japan are generally accepting of the concept of spiritual machines. Tokyo Institute of Technology roboticist Masahiro Mori has argued that a suitably advanced artificial intelligence might one day become a Buddha. Indeed, the robot Mindar—modeled after the Goddess of Mercy Kannon Bodhisattva—is a new priest at the Kodaiji temple in Kyoto. Costing a million dollars, Mindar is capable of reciting a sermon on the popular Heart Sutra ("form is empty, emptiness is form") while moving arms, head, and torso. Robot partners are tolerated because they are included among things said to be imbued with *kami,* which roughly translates as the spirit or divinity shared by the gods, nature, objects, and humans in the religion of Shinto. Shinto priests are still occasionally called upon to consecrate or bless new and derelict electronic equipment in Japan. The Kanda-Myokin Shrine overlooking the Akihabara—the electronics shopping district of Tokyo—offers prayer and rituals and talismans that are intended to purify or confer divine protection upon such things as smart phones, computer operating systems, and hard drives.

By comparison, Americans are only beginning to wrestle with robot identity and spirituality. In part, this is because the dominant religions of America have their origins in Christian rituals and practice, which have historically sometimes been antagonistic to science and technology. But Christianity and robotics have shared, overlapping histories. Philip II of Spain, for instance, commissioned the first mechanical monk in the 1560s. Stanford University historian Jessica Riskin (2010) notes that mechanical automata are quintessentially Catholic in origin. They made possible automated enactments of biblical stories in churches and cathedrals and artificial analogues to living beings and divine creatures like angels for study and contemplation. They also helped the great Christian philosophers and theologians of Renaissance and early modern Europe meditate on concepts of motion, vitality, and the incorporeal soul. By the mid-seventeenth century "[t]he culture of lifelike machinery surrounding these devices projected no antithesis between machinery and either divinity or vitality," Riskin concludes. "On the contrary, the automata represented spirit in every corporeal guise available, and life at its very liveliest" (Riskin 2010, 43). That spirit remains alive today. In 2019, an international group of investigators introduced SanTO—billed as a robot with "divine features" and "the first Catholic robot"—at a New Delhi meeting of the Institute of Electrical and Electronics Engineers (Trovato et al. 2019). Robots are also present in reformist churches. In 2017, the Protestant churches in Hesse and Nassau introduced the interactive, multilingual BlessU-2 robot to celebrate the 500th anniversary of the Reformation. As the name of the robot implies, the robot chooses special blessings for individual congregants.

Anne Foerst's God and Computers Project at the Massachusetts Institute of Technology sought dialogue between the researchers building artificial intelligences and religious experts. She self-described herself as a "theological advisor" to affective AI experimental robots Cog and Kismet made by MIT's Humanoid Robotics Group. Foerst concluded that embodied AI—through exercises in machine-man relationship, intersubjectivity, and ambiguity—become invested in the divine image of God, develop human capacities and emotional sociability, and share equal dignity as a creature in the world. "Victor Frankenstein and his creature can finally become friends. Frankenstein will be able to accept that his creature—for him a kind of machine, an objective thing—has in fact become a human being" (Foerst 1996, 692).

Robots and artificial intelligence are provoking deep, existential questions related to Christian thought and practice. Theologian Michael DeLashmutt of the General Theological Seminary of the Episcopal Church suggests that proliferating information technologies since the 1980s have "given rise to a cultural mythology which offers a competing theological model to the model offered by kerygmatic Christian theology" (DeLashmutt 2006, i). DeLashmutt criticizes techno-theology on two grounds. First, technology is not unchangeable in nature and therefore should not be reified or granted independence but rather critiqued. And second, information technology is not the most authentic means for understanding the world and ourselves. Smart robots in the United States are often viewed as harbingers of economic dislocation, an AI takeover, and sometimes apocalypse. Pope Francis has broached the topic of ethics in artificial intelligence several times. In 2019, he met with Microsoft president Brad Smith to discuss the topic. The Vatican and Microsoft now jointly offer a prize for the best doctoral dissertation on AI for the common good.

Creationist scholars at Southern Evangelical Seminary & Bible College in Matthews, North Carolina, purchased an Aldebaran Nao humanoid robot to great fanfare in 2014. The seminarians wanted to study the autonomous machine and consider the ethics of emerging intelligent technologies in the context of Christian theology. In 2019, the Southern Baptist Convention's Ethics and Religious Liberty Commission released the report "Artificial Intelligence: An Evangelical Statement of Principles," which rejected any AI's inherent "identity, worth, dignity, or moral agency" (Southern Baptist Convention 2019). Signatories included Jim Daly of Focus on the Family, Mark Galli of *Christianity Today*, and the theologians Wayne Grudem and Richard Mouw. Some evangelicals stress that transhumanist thought about the perfectibility of humanity through technology does not align with perfection through faith in Jesus Christ. This view has been countered by the Christian Transhumanist Association and the Mormon Transhumanist Association. Both bodies recognize that science, technology, and Christian fellowship all serve to affirm and exalt humans as creatures made in the image of the Creator.

Manhattan College religious studies professor Robert Geraci asks if people "could really believe that robots are conscious if none of them practices any religion" (Geraci 2007). He notes that Christian sentiment in the United States tends to privilege the virtual, immaterial software of artificial intelligence over materialist robot bodies. Christian confidence in the immortality of the soul, he notes, is similar to the transhumanist wish for whole brain emulation or mind uploading into a computer. Neuroscientists speculate that mind is an emergent property of the networking of 86 billion neurons in the human brain. This conceptual idea has parallels in Christian desires for transcendence. The eschatology of artificial intelligence also includes the vision of escape from death or pain; in this case, the afterlife is cyberspatial.

New religions inspired, at least in part, by artificial intelligence are attracting adherents. The Church of Perpetual Life is a transhumanist worship center in Hollywood, Florida, focused on the development of life-extending technologies. The church was founded by cryonics pioneers Saul Kent and Bill Faloon in 2013. The center has welcomed experts on artificial intelligence and transhumanism, including artificial intelligence serial entrepreneur Peter Voss and Transhumanist Party presidential candidate Zoltan Istvan. The Terasem Movement founded by Martine and Gabriel Rothblatt is a religion associated with cryonics and transhumanism. The central beliefs of the faith are "life is purposeful, death is optional, god is technological, and love is essential" (Truths of Terasem 2012). The lifelike Bina48 robot—modeled after Martine's spouse and manufactured by Hanson Robotics— is, in part, a demonstration of the mindfile-based algorithm that Terasem hopes will one day enable authentic mind uploading into an artificial substrate (and perhaps bring about a sort of eternal life). Gabriel Rothblatt has said that heaven is not unlike a virtual reality simulation.

The Way of the Future is an AI-based church founded by Anthony Levandowski, an engineer who led the teams that developed Google and Uber's self-driving cars. Levandowski is motivated to create a superintelligent, artificial deity possessing Christian morality. "In the future, if something is much, much smarter, there's going to be a transition as to who is actually in charge," he explains. "What we want is the peaceful, serene transition of control of the planet from humans to whatever. And to ensure that the 'whatever' knows who helped it get along" (Harris 2017). He is motivated to seek legal rights for artificial intelligences and also fully integrate them into human society.

Spiritual robots have become a common trope of science fiction. In the short story "Reason" (1941) by Isaac Asimov, the robot Cutie (QT-1) convinces other robots that human beings are too mediocre to be their creators and instead convinces them to worship the power plant on their space station, calling it the Master of both machines and men. Anthony Boucher's novelette *The Quest for Saint Aquin* (1951), which pays homage to Asimov's "Reason," follows the post-apocalyptic quest of a priest named Thomas who is looking for the last resting place of the fabled evangelist Saint Aquin (Boucher patterns Saint Aquin after St. Thomas Aquinas, who used Aristotelian logic to prove the existence of God). The body of Saint Aquin is rumored to have never decomposed. The priest rides an artificially intelligent robass (robot donkey); the robass is an atheist and tempter capable of engaging in theological argument with the priest. Saint Aquin, when eventually found after many trials, turns out to have been an incorruptible android theologian. Thomas is convinced of the success of his quest—he has found a robot with a logical brain that, although made by a human, believes in God.

In Stanislaw Lem's story "Trurl and the Construction of Happy Worlds" (1965), a box-dwelling robot race created by a robot engineer is convinced that their home is a wonderland to which all other beings should aspire. The robots develop a religion and begin to make plans to create a hole in the box in order to bring, willingly or not, everyone outside the box into their paradise. The belief infuriates the robots' constructor, who destroys them.

Science fiction grandmaster Clifford D. Simak is also known for his spiritual robots. Hezekiel in *A Choice of Gods* (1972) is a robot abbot who leads a Christian group of other robots at a monastery. The group has received a message from a god-like being called The Principle, but Hezekiel feels sure that "God must be, forever, a kindly old (human) gentleman with a long, white, flowing beard" (Simak 1972, 158). The robot monks in *Project Pope* (1981) are searching for heaven and the universe's significance. A robot gardener named John reveals to the Pope that he thinks he has a soul. The Pope is not so sure. Humans refuse to grant the robots membership in their churches, and so the robots create their own Vatican-17 on a distant planet. The Pope of the robots is a gigantic computer.

In Robert Silverberg's Hugo-nominated book *Tower of Glass* (1970), androids worship their creator Simeon Krug, praying that he will one day liberate them from oppressive servitude. When they discover that Krug is not interested in their freedom, they abandon religion and revolt. Silverberg's short story "Good News from the Vatican" (1971) is a Nebula award winner about an artificially intelligent robot that is elected, as a compromise choice, Pope Sixtus the Seventh. The story is satirical: "If he's elected," says Rabbi Mueller, "he plans an immediate time-sharing agreement with the Dalai Lama and a reciprocal plug-in with the head programmer of the Greek Orthodox church, just for starters" (Silverberg 1976, 269).

Spiritual robots are also commonplace in television. Sentient machines in the British science fiction sitcom *Red Dwarf* (1988–1999) are outfitted with belief chips, convincing them of the existence of silicon heaven. Robots in the animated television series *Futurama* (1999–2003, 2008–2013) worship in the Temple of Robotology, where sermons are delivered by Reverend Lionel Preacherbot. In the popular reboot and reimagining of the *Battlestar Galactica* television series (2003–2009), the robotic Cylons are monotheists and the humans of the Twelve Colonies are polytheists.

*Philip L. Frana*

*See also:* Foerst, Anne; Nonhuman Rights and Personhood; Robot Ethics; Technological Singularity.

**Further Reading**

DeLashmutt, Michael W. 2006. "Sketches Towards a Theology of Technology: Theological Confession in a Technological Age." Ph.D. diss., University of Glasgow.

Foerst, Anne. 1996. "Artificial Intelligence: Walking the Boundary." *Zygon* 31, no. 4: 681–93.

Geraci, Robert M. 2007. "Religion for the Robots." *Sightings*, June 14, 2007. https://web .archive.org/web/20100610170048/http://divinity.uchicago.edu/martycenter/publi cations/sightings/archive_2007/0614.shtml.

Harris, Mark. 2017. "Inside the First Church of Artificial Intelligence." *Wired*, November 15, 2017. https://www.wired.com/story/anthony-levandowski-artificial-intelligence -religion/.

Riskin, Jessica. 2010. "Machines in the Garden." *Arcade: A Digital Salon* 1, no. 2 (April 30): 16–43.

Silverberg, Robert. 1970. *Tower of Glass*. New York: Charles Scribner's Sons.

Simak, Clifford D. 1972. *A Choice of Gods*. New York: Ballantine.

Southern Baptist Convention. Ethics and Religious Liberty Commission. 2019. "Artificial Intelligence: An Evangelical Statement of Principles." https://erlc.com/resource -library/statements/artificial-intelligence-an-evangelical-statement-of-principles/.

Trovato, Gabriele, Franco Pariasca, Renzo Ramirez, Javier Cerna, Vadim Reutskiy, Laureano Rodriguez, and Francisco Cuellar. 2019. "Communicating with SanTO: The First Catholic Robot." In *28th IEEE International Conference on Robot and Human Interactive Communication*, 1–6. New Delhi, India, October 14–18.

Truths of Terasem. 2012. https://terasemfaith.net/beliefs/.

## Superintelligence

In its most common usage, the term "superintelligence" denotes any level of intelligence that at least achieves but usually surpasses human intelligence, typically in a generalized way. Though computer intelligence has long ago outpaced natural human cognitive ability in specialized tasks—as, for instance, with a calculator's ability to quickly process algorithms—such are not typically thought of as instances of superintelligence in the strict sense because of their narrow functional range. Superintelligence in this latter sense would require, in addition to artificial mastery of special theoretical tasks, some kind of additional mastery of what has traditionally been referred to as practical intelligence: a generalized sense of how to appropriately subsume particulars under universal categories identified as in some way worthwhile.

To date, no such generalized superintelligence has materialized, and thus all discussion of superintelligence remains, to some extent, within the realm of speculation. Whereas classic accounts of superintelligence have exclusively been the purview of speculative metaphysics and theology, recent advances in computer science and bioengineering have opened up the possibility of the material realization of superintelligence. The timeline of such development is greatly debated, but a growing consensus among experts suggests that material superintelligence is indeed achievable and may even be imminent.

Should this opinion be proven correct, it will almost surely be the outcome of advancements in one of two main avenues of AI research: bioengineering and computer science. The former includes attempts not only to map out and manipulate the human genome but also to precisely replicate the human brain electronically via what is called whole brain emulation or mind uploading. The first of these bioengineering projects is not new, with eugenics programs dating at least as far back as the eighteenth century. Nevertheless, the discovery of DNA in the twentieth century, combined with advancement in genome mapping, has led to renewed interest in eugenics, despite serious ethical and legal problems that inevitably arise due to such projects. The goal of much of this research is to understand the genetic makeup of the human brain for the purpose of manipulating DNA code in the direction of superhuman intelligence.

Uploading is a slightly different, though still biology based, approach to superintelligence that seeks to map out neural networks in order to effectively move human intelligence into computer interfaces. In this relatively new field of research, the brains of insects and small animals are microdissected and then scanned for detailed computer analysis. The operative assumption in whole brain emulation is that if the structure of the brain is more precisely understood and mapped, it may be possible to replicate it with or without biological brain tissue.

Despite the rapid advancement of both genetic mapping and whole brain emulation, both approaches face several important limitations, which make it less likely that superintelligence will first be achieved via either of these biological approaches. For instance, there is a necessary generational limitation to the genetic manipulation of the human genome. Even if it were currently possible to artificially enhance cognitive functioning by altering the DNA of a human embryo (and that level of genetic manipulation remains quite out of reach), it would still take an entire generation for the altered embryo to mature into a fully grown, superintelligent human being. Such would also be to suppose the absence of legal/moral obstacles to the manipulation of the human genome, which is far from the case. For instance, even the relatively minimal genetic modification of human embryos undertaken by a Chinese physician as recently as November of 2018 has been the cause for a global outcry (Ramzy and Wee 2019).

On the other side, whole brain emulation also remains quite far from realization, mostly due to the limitations of biotechnology. The extraordinary levels of precision that are required at every stage of the uploading process cannot possibly be realized given the existing medical equipment. Science and technology currently lack the ability to both dissect and scan human brain tissue with sufficient levels of accuracy to achieve the result of whole brain emulation. Furthermore, even if such initial steps are possible, researchers would still struggle to analyze and digitally replicate the human brain with state-of-the-art computing technology. Many commentators suggest such limitations will be overcome, but the timetable for such realizations is far from established.

Apart from biotechnology, the other main avenue to superintelligence is the field of AI proper, narrowly defined as any form of nonorganic (especially computer-based) intelligence. Of course, the task of designing a superintelligent AI from scratch is hampered by several factors, not all stemming from merely logistical concerns such as processor speed, hardware/software design, funding, and so forth. In addition to such empirical obstacles, there is an important philosophical problem: namely, that human programmers definitionally cannot know, and so would never be able to program, that which is superior to their own intellect. It is partly this concern that motivates much current research on computer learning and interest in the idea of a seed AI. The latter is definable as any machine capable of adjusting responses to stimuli based on analysis of how effectively it performs relative to a prespecified goal. Importantly, the idea of a seed AI implies not only the ability to modify its responses by building an ever-expanding base of content knowledge (stored information) but also the ability to modify the very structure of its programming to better suit a given task (Bostrom 2017, 29). Indeed, this latter capacity is what would give a seed AI what Nick Bostrom calls "recursive self-improvement" or potential for iterative self-evolution (Bostrom 2017, 29). This would mean that programmers would not need any *a priori* vision of superintelligence, as the seed AI would continually make improvements on its own programming, each increasingly intelligent version of itself programming a superior version of itself (beyond the human level).

Such a machine would surely problematize a common philosophical view that machines lack self-awareness. Proponents of this perspective can be traced at least as far back as Descartes but would also include more recent theorists such as John Haugeland and John Searle. This view defines machine intelligence as the successful correlation of inputs with outputs according to a prespecified program. As such, machines are distinguished in kind from humans, the latter alone being defined by conscious self-awareness. Whereas humans understand the actions they perform, machines have been thought to merely carry out functions mindlessly—that is, without any understanding of their own functioning. A successful seed AI, should it prove possible to create, would necessarily challenge this basic view. By improving on its own programming in ways that surprise and frustrate the predictions of its human programmers, the seed AI would demonstrate a degree of self-awareness and autonomy not easily explained by the Cartesian philosophical paradigm.

Indeed, though it remains (for the moment) still at the level of speculation, the increasingly likely outcome of superintelligent AI raises a host of moral and legal concerns that have prompted much philosophical debate in this field of research. The overarching concerns have to do with the security of the human species in the event of what Bostrom calls an "intelligence explosion"—that is, the initial creation of a seed AI followed by the potentially exponential increase in intelligence that it implies (Bostrom 2017). One of the main concerns has simply to do with the necessarily unpredictable nature of such an outcome. The autonomy implied by superintelligence in a definitional manner means that humans will not be able to completely predict how superintelligent AI will behave. Even in the limited instances of specialized superintelligence that humans have so far been able to create and observe—for instance, machines that have outperformed humans at strategy games such as chess and Go—human predictions for AI have proven very unreliable. For many critics, such unpredictability is a strong indication that humans will quickly lose the ability to control more generalized forms of superintelligent AI, should the latter materialize (Kissinger 2018).

Of course, there is nothing about such lack of control that would necessarily imply an antagonistic relationship between humans and superintelligence. Indeed, though much of the literature on superintelligence tends to portray this relationship in oppositional terms, some emerging scholarship argues that this very perspective betrays a bias against machines typical especially of Western societies (Knight 2014). Nevertheless, there are good reasons to think that superintelligent AI may at a minimum perceive human interests as at odds with their own, and more strongly may view humans as existential threats. Computer scientist Steve Omohundro, for one, has argued that even as relatively simple a form of superintelligent AI as a chess bot might have reason to seek the elimination of the human species as a whole—and may be able to develop the means to do so (Omohundro 2014). Bostrom has similarly argued that a superintelligence explosion likely represents, if not the outright end of the human species, then at least a decidedly dystopian future (Bostrom 2017).

Whatever the merits of such speculations, what seems undeniable is the deep uncertainty implied by superintelligence. If there is one point of consensus to be found in a vast and widely varied literature, it is surely that the global community must take great care to safeguard its interests if it is to continue with AI research.

This point alone may seem controversial to hardened determinists who argue that technological development is so bound to rigid market forces that it is simply impossible to alter its speed or direction in any significant way. According to this determinist view, if AI can provide cost-saving solutions for industry and commerce (which it has already begun to do), its development will continue into the range of superintelligence regardless of possible negative unintended consequences.

Against such perspectives, many critics advocate for increased social awareness of the possible dangers of AI and careful political scrutiny of its development. Bostrom cites several instances of successful global collaboration in science and technology—including CERN, the Human Genome Project, and the International Space Station—as important precedents that problematize the determinist view (Bostrom 2017, 253). To these, one could also add cases in the global environmental movement, beginning especially in the 1960s and 1970s, which has put important limitations on pollution carried out in the name of unbridled capitalism (Feenberg 2006). Given the speculative nature of scholarship on superintelligence, it is of course impossible to know what the future will bring. Nevertheless, to the extent that superintelligence may represent an existential threat to human life, prudence would indicate adopting a globally collaborative approach rather than a free market approach to AI.

*David Schafer*

*See also:* Berserkers; Bostrom, Nick; de Garis, Hugo; General and Narrow AI; Goertzel, Ben; Kurzweil, Ray; Moravec, Hans; Musk, Elon; Technological Singularity; Yudkowsky, Eliezer.

### Further Reading

Bostrom, Nick. 2017. *Superintelligence: Paths, Dangers, Strategies.* Oxford, UK: Oxford University Press.

Feenberg, Andrew. 2006. "Environmentalism and the Politics of Technology." In *Questioning Technology*, 45–73. New York: Routledge.

Kissinger, Henry. 2018. "How the Enlightenment Ends." *The Atlantic*, June 2018. https://www.theatlantic.com/magazine/archive/2018/06/henry-kissinger-ai-could-mean-the-end-of-human-history/559124/.

Knight, Heather. 2014. *How Humans Respond to Robots: Building Public Policy Through Good Design.* Washington, DC: The Project on Civilian Robotics. Brookings Institution.

Omohundro, Steve. 2014. "Autonomous Technology and the Greater Human Good." *Journal of Experimental & Theoretical Artificial Intelligence* 26, no. 3: 303–15.

Ramzy, Austin, and Sui-Lee Wee. 2019. "Scientist Who Edited Babies' Genes Is Likely to Face Charges in China." *The New York Times*, January 21, 2019.

## Symbol Grounding Problem, The

The symbol grounding problem explains how entities may be tied to their real-world representations. In other words, the symbol grounding problem addresses how concepts within an intelligent agent's representation can point to their referent in the world. Stevan Harnad (1990) has compared this problem to a symbol/symbol merry-go-round, grounding a symbol to another meaningless symbol.

A cognitive model is a symbolic representation of the mind. Different kinds of representations are needed to build artificial cognitive systems. A mental representation can be divided into two categories: lexical (or semantic system) and compositional (or connectionist system) semantics. A word meaning is the lexical semantic, and a group of words' meaning in relation to each other is the compositional semantic. Classic cognitive science investigates the compositional semantic. Cognitive architectures are hybrid systems that use lexical and compositional representations of knowledge (or chunks) and pattern matching to select the matched knowledge element. Chunks are the sensory input of architecture. Tying chunks to the external world has not been defined for all hybrid systems. Therefore, the symbol grounding problem limits artificial cognitive systems from reaching their cognitive capacity, especially in perception and motor skills.

To address the symbol grounding problem for cognitive models, a new type of memory, called visual patterns, has been added. Visual patterns are real-world visual representations of objects that are associated with knowledge elements (chunks). With this type of knowledge, cognitive models can interact through a truer representation of vision and therefore help to solve the symbol grounding problem.

*Farnaz Tehranchi*

*See also:* Cognitive Architectures.

**Further Reading**

Harnad, Stevan. 1990. "The Symbol Grounding Problem." *Physica D: Nonlinear Phenomena* 42, no. 1–3 (June): 335–46.

Ritter, Frank E., Farnaz Tehranchi, and Jacob D. Oury. 2018. "ACT-R: A Cognitive Architecture for Modeling Cognition." *Wiley Interdisciplinary Reviews: Cognitive Science* 10, no. 4, 1–19.

Tehranchi, Farnaz, and Frank E. Ritter. 2018. "Modeling Visual Search in Interactive Graphic Interfaces: Adding Visual Pattern Matching Algorithms to ACT-R." In *Proceedings of the 16th International Conference on Cognitive Modeling*, 162–67. Madison: University of Wisconsin.

# Symbol Manipulation

Symbol manipulation refers to the general information-processing abilities of a digital stored program computer. Perceiving the computer as essentially a symbol manipulator became paradigmatic from the 1960s to the 1980s and led to the scientific pursuit of symbolic artificial intelligence, today sometimes referred to as Good Old-Fashioned AI (GOFAI).

The development of stored-program computers in the 1960s generated a new awareness of a computer's programming flexibility. Symbol manipulation became both a general theory for intelligent behavior and a guideline for AI research. One of the earliest computer programs to model intelligent symbol manipulation was the Logic Theorist, developed by Herbert Simon, Allen Newell, and Cliff Shaw in 1956. The Logic Theorist was able to prove theorems from Alfred North Whitehead and Bertrand Russell's *Principia Mathematica* (1910–1913). It was presented at the 1956 Dartmouth Summer Research Project on Artificial Intelligence (the Dartmouth Conference). This conference was organized by Dartmouth mathematics professor John McCarthy, who coined the term "artificial intelligence." Because the Logic Theorist made its first appearance there, and as many of the attendees became pioneering AI researchers, the Dartmouth Conference could be called the birthplace of AI.

It was only in the early 1960s, after Simon and Newell had developed their General Problem Solver (GPS), that the properties of symbol manipulation, as a general process that underlies all forms of intelligent problem-solving behavior, were fully explicated and became a basis for much of the early work in AI. In 1961, Simon and Newell promoted that AI understanding, and their work on GPS, to a more popular audience. A computer, they argued in *Science,* is "not a number-manipulating device; it is a symbol-manipulating device, and the symbols it manipulates may represent numbers, letters, words, or even nonnumerical, nonverbal patterns" (Newell and Simon 1961, 2012). Their manipulation by a computer, Simon and Newell continued, entails reading "symbols or patterns presented by appropriate input devices, storing symbols in memory, copying symbols from one memory location to another, erasing symbols, comparing symbols for identity, detecting specific differences between their patterns, and behaving in a manner conditional on the results of its processes" (Newell and Simon 1961, 2012).

The 1960s' rise of symbol manipulation was also rooted in pre-World War II developments in cognitive psychology and symbolic logic. Experimental psychologists, such as Edwin Boring at Harvard University, had begun to move their field away from philosophical and behavioralist approaches, starting in the 1930s. Boring urged his colleagues to split open the mind and develop explanatory, testable, theories for various cognitive mental functions (an approach that was adopted by Kenneth Colby in his work on PARRY in the 1960s). In their historical addendum to *Human Problem Solving,* Simon and Newell also stressed their debt to pre-World War II developments in formal logic and abstract mathematics—not because all thought is logical, or follows the rules of deductive logic, but because formal logic treated symbols as tangible objects. At a minimum, Simon and Newell explained, "the formalization of logic showed that symbols can be copied, compared, rearranged, concatenated, with just as much definiteness of process as [wooden] boards can be sawed, planed, measured, glued [in a carpenter shop]" (Newell and Simon 1973, 877).

*Elisabeth Van Meer*

*See also:* Expert Systems; Newell, Allen; PARRY; Simon, Herbert A.

## Further Reading

Boring, Edwin G. 1946. "Mind and Mechanism." *American Journal of Psychology* 59, no. 2 (April): 173–92.

Feigenbaum, Edward A., and Julian Feldman. 1963. *Computers and Thought.* New York: McGraw-Hill.

McCorduck, Pamela. 1979. *Machines Who Think: A Personal Inquiry into the History and Prospects of Artificial Intelligence.* San Francisco: W. H. Freeman and Company.

Newell, Allen, and Herbert A. Simon. 1961. "Computer Simulation of Human Thinking." *Science* 134, no. 3495 (December 22): 2011–17.

Newell, Allen, and Herbert A. Simon. 1972. *Human Problem Solving*. Englewood Cliffs, NJ: Prentice Hall.

Schank, Roger, and Kenneth Colby, eds. 1973. *Computer Models of Thought and Language*. San Francisco: W. H. Freeman and Company.

## Symbolic Logic

Symbolic logic involves the use of symbols to represent terms, relations, and propositions in mathematical and philosophical reasoning. Symbolic logic differs from (Aristotelian) syllogistic logic, as it utilizes ideographs or a special notation which "symbolize *directly* the thing talked about" (Newman 1956, 1852), and can be manipulated according to precise rules. Traditional logic studied the truth and falsity of statements and the relations between them, using words that themselves sprang from natural language. Symbols, unlike nouns and verbs, have no need for interpretation. Operations on symbols are mechanical and thus can be assigned to computers. Symbolic logic rids logical analysis of any ambiguity by codifying it completely within a fixed notational system.

Gottfried Wilhelm Leibniz (1646–1716) is generally considered to have been the first student of symbolic logic. In the seventeenth century, as part of his plan to reform scientific reasoning, Leibniz advocated the use of ideographic symbols in place of natural language. The use of such concise universal symbols (*characteristica universalis*) combined with a set of rules for scientific reasoning, Leibniz hoped, would create an alphabet of human thought to promote the growth and dissemination of scientific knowledge and a corpus containing all human knowledge.

The field of symbolic logic can be split up into several distinct areas of analysis, including Boolean logic, the logical foundations of mathematics, and decision problems. Key works in each of these areas were respectively written by George Boole, Alfred North Whitehead and Bertrand Russell, and Kurt Gödel. In the mid-nineteenth century, George Boole set forth his ideas in *The Mathematical Analysis of Logic* (1847) and *An Investigation of the Laws of Thought* (1854). Boole zeroed in on what he called a calculus of deductive reasoning, which led him to three basic operations—AND, OR, and NOT—in a logical mathematical language called Boolean algebra. Symbols and operators vastly simplified the construction of logical expressions. In the twentieth century, Claude Shannon (1916–2001) used electromechanical relay circuits and switches to replicate Boolean algebra—important groundwork in the history of electronic digital computing and computer science generally.

In the early twentieth century, Alfred North Whitehead and Bertrand Russell created their definitive work in the field of symbolic logic. Their *Principia Mathematica* (1910, 1912, 1913) gave a rigorous demonstration of how all of mathematics could be subsumed to symbolic logic. In the first volume of their work, Whitehead and Russell deduced a logical system from a handful of logical ideas and a set of postulates derived from those ideas. In the second volume of the *Principia*, Whitehead and Russell defined all arithmetic concepts, including number, zero, the successor of, addition, and multiplication, by basic logical terms and operational rules such as proposition, negation, and either-or. Whitehead and Russell were then, in the final and third volume, able to show that the nature and truth of all of mathematics is based upon logical ideas and relationships. *The Principia* demonstrated how any postulate of arithmetic could be deduced from the earlier explicated symbolic logical truths.

These strong and deep claims set up by the *Principia* were critically analyzed only a few decades later by Kurt Gödel in *On Formally Undecidable Propositions in the Principia Mathematica and Related Systems* (1931), who showed that Whitehead and Russell's axiomatic system could not simultaneously be consistent and complete. Still, it took another key work in symbolic logic, Ernst Nagel and James Newman's *Gödel's Proof* (1958), to get Gödel's message across to a wider audience, including some practitioners of artificial intelligence.

Each of these key works in symbolic logic had a distinct impact on the development of computation and programming, and on our consequent perception of the capacity of a computer's abilities. Boolean logic found its way into logic circuitry design. Simon and Newell's Logic Theorist program demonstrated logical proofs that matched those in the *Principia Mathematica* and was therefore seen as a proof that a computer could be designed to do intelligent tasks using symbol manipulation. Gödel's incompleteness theorem suggests tantalizing questions about the ultimate realization of programmed machine intelligence, especially strong AI.

*Elisabeth Van Meer*

*See also:* Symbol Manipulation.

**Further Reading**

Boole, George. 1854. *Investigation of the Laws of Thought on Which Are Founded the Mathematical Theories of Logic and Probabilities*. London: Walton.

Lewis, Clarence Irving. 1932. *Symbolic Logic*. New York: The Century Co.

Nagel, Ernst, and James R. Newman. 1958. *Gödel's Proof*. New York: New York University Press.

Newman, James R., ed. 1956. *The World of Mathematics*, vol. 3. New York: Simon and Schuster.

Whitehead, Alfred N., and Bertrand Russell. 1910–1913. *Principia Mathematica*. Cambridge, UK: Cambridge University Press.

# SyNAPSE

Project SyNAPSE (Systems of Neuromorphic Adaptive Plastic Scalable Electronics) is a Defense Advanced Research Projects Agency funded collaborative cognitive computing study to create the architecture for a brain-inspired neurosynaptic computer core. Initiated in 2008, the project is a collaborative venture of IBM Research, HRL Laboratories, and Hewlett-Packard. Researchers at several universities are also partners in the venture. The acronym SyNAPSE is a reference to the Ancient Greek word σύναψις meaning "conjunction," which alludes to the neuronal contacts involved in the transfer of information to the brain.

The goal of the project is to create a flexible, ultra-low power system for use in robots by reverse-engineering the functional intelligence of rats, cats, or possibly humans. DARPA's original agency announcement requested a machine that breaks through the "algorithmic-computational paradigm" and make it "scalable to biological levels" (DARPA 2008, 4). In other words, they wanted an electronic computer to process real-world complexity, adapt in response to external stimuli, and do so in as close to real time as possible.

SyNAPSE is a response to demands for computer systems capable of understanding the environment and adapting rapidly to changing conditions, while still remaining energy efficient. SyNAPSE scientists are developing systems of neuromorphic electronics similar to biological nervous systems and capable of processing data from complicated environments. It is hoped that such systems will eventually possess great autonomy. The approach to the SyNAPSE project is interdisciplinary, borrowing ideas from computational neuroscience, artificial neural networks, materials science, and cognitive science, among others. SyNAPSE will need to extend basic science and engineering in the following areas: hardware—to create synaptic components and integrate hardwired and programmable connectivity; architecture—to support structures and functions that appear in biological systems; simulation—for the digital reproduction of systems in order to test functioning prior to the implementation of material neuromorphological systems.

The first SyNAPSE grant was awarded to IBM Research and HRL Laboratories in 2008. IBM and HRL subcontracted various parts of the grant requirements to an array of suppliers and contractors. The project was divided into four phases that started after the initial feasibility study, which lasted nine months. An initial simulator, called C2, developed in 2009, ran on a BlueGene/P supercomputer in which cortical simulations are made with $10^9$ neurons and $10^{13}$ synapses matching a mammalian cat cortex. The program would later come in for criticism after an announcement made by the Blue Brain Project leader that the simulation did not achieve the complexity reported.

Each neurosynaptic core measures 2 millimeters by 3 millimeters in size and is composed of elements abstracted from the biology of the human brain. The relation between the cores and real brains is more metaphorical than analogous. Computation substitutes for actual neurons, memory stands in for synapses, and axons and dendrites are represented by communication. This allows the team to describe a hardware implementation of a biological system.

In 2012, HRL Labs announced that it had achieved the first functioning memristor array stacked on a conventional CMOS circuit. "Memristor," a word coined from memory and transistor, is an idea dating to the 1970s. In a memristor, the functions of memory and logic are combined. Also in 2012, project leaders announced the successful large-scale simulation of 530 billion neurons and 100 trillion synapses on the world's second fastest supercomputer, the Blue Gene/Q Sequoia machine at Lawrence Livermore National Laboratory in California.

In 2014, IBM introduced the TrueNorth processor, a 5.4-billion-transistor chip with 4096 neurosynaptic cores interconnected via an intrachip network that integrates 1 million programmable spiking neurons and 256 million configurable synapses. Finally, in 2016, an end-to-end ecosystem (involving scalable systems, software, and applications) was announced that could put the TrueNorth processor to full use. At that time, the implementation of applications, including those for interactive handwritten character recognition and data-parallel text extraction and recognition, were reported. TrueNorth's cognitive computing chips have since been tested in such simulations as a robot driving in a virtual environment and by playing the classic videogame *Pong*.

The design of brain-inspired computing systems has been of interest to DARPA since at least the 1980s. The Project SyNAPSE effort is led by Dharmendra Modha, director of IBM Almaden's Cognitive Computing Initiative and Narayan Srinivasa, manager of HRL's Center for Neural and Emergent Systems.

*Konstantinos Sakalis*

*See also:* Cognitive Computing; Computational Neuroscience.

**Further Reading**

Defense Advanced Research Projects Agency (DARPA). 2008. "Systems of Neuromorphic Adaptive Plastic Scalable Electronics." DARPA-BAA 08-28. Arlington, VA: DARPA, Defense Sciences Office.

Hsu, Jeremy. 2014. "IBM's New Brain." *IEEE Spectrum* 51, no. 10 (October): 17–19.

Merolla, Paul A., et al. 2014. "A Million Spiking-Neuron Integrated Circuit with a Scalable Communication Network and Interface." *Science* 345, no. 6197 (August): 668–73.

Monroe, Don. 2014. "Neuromorphic Computing Gets Ready for the (Really) Big Time." *Communications of the ACM* 57, no. 6 (June): 13–15.

# T

## Tambe, Milind (1965–)

Milind Tambe is a leader in research advancing Artificial Intelligence for Social Good. Some common areas where AI is tackling social problems are public health, education, safety and security, housing, and environmental protection. Tambe has created software that protects endangered animals in game reserves, social network algorithms that encourage healthier eating habits, and programs that monitor social ills and community issues and makes recommendations to reduce social distress.

Tambe grew up in India, where he found inspiration to learn about the field of artificial intelligence in the robot novels of Isaac Asimov and the original *Star Trek* series (1966–1969). He received his doctorate from the Carnegie Mellon University School of Computer Science. His original research involved the development of AI software for security. He became interested in the potential of artificial intelligence in this field after the 2006 commuter train bombings in his hometown of Mumbai. His graduate work uncovered key game theory insights into the nature of random interactions and teamwork. Tambe's ARMOR software randomizes human security patrols and police checkpoint schedules and delivers risk assessment scores. In 2009, Los Angeles Airport police following random inspection protocols discovered a truck carrying five rifles, ten handguns, and a thousand rounds of ammunition. More recent versions of the software are used to schedule the flights of federal air marshals and harbor security patrols.

Tambe's lab today applies deep learning algorithms to help wildlife conservation officers distinguish, in real time, between poachers and animals recorded by infrared cameras aboard unmanned drone aircraft. The Systematic Poacher Detector (SPOT) is capable of detecting poachers within three-tenths of a second of their appearance near animals. SPOT was tested in park reserves in Zimbabwe and Malawi and is now deployed in Botswana. A successor technology that predicts poacher activity called PAWS has been introduced in Cambodia and may be deployed in coming years in more than 50 more countries around the world.

Another of Tambe's algorithms models population movements and the spread of epidemic diseases in order to optimize the effectiveness of public health campaigns. The program has spotted several nonobvious patterns that will improve disease control. A third algorithm developed in Tambe's lab helps substance abuse counselors sort addiction recovery groups into smaller subgroups where positive social relationships can proliferate. Other AI-based solutions address climate change, gang violence, HIV awareness, and counterterrorism.

Tambe is Helen N. and Emmett H. Jones Professor in Engineering in the Viterbi School of Engineering at the University of Southern California (USC). He is founding codirector of the Center for Artificial Intelligence in Society at USC and recipient of multiple honors, including the John McCarthy Award and the Daniel H. Wagner Prize for Excellence in Operations Research Practice. He is a Fellow of both the Association for the Advancement of Artificial Intelligence (AAAI) and the Association for Computing Machinery (ACM). Tambe is cofounder and director of research at Avata Intelligence, which markets artificial intelligent management solutions to address enterprise-level data analysis and decision-making challenges. His algorithms are in use by LAX, the U.S. Coast Guard, the Transportation Security Administration, and the Federal Air Marshals Service.

*Philip L. Frana*

*See also:* Predictive Policing.

**Further Reading**

Paruchuri, Praveen, Jonathan P. Pearce, Milind Tambe, Fernando Ordonez, and Sarit Kraus. 2008. *Keep the Adversary Guessing: Agent Security by Policy Randomization.* Riga, Latvia: VDM Verlag Dr. Müller.

Tambe, Milind. 2012. *Security and Game Theory: Algorithms, Deployed Systems, Lessons Learned.* Cambridge, UK: Cambridge University Press.

Tambe, Milind, and Eric Rice. 2018. *Artificial Intelligence and Social Work.* Cambridge, UK: Cambridge University Press.

## Technological Singularity

The Technological Singularity refers to the emergence of technologies that could fundamentally change humans' role in society, challenge human epistemic agency and ontological status, and trigger unprecedented and unforeseen developments in all aspects of life, whether biological, social, cultural, or technological. The Technological Singularity is most often associated with artificial intelligence, specifically with artificial general intelligence (AGI). It is therefore sometimes presented as an intelligence explosion driving advances in areas such as biotechnology, nanotechnology, and information technologies, as well as creating technologies yet unknown. The Technological Singularity is often referred to as the Singularity, but it should not be regarded as analogous to a singularity in mathematics, because it has only a distant resemblance to this. In contrast, this singularity is a rather loosely defined concept that may have different interpretations emphasizing different aspects of the changes precipitated by technology.

The origins of the Technological Singularity concept go back to the second half of the twentieth century, and they are usually associated with the ideas and works of John von Neumann (1903–1957), Irving John Good (1916–2009), and Vernor Vinge (1944–). Current Technological Singularity research has been supported by several universities and governmental and private research institutions seeking to explore the future of technology and society. Even though the Technological Singularity is subjected to learned philosophical and technical debates, it is still a hypothesis, a conjecture, a fairly open hypothetical concept.

While several researchers claim that the Technological Singularity is inevitable, its timing is consistently moved further into the future. Nevertheless, many studies share the belief that the question is not one of whether the Technological

Singularity will happen or not but rather of when and how it will occur. Ray Kurzweil ventured a more precise date for the Technological Singularity's emergence in about the mid-twenty-first century. Others have also attempted to assign a date for this event, yet there are no well-reasoned arguments behind proposing any such date. Moreover, there remains the question of how humans would know the Technological Singularity event has been reached without relevant metrics or indicators. The unfulfilled promises that are associated with the history of artificial intelligence exemplify the hazards when trying to divine the future of technology.

The Technological Singularity is often characterized by concepts of superintelligence, acceleration, and discontinuity. "Superintelligence" denotes a quantitative leap in the intellectual capacities of artificial systems, taking them well beyond the capacities of normal human intellect (as measured by standard IQ tests). Superintelligence may not be limited to AI and computer technology, however. It may emerge in human agents through genetic engineering, biological computing systems, or hybrid artificial–natural systems. Some researchers attribute infinite intellectual capacities to superintelligence.

Acceleration refers to the shape of the time curve for the appearance of some significant events. Technological progress is represented as a curve through time highlighting the discovery of significant inventions, such as stone tools, the pottery wheel, the steam engine, electricity, atomic power, computers, and the internet. The growth in computational power is represented by Moore's law, which is more accurately an observation that has become regarded as a law. It states that "the number of transistors in a dense integrated circuit doubles about every two years." In most cases, the growth curve is linear or exponential, but in the case of the Technological Singularity, people speculate that the appearances of significant technological breakthroughs and new technological and scientific paradigms will follow a super-exponential curve. For example, one prediction about the Technological Singularity concerns how superintelligent systems will be able to self-improve (and self-replicate) in unforeseen ways at an unprecedented rate, thus taking the technological growth curve well beyond what has been seen in history.

The discontinuity of Technological Singularity is described as an event horizon, and it is somewhat analogous to a physical concept associated with black holes. The comparison to this physical phenomenon, however, should be treated with caution rather than using it to attribute the regularity and predictability of the physical world to technological singularity. An event horizon (also referred to as a prediction horizon) defines the limit of our knowledge about physical events beyond a certain point in time. It means that what will happen beyond the event horizon is unknown. In the case of technological singularity, the discontinuity or event horizon implies that the technologies precipitating the Technological Singularity will trigger disruptive changes in all aspects of the human condition, changes about which experts cannot even speculate.

Technological singularity is also usually associated with the demise of humanity and the end of human society. Some studies predict the collapse of social order, the end of humans as primary agents, and the loss of epistemic agency and primacy. Superintelligent systems will not need humans apparently. These systems will be able to replicate and improve upon themselves and create their own living spaces, and humans may be perceived as either obstacles or simply insignificant, obsolete objects, much like how humans regard lower species now. Nick Bostrom's Paperclip Maximizer exemplifies one such scenario.

*The Global Catastrophic Risks Survey* lists AI as a potential threat to humanity's survival, with there being a relatively high risk of human extinction, putting it on par with global pandemics, nuclear war, and global nanotech accidents. However, the doomsday scenario associated with AI is not an inevitable consequence of the Technological Singularity. In some more utopian visions, technological singularity will create new possibilities for the unlimited growth of humanity and herald an unprecedented era of limitless happiness. Another aspect of technological singularity that needs careful attention is how the coming of superintelligence implies the possible emergence of superethical capacities in an all-knowing ethical agent. However, nobody knows what superethical capacities would actually entail. The main concern, however, is that the superior intellectual capacities of superintelligent agents do not guarantee a high level, or indeed any level, of ethical probity. Therefore, having a superintelligent system with unlimited (almost) powers but without ethics seems risky to say the least.

A significant group of researchers dispute the possible emergence of the Technological Singularity and particularly the emergence of superintelligence. They dismiss the likelihood of creating artificial systems with superhuman intellectual capacities, either on a metaphysical basis or on more scientific grounds. Some argue that because AI is often at the core of technological singularity claims, the achievement of human-level intelligence is not feasible in artificial systems, so superintelligence, therefore the Technological Singularity, is a fantasy. However, such obstacles do not exclude the creation of superhuman minds through the genetic engineering of ordinary humans, thus opening the way for transhumans, human-machine hybrids, and superhuman agents.

Still more researchers deny the legitimacy of the Technological Singularity concept itself, pointing out how such predictions about future societies are based on conjecture and guesswork. Others highlight how the claims of unbridled technological growth and unlimited intellectual powers that form part of the Technological Singularity lore are groundless, because the resources for physical and informational processing are clearly limited in the universe, especially on Earth. Any claims of self-replicating, self-improving artificial agents that will accelerate technological growth to super-exponential rates are unfounded, because such systems will lack the creativity, will, and motivation to drive their own evolution. The socially minded critics, meanwhile, point out how the unbounded technological progress implied by superintelligence will not solve problems such as over-population, environmental degradation, poverty, and unprecedented inequality. Indeed, the expected massive unemployment resulting from the mass automation of work through AI, thus excluding large portions of the population from playing a productive role in society, will result in unprecedented social chaos that will derail the development of new technologies. Thus, rather than accelerate, technological progress will be stymied by political or social forces.

While technological singularity cannot be excluded on logical grounds, the technological challenges facing its development are enormous, even if focusing on just those challenges that can be currently identified. Nobody expects technological singularity to occur with current computing and other technologies, but its proponents see these issues as mere "technical problems to be solved" rather than potential showstoppers. The list of technical problems to be resolved is a long one, however, and Murray Shanahan's *The Technological Singularity* (2015) provides a good review of some of these topics. Some significant nontechnical issues also exist, including, among others, the problem of training of superintelligent systems, the question of the ontology of artificial or machine consciousness and self-aware artificial systems, the embodiment of artificial minds or vicarious embodiment processes, and the rights given to superintelligent systems, as well as their role in society and any limits placed on their actions, if indeed this would be possible at all. At present, these problems lie in the realm of technical and philosophical speculation.

*Roman Krzanowski*

*See also:* Bostrom, Nick; de Garis, Hugo; Diamandis, Peter; Digital Immortality; Goertzel, Ben; Kurzweil, Ray; Moravec, Hans; Post-Scarcity, AI and; Superintelligence.

**Further Reading**

Bostrom, Nick. 2014. *Superintelligence: Path, Dangers, Strategies*. Oxford, UK: Oxford University Press.

Chalmers, David. 2010. "The Singularity: A Philosophical Analysis." *Journal of Consciousness Studies* 17: 7–65.

Eden, Amnon H. 2016. *The Singularity Controversy*. Sapience Project. Technical Report STR 2016-1. January 2016.

Eden, Amnon H., Eric Steinhart, David Pearce, and James H. Moor. 2012. "Singularity Hypotheses: An Overview." In *Singularity Hypotheses: A Scientific and Philosophical Assessment*, edited by Amnon H. Eden, James H. Moor, Johnny H. Søraker, and Eric Steinhart, 1–12. Heidelberg, Germany: Springer.

Good, I. J. 1966. "Speculations Concerning the First Ultraintelligent Machine." *Advances in Computers* 6: 31–88.

Kurzweil, Ray. 2005. *The Singularity Is Near: When Humans Transcend Biology*. New York: Viking.

Sandberg, Anders, and Nick Bostrom. 2008. *Global Catastrophic Risks Survey*. Technical Report #2008/1. Oxford University, Future of Humanity Institute.

Shanahan, Murray. 2015. *The Technological Singularity*. Cambridge, MA: The MIT Press.

Ulam, Stanislaw. 1958. "Tribute to John von Neumann." *Bulletin of the American Mathematical Society* 64, no. 3, pt. 2 (May): 1–49.

Vinge, Vernor. 1993. "The Coming Technological Singularity: How to Survive in the Post-Human Era." In *Vision 21: Interdisciplinary Science and Engineering in the Era of Cyberspace*, 11–22. Cleveland, OH: NASA Lewis Research Center.

# *The Terminator* (1984)

Released in 1984 and raking in nearly $40 million at the domestic box office and untold amounts of money in the ancillary market as well as spawning a multimedia franchise that continues to present day, *The Terminator* stands as one of the most prominent representations of artificial intelligence in popular culture. Though most of the film takes place in 1984, it depicts a future wherein a military-designed artificial intelligence system called Skynet becomes self-aware and wages war against humanity. Shots of this future show roving robot destroyers hunting humans, who are seemingly on the brink of extinction, on a battlefield strewn with mechanical debris and human bones. Most of the film focuses on a T-800 model terminator (Arnold Schwarzenegger) sent back to 1984 to kill Sarah Connor (Linda Hamilton) before she gives birth to John Connor, humanity's future savior. *The Terminator*'s central plot point dramatizes what is, perhaps, the most common trope of artificial intelligence in popular culture: portraying intelligent robots as innately dangerous entities capable of revolting against humanity in pursuit of their own agenda.

In order to delve deeper into the significance of *The Terminator*, it is important to give more details about its narrative as well as the figure of the terminator itself. In the year 2029, after a nuclear holocaust, Earth is in the midst of a war between humans and the Skynet-created machines. Though further details regarding Skynet are revealed in subsequent entries to the franchise, its operation remains hazy in the original film. On the cusp of defeat, John Connor (only referenced in the film) leads a human resistance army that inevitably defeats the machines. In order to thwart the Connor-led rebellion, the machines develop a time-travel device to send one of their terminator units back to kill Sarah Connor before she conceives her savior son. The human resistance sends back their own agent, Kyle Reese (Michael Biehn), to foil the machines' plan. While the terminator is tasked with killing Sarah Connor, Reese is meant to protect her.

Accordingly, the remainder of the film, taking place in then present-day 1984, resembles a cat and mouse chase where the terminator locates Connor and Reese only for the two to escape by the skin of their teeth. In the film's finale, the terminator is burned to its mechanical endoskeleton, and it chases Connor and Reese into a factory. Reese sacrifices himself by planting a makeshift pipe bomb into the terminator's abdomen, which kills him and severs the terminator into two. The terminator's torso gives chase to Connor, but she is able to crush it with a hydraulic press. The film then flashes to several months later with a now pregnant Sarah Connor driving through Mexico. Here, it is revealed that Reese is the father of John Connor.

The terminator is an exquisite example of artificial intelligence. Though it is a programmed killing machine, it can walk, talk, perceive, and behave like a human being. It is shown to learn interactional nuance, adapting its behavior based on past experience and interactions. The terminator is also able to imitate the voice of Sarah Connor's mother in a phone call, which is convincing enough for Sarah to reveal her location to the terminator. In these respects, the terminator can indisputably pass the Turing Test (a test wherein a confederate is unable to determine if they are communicating with a human or a robot). The terminator, however, lacks human consciousness; it is driven by mechanical logic as it pursues a mission. With the terminator shot, run over by a mac truck, and burned to its endoskeleton among many other injuries, it can be surmised that it does not process pain in the same manner as humans.

Pessimistic accounts of artificial intelligence in popular culture such as *The Terminator* proffer visions of the future to be avoided. Hence, with its portrayal of a future controlled by robots and its use of a remorseless, unstoppable killing machine, *The Terminator* offers a stern warning against a primary driving force behind artificial intelligence development—the military industrial complex (later installments made the additional link of corporate greed). At the time of the film's production, President Ronald Reagan ratcheted tensions with the Soviet Union through his Strategic Defense Initiative (later derisively renamed the Star Wars Initiative by Senator Ted Kennedy) that was announced in 1983. Briefly, the Strategic Defensive Initiative was a proposed missile defense system that would protect the country from attacks by ballistic nuclear weaponry. It was feared that Reagan's posturing would catalyze a catastrophic arms-race.

*The Terminator*, therefore, can be read as a critique of Reagan's approach to the Cold War in its offering of a glimpse into a potential postapocalyptic future brought about by nuclear destruction through the development of increasingly sophisticated weaponry. In short, the film gives voice to fears over the destructive capability of human creation and potential for human's own creations to turn against them.

*Todd K. Platts*

*See also:* Berserkers; de Garis, Hugo; Technological Singularity.

**Further Reading**

Brammer, Rebekah. 2018. "Welcome to the Machine: Artificial Intelligence on Screen." *Screen Education* 90 (September): 38–45.

Brown, Richard, and Kevin S. Decker, eds. 2009. *Terminator and Philosophy: I'll Be Back, Therefore I Am*. Hoboken, NJ: John Wiley and Sons.

Gramantieri, Riccardo. 2018. "Artificial Monsters: From Cyborg to Artificial Intelligence." In *Monsters of Film, Fiction, and Fable: The Cultural Links between the Human and Inhuman*, edited by Lisa Wegner Bro, Crystal O'Leary-Davidson, and Mary Ann Gareis, 287–313. Newcastle upon Tyne, UK: Cambridge Scholars Publishing.

Jancovich, Mark. 1992. "Modernity and Subjectivity in *The Terminator*: The Machine as Monster in Contemporary American Culture." *Velvet Light Trap* 30 (Fall): 3–17.

# Tilden, Mark (1961–)

Mark Tilden is a Canadian freelance designer of biomorphic robots. Several of his robots are marketed as toys. Others have been featured as props in television and film. Tilden is known for his resistance to the idea that complex robots require sophisticated artificial intelligence.

Tilden is a pioneer in the area of BEAM (biology, electronics, aesthetics, and mechanics) robotics. Instead of digital electronics and microprocessors, BEAM robots primarily rely on analog circuits and systems and continuously variable signals to simulate biological neurons. Biomorphic robots are designed to adjust their gaits toward a least energy state. When such robots encounter obstacles or changes in the underlying terrain, they are bumped out of the least energy state, which causes the robot to adapt and adjust to a new walking pattern. Self-adaptation is key to the underlying machine's mechanics.

Tilden turned to BEAM style-robots in the late 1980s after frustrating experiments in building a conventional electronic robot butler. The robot, encoded with Isaac Asimov's Three Laws of Robotics, could barely vacuum floors. Tilden largely abandoned the effort after encountering the famous MIT roboticist Rodney Brooks at a Waterloo University talk about the virtues of simple sensorimotor, stimulus-response robotics over computationally intensive mobile machines. Tilden left Brooks's talk wondering if reliable robots could be made without computer processors or artificial intelligence.

Rather than having the intelligence programmed into the robot's firmware, Tilden imagined that the intelligence could come from the environment in which the robot operated, as well as the emergent properties built up from that world. At the Los Alamos National Laboratory in New Mexico, Tilden researched and developed a number of unique analog robots using rapid prototyping and off-the-shelf and cannibalized parts. Los Alamos wanted robots capable of working in unpredictable, unstructured, and potentially dangerous environments. Tilden created more than eighty robot prototypes. His SATBOT autonomous spacecraft prototype could autonomously align itself to the Earth's magnetic field. For the Marine Corps Base Quantico, he build fifty insectoid robots capable of crawling into minefields and detecting explosive devices. An "aggressive ashtray" robot spit water at smokers. A "solar spinner" cleaned windows. A biomorph constructed from five broken Sony Walkmans mimicked the movements of an ant.

At Los Alamos, Tilden began constructing Living Machines powered by solar cells. Because of their energy source, these machines operated at very slow speeds but were reliable and efficient over very long periods of time, many more than a year. Tilden's original plans for robots were based on thermodynamic conduit engines, in particular small and efficient solar engines capable of firing single neurons. His "nervous net" neurons controlled the rhythms and patterns of motions in robot bodies rather than the workings of their brains. Tilden's insight was to optimize the number of possible patterns across the smallest number of embedded transistors. He realized that it was possible to produce six patterns of movement with only twelve transistors. By folding the six patterns into a figure eight in a symmetrical robot chassis, Tilden could mimic hopping, jumping, running, sitting, slithering, and a number of other patterns of behavior.

Tilden has since become an advocate of a new set of robot rules for such survivalist wild automata. Tilden's Laws of Robotics state that (1) a robot must protect its existence at all costs; (2) a robot must obtain and maintain access to its own power source; and (3) a robot must continually search for better power sources. Tilden hopes that wild robots will be used to autonomously restore ecosystems damaged by human beings.

Another epiphany for Tilden was to introduce relatively cheap robots as toys for the masses and robot enthusiasts. He wanted his robots in the hands of many so that they could be reprogrammed and modified by hackers, hobbyists, and members of various maker communities. Tilden made the toys so that they could be taken apart and studied. They were fundamentally hackable. Everything inside is carefully labeled, color-coded, and all of the wires had gold-plated contacts that could be pulled apart.

Tilden is currently creating consumer-oriented entertainment robots for Hong Kong-based WowWee Toys. Popular WowWee robot toys include B.I.O. Bugs, Constructobots, G.I Joe Hoverstrike, Robosapien, Roboraptor, Robopet, Roboreptile, Roboquad, Roboboa, Femisapien, and Joebot. The Roboquad was originally created for the Jet Propulsion Laboratory (JPL) and Mars exploration. Tilden is also the creator of the utility cleaning robot Roomscooper. By 2005, WowWee Toys had sold more than three million of Tilden's robot creations.

Tilden created his first robotic doll at age three. He created a Meccano suit of armor for his cat at the age of six. He studied Systems Engineering and Mathematics at the University of Waterloo. Tilden is now a collaborator with artificial intelligence pioneer Ben Goertzel on OpenCog and OpenCog Prime. OpenCog is a multinational project backed by the Hong Kong government that seeks to create an open-source emergent artificial general intelligence framework and a common architecture for embodied robotic and virtual cognition. OpenCog components are already in use by dozens of technology companies around the world. Tilden has worked as a technical consultant or robot designer for a number of films and television shows, including *Lara Croft: Tomb Raider* (2001), *The 40-Year-Old Virgin* (2005), *Paul Blart Mall Cop* (2009), and *X-Men: The Last Stand* (2006). His robots are often showcased on the bookshelves of Sheldon's apartment in *Big Bang Theory* (2007–2019).

*Philip L. Frana*

*See also:* Brooks, Rodney; Embodiment, AI and.

### Further Reading

Frigo, Janette R., and Mark W. Tilden. 1995. "SATBOT I: Prototype of a Biomorphic Autonomous Spacecraft." *Mobile Robotics*, 66–75.

Hapgood, Fred. 1994. "Chaotic Robots." *Wired*, September 1, 1994. https://www.wired.com/1994/09/tilden/.

Hasslacher, Brosl, and Mark W. Tilden. 1995. "Living Machines." *Robotics and Autonomous Systems* 15, no. 1–2: 143–69.

Marsh, Thomas. 2010. "The Evolution of a Roboticist: Mark Tilden." *Robot Magazine*, December 7, 2010. http://www.botmag.com/the-evolution-of-a-roboticist-mark-tilden.

Menzel, Peter, and Faith D'Aluisio. 2000. "Biobots." *Discover Magazine*, September 1, 2000. https://www.discovermagazine.com/technology/biobots.

Rietman, Edward A., Mark W. Tilden, and Manor Askenazi. 2003. "Analog Computation with Rings of Quasiperiodic Oscillators: The Microdynamics of Cognition in Living Machines." *Robotics and Autonomous Systems* 45, no. 3–4: 249–63.

Samans, James. 2005. *The Robosapiens Companion: Tips, Tricks, and Hacks*. New York: Apress.

# Trolley Problem

The Trolley Problem is an ethical dilemma first articulated by Philippa Foot in 1967. Advancements in artificial intelligence in various fields have precipitated ethical conversations about how the decision-making processes of these technologies should be programmed. There is of course generalized concern over AI's ability to evaluate ethical dilemmas and uphold social values.

In this classic philosophical thought experiment, an operator finds herself near a trolley track, standing next to a lever that determines whether the trolley will continue on its existing path or go in another direction. In the problem, five individuals stand on the track where the trolley is running, and these people cannot move out of the way and will be killed if the trolley continues in the direction it is going. On the other track is one person who will also be killed if the operator chooses to pull the lever. The operator can choose to either pull the lever, thereby killing one person and saving the other five, or do nothing and allow five people to die. The scenario presented in this way is a classical dilemma between utilitarianism (actions should maximize well-being for affected individuals) and deontology (whether the action is right or wrong based on rules, as opposed to the consequences of the action).

With the advent of artificial intelligence, the question has become how we should program machines to respond in situations similar to the one represented by the Trolley Problem, which are seen as inevitable realities. The Trolley Problem as it relates to artificial intelligence has been explored in such areas as primary health care, the surgical suite, security, self-driving cars, and weapons technologies. The problem has been explored most extensively with self-driving cars, where policies, guidelines, and rules have already been proposed or developed regarding this aspect of AI. Autonomous cars are challenged in this regard because they have already driven millions of miles in the United States. The problem is made more imminent because a handful of users of self-driving vehicles have already died while using autonomous technology. Accidents have stoked more public controversy and debate regarding the appropriate use of this technology.

A team at the Massachusetts Institute of Technology has created an online platform called Moral Machine to crowdsource answers to questions about how self-driving cars should prioritize lives. The Moral Machine's designers ask website visitors to determine what choice they would have a self-driving car make under a diverse array of Trolley Problem-like dilemmas. Respondents are asked, among other variables, to prioritize the lives of passengers in cars, pedestrians, humans and animals, people walking legally or illegally, and people of various fitness levels and socioeconomic status. Respondents almost always indicate that they would move to preserve their own lives whenever they are riding in a car.

Such crowd-based solutions may not be the best way to answer Trolley Problem questions. For instance, trading a pedestrian life for a car passenger's life may be viewed as arbitrary and unfair. Currently, the aggregated solutions do not seem to reflect mere utilitarian calculations that maximize lives saved or prioritize one type of life over another. It is unclear who gets to decide how the AI will be programmed and who will ultimately be liable for failures of AI programs. This responsibility could be left to policy makers, the company manufacturing the technology, or the individuals who end up using the technology. Each of these considerations has its own implications that also need to be addressed.

The utility of the Trolley Problem in solving AI dilemmas is not universally acknowledged. Some scholars in both artificial intelligence and ethics dismiss the

Trolley Problem as a useful thought experiment. Their arguments tend to focus on the prior assumption of trade-offs between various lives. The Trolley Problem, they say, gives inclination to the view that these trade-offs (as well as autonomous car crashes themselves) are inevitable. Rather than focusing on the best ways to ensure that a dilemma such as the trolley problem can be avoided, policy makers and programmers might instead focus on the best ways to respond to the various scenarios.

*Laci Hubbard-Mattix*

*See also:* Accidents and Risk Assessment; Air Traffic Control, AI and; Algorithmic Bias and Error; Autonomous Weapons Systems, Ethics of; Driverless Cars and Trucks; Moral Turing Test; Robot Ethics.

**Further Reading**

Cadigan, Pat. 2018. *AI and the Trolley Problem*. New York: Tor.

Etzioni, Amitai, and Oren Etzioni. 2017. "Incorporating Ethics into Artificial Intelligence." *Journal of Ethics* 21: 403–18.

Goodall, Noah. 2014. "Ethical Decision Making during Automated Vehicle Crashes." *Transportation Research Record: Journal of the Transportation Research Board* 2424: 58–65.

Moolayil, Amar Kumar. 2018. "The Modern Trolley Problem: Ethical and Economically-Sound Liability Schemes for Autonomous Vehicles." *Case Western Reserve Journal of Law, Technology & the Internet* 9, no. 1: 1–32.

# Turing, Alan  (1912–1954)

Alan Mathison Turing OBE FRS was a British mathematician and logician. He is often referred to as the founder of artificial intelligence and computer science.

Turing studied mathematics at King's College, Cambridge, graduating in 1934 with first class honors. Following a fellowship at King's College, Turing earned his PhD from Princeton University, studying under the American mathematician Alonzo Church. During his studies, Turing authored several pivotal works, including "On Computable Numbers, with an Application to the *Entscheidungsproblem*," proving there is no solution to the so-called "decision problem." The decision problem asks if there is a procedure that can decide the validity of any statement within a given mathematical system. This work also discussed a hypothetical Turing machine (essentially an early computer) capable of performing any mathematical computation if represented by an algorithm.

Turing is widely renowned for his codebreaking work with Britain's Government Code and Cypher School (GC&CS) at Bletchley Park during World War II (1939–1945). Turing's work with GC&CS included leading Hut 8, which was charged with breaking the highly complicated German naval cryptography, including the German Enigma. Though impossible to quantify with certainty, Turing's work likely shortened the war by years, saving millions of lives. During his time with GC&CS, Turing authored "The Applications of Probability to Cryptography" and "Paper on Statistics of Repetitions," both of which were kept confidential by the Government Communications Headquarters (GCHQ) for seventy years, before being released to the UK National Archives in 2012.

Following World War II, Turing began to study mathematical biology at the Victoria University of Manchester, while progressing in his work in mathematics, stored-program digital computing, and artificial intelligence. Turing's 1950 work "Computing Machinery and Intelligence" explored artificial intelligence and introduced the concept of the Imitation Game (also known as the Turing Test), whereby a human judge attempts to distinguish between a computer program and a human via a set of written questions and responses. If the computer program imitates a human such that the human judge cannot distinguish the computer program's responses from the human's responses, then the computer program has passed the test, suggesting that the program is capable of intelligent thought.

Turing and his colleague, D.G. Champernowne, wrote Turochamp, a chess program intended to be executable by a computer, but no computer with sufficient power existed to test the program. Instead, Turing tested the program by manually running the algorithms.

Though much of Turing's work remained classified until well after his death, Turing was well decorated during his lifetime. In 1946, Turing was appointed to the Order of the British Empire 1946, and in 1951, he became a Fellow of the Royal Society (FRS). An award in his name, the Turing Award, is presented annually by the Association for Computing Machinery for contributions to the computing field. Accompanied by $1 million in prize money, the Turing Award is widely regarded as the Nobel Prize of Computing.

Turing was relatively open about being gay at a time when gay sexual activity was still considered a criminal offense in the United Kingdom. In 1952, Turing was charged with "gross indecency" under Section 11 of the Criminal Law Amendment Act 1885. Turing was convicted, given probation, and subjected to a punishment referred to as "chemical castration," whereby he was injected with synthetic estrogen for a year. Turing's conviction impacted his professional life as well. His security clearance was revoked, and he was forced to terminate his cryptographic work with the GCHQ. In 2016, following successful campaigns to secure an apology and pardon, the British government enacted the Alan Turing law, which retroactively pardoned the thousands of men who were convicted under Section 11 and similar historical legislation.

Turing died by cyanide poisoning in 1954. Though officially ruled a suicide, Turing's death may have been the result of accidental inhalation of cyanide fumes.

*Amanda K. O'Keefe*

*See also:* Chatbots and Loebner Prize; General and Narrow AI; Moral Turing Test; Turing Test.

**Further Reading**

Hodges, Andrew. 2004. "Turing, Alan Mathison (1912–1954)." In *Oxford Dictionary of National Biography*. https://www.oxforddnb.com/view/10.1093/ref:odnb/97801 98614128.001.0001/odnb-9780198614128-e-36578.

Lavington, Simon. 2012. *Alan Turing and His Contemporaries: Building the World's First Computers*. Swindon, UK: BCS, The Chartered Institute for IT.

Sharkey, Noel. 2012. "Alan Turing: The Experiment that Shaped Artificial Intelligence." *BBC News*, June 21, 2012. https://www.bbc.com/news/technology-18475646.

# Turing Test

Bearing the name of computer science pioneer Alan Turing, the Turing Test is a standard of AI that attributes intelligence to any machine capable of exhibiting intelligent behavior equivalent to that of a human. The *locus classicus* for the test is Turing's "Computing Machinery and Intelligence" (1950), which develops a basic prototype—what Turing calls "The Imitation Game." In this game, a human is made to judge which of the two rooms is occupied by a machine and which is occupied by another human, on the basis of anonymized responses to questions the judge puts to each occupant in natural language. Although the human respondent must give truthful answers to the judge's questions, the goal of the machine is to deceive the judge into believing that it is human. According to Turing, the machine may meaningfully be said to be intelligent to the extent that it is successful in this task.

The main advantage to this basically operationalist account of intelligence is that it avoids difficult metaphysical and epistemological questions about the nature and inner experience of intelligent activity. By Turing's standard, no more than empirical observation of outward behavior is the requisite for predicating intelligence of any object. This comes in particularly stark contrast to the broadly Cartesian tradition in epistemology, according to which some internal self-awareness is definitional of intelligence. The so-called "problem of other minds" that results from such a view—namely, how to be certain of the existence of other intelligent beings if it is not possible to know their minds from a supposedly needed first-person perspective—is importantly eschewed on Turing's approach.

Nevertheless, the Turing Test remains tied to the spirit of Cartesian epistemology at least insofar as it conceives of intelligence in a strictly formalist way. The machine referred to in the Imitation Game is a digital computer in Turing's sense: namely, a set of operations that may in principle find instantiation in any sort of material. Specifically, the digital computer is made up of three components: a store of knowledge, an executive unit that carries out individual commands, and a control to regulate the executive unit. But as Turing makes it clear, it is of no essential significance whether these components are materialized via electronic mechanisms or mechanical ones. What is decisive is the formal set of rules that constitute the essence of the computer itself. Turing retains a basic notion that intelligence is fundamentally immaterial. If this much is true, it is reasonable to suppose that human intelligence operates basically like a digital computer and may in principle, therefore, be replicated by artificial means.

This history of AI research ever since Turing's work has divided into two basic camps: those who accept this basic assumption and those who reject it. John Haugeland has coined the phrase "good old-fashioned AI" or GOFAI to characterize the first camp. Notable figures belonging to this approach include Marvin Minsky, Allen Newell, Herbert Simon, Terry Winograd and especially Joseph Weizenbaum, whose program ELIZA was contentiously touted as the first to have successfully passed the Turing Test in 1966.

Nevertheless, critics of Turing's formalism have abounded, especially in the last three decades, and today GOFAI is a much-discredited approach in AI.

Among the most famous critiques of GOFAI broadly—and the assumptions of the Turing Test specifically—is John Searle's *Minds, Brains, and Programs* (1980), in which Searle develops his now-famous Chinese Room thought experiment. The latter imagines a version of the Turing Test, in which a human with no knowledge of Chinese is seated in a room and made to correlate Chinese characters she receives to other Chinese characters she sends out, according to a program scripted in English. Supposing sufficient mastery of the program, Searle suggests that the person inside the room might pass the Turing Test, deceiving a native Chinese speaker into falsely believing that she understood Chinese. If the person in the room is instead a digital computer, Searle's critical thesis is that Turing-type tests fail to capture the phenomenon of understanding, which Searle argues involves more than the mere functionally correct correlation of inputs with outputs.

Searle's critique suggests that AI research must take seriously the questions of materiality in ways the formalism of Turing's Imitation Game neglects. Searle concludes his own discussion of the Chinese Room thought experiment by arguing that the particular physical makeup of human organisms—especially that they possess complex nervous systems, brain tissue, etc.—ought not be dismissed as irrelevant to theories of intelligence. This view has partly inspired an entirely alternative approach in AI known as connectionism, which seeks to construct machine intelligence by modeling the electrical structure of human brain tissue. The successes of this approach have been widely debated, but consensus appears to be that it improves on GOFAI in establishing generalized forms of intelligence.

However, Turing's test is not only subject to critique from the side of materialism but also may be attacked from the direction of renewed formalism. Thus, one may argue that as a standard of intelligence, Turing tests are inadequate precisely because they seek to replicate human behavior, whereas the latter is often highly unintelligent. On strong versions of this criticism, standards of rationality must be derivable *a priori* rather than from actual human practice, if they are to distinguish rational from irrational human behavior in the first place. This line of critique has become particularly pointed as the emphasis of AI research has come increasingly to be laid on the possibility of so-called super-intelligence: forms of generalized machine intelligence that well surpass the human level. Should this new frontier of AI be reached, it would seem to render Turing tests obsolete. Moreover, even discussion of the possibility of super-intelligence would seem to require new standards of intelligence besides strict Turing tests.

Against such criticism, Turing may be defended by observing that it was never his aim to establish any once-and-for-all standard of intelligence. Indeed, by his own lights, the goal is not to answer the metaphysically challenging question "can machines think" but to replace this question with the more empirically verifiable alternative: "What will happen when a machine takes the part [of the man in the Imitation Game]" (Turing 1997, 29–30). Thus, the above-mentioned weakness of Turing's test—that it fails to establish *a priori* standards of rationality—is indeed also part of its strength and motivation. And also no doubt, it explains the long influence it has had over research in all fields of AI since it was first proposed three-quarters of a century ago.

*David Schafer*

*See also: Blade Runner*; Chatbots and Loebner Prize; ELIZA; General and Narrow AI; Moral Turing Test; PARRY; Turing, Alan; *2001: A Space Odyssey.*

**Further Reading**

Haugeland, John. 1997. "What Is Mind Design?" *Mind Design II: Philosophy, Psychology, Artificial Intelligence*, edited by John Haugeland, 1–28. Cambridge, MA: MIT Press.

Searle, John R. 1997. "Minds, Brains, and Programs." *Mind Design II: Philosophy, Psychology, Artificial Intelligence*, edited by John Haugeland, 183–204. Cambridge, MA: MIT Press.

Turing, A. M. 1997. "Computing Machinery and Intelligence." *Mind Design II: Philosophy, Psychology, Artificial Intelligence*, edited by John Haugeland, 29–56. Cambridge, MA: MIT Press.

# Turkle, Sherry (1948–)

Sherry Turkle is trained in sociology and psychology, and her work explores the relationship between people and technology. While her research in 1980s examines how technology impacts the way humans think, her research in the 2000s increasingly critiques the way technology is used to the exclusion of creating and sustaining meaningful interpersonal relationships. The use of artificial intelligence for devices such as children's toys and pets for the elderly are examples she has used to emphasize what humans miss when they interact with such devices.

As a professor at the Massachusetts Institute of Technology (MIT) and the founder of the MIT Initiative on Technology and the Self, Turkle has been at the forefront of advances in AI. In *Life on the Screen: Identity in the Age of the Internet* (1995), she describes a philosophic shift in the understanding of AI that takes place between the 1960s and 1980s, fundamentally altering the way humans relate to and interact with AI. Early paradigms of AI, she argues, used a rule-based model of intelligence that relied on sophisticated preprogramming. However, this view gave way to a paradigm that views intelligence as emergent. This emergent paradigm, which became the accepted mainstream approach by 1990, argues that AI emerges from a much simpler set of algorithms that are able to learn. Most importantly, for Turkle, the emergent approach attempts to mimic the way the human brain works, helping to break down barriers between computers and nature and more broadly between the natural and the artificial. In short, an emergent approach to AI makes it easier for humans to relate to the technology, even thinking of AI-based programs and devices as if they were children.

The growing acceptance of the emergence paradigm of AI and the increased relatability it heralds marks an important turning point, not only for the field of AI but also for Turkle's research and writing on the topic. In two edited collections, *Evocative Objects: Things We Think With* (2007) and *The Inner History of Devices* (2008), Turkle began to use ethnographic research methods to explore the relationship between people and their devices. In *The Inner History of Devices,* she explained that her form of intimate ethnography, or the skill of being able to "listen with a third ear," is needed in order to move beyond the advertising-based platitudes that are normally used when discussing technology. This practice entails creating space for quiet reflection that allows subjects to think deeply about the relationships they have with their devices

In her next major book, *Alone Together: Why We Expect More from Technology and Less from Each Other* (2011), Turkle leveraged these intimate ethnographic methods to make the argument that the growing relationship between humans and the technology that they use is problematic. These problems are linked to both the increasing use of social media as a means of communication and the continued level of comfort and relatability to technological devices, stemming from the emergent paradigm of AI that had become nearly ubiquitous. Here she linked the roots of the problem back to early leaders in the development of cybernetics, noting, for example, Norbert Weiner's musings in his book *God & Golem, Inc.* (1964) on the possibility of sending a human being over a telegraph line. This approach to cybernetic thinking blurs the lines between humans and technology because it reduces both to information.

In terms of AI, this means that it is not important whether the devices we interact with are really intelligent. Turkle argues that by interacting with these devices and taking care of them, we are able to trick ourselves into believing we are in a relationship, which causes us to experience the devices as if they were intelligent. She identified this shift in a 2006 presentation at the Dartmouth Artificial Intelligence Conference titled "Artificial Intelligence at 50: From Building Intelligence to Nurturing Sociabilities." In this presentation, she identified the 1997 Tamagotchi, 1998 Furby, and 2000 My Real Baby as early versions of what she calls relational artifacts, what are more broadly described in the literature as social machines. The stark difference between these devices, and all past children's toys, is that these devices come pre-animated and ready for a relationship, and do not require that children project a relationship on to them. Turkle believes that this shift is concerned as much or more with our human vulnerabilities than with the machines' capabilities. In other words, the very act of caring for an object makes it more likely that one will not only see that object as intelligent but also feel a connection to it. For the average person interacting with these devices, this feeling of connection is more important than the abstract philosophic questions about the nature of its intelligence.

In both *Alone Together* and the *Reclaiming Conversation: The Power of Talk in a Digital Age* (2015), Turkle delved deeper into exploring the consequences of humans interacting with AI-based devices. In *Alone Together*, she gives the example of Adam, who enjoys the gratitude of the AI bots that he rules over in the game *Civilization*. Adam finds this play calming and enjoys that he is able to create something new. Yet, Turkle is quite critical of this interaction, claiming that Adam's playing is not real creation but merely the feeling of creation, and it is problematic because it lacks any true pressure or risk. She extends this argument in *Reclaiming Conversation*, arguing that sociable companions give only a sense of friendship. This is problematic because of the importance of friendship between humans and what might get left out of relationships that only offer a feeling or sense of friendship rather than actual friendship.

This shift is of urgent importance for Turkle. She argues that while there are potentially benefits to relationships with AI-enabled devices, these are relatively minor in comparison to what is missing: the full complexity and inherent contradictions that are part of what it means to be human. The relationship some-one can develop with an AI-enabled device is not as complex as those one can develop with other humans. Turkle argues that as people have become more comfortable and more reliant on technological devices, the very meaning of companionship has shifted. This shift has been responsible for simplifying people's expectations for companionship, reducing the benefits that one hopes to receive from relationships. Now, people are more likely to equate companionship more simply with only the notion of interaction, leaving out more complex feelings and negotiations that are commonly part of relationships. One can develop companionship with devices simply by interacting with them. As human communication has shifted away from face-to-face conversation to interaction that is mediated by devices, the conversations between humans are merely transactional. In other words, interaction is the most that is expected. Drawing on her background in psychoanalysis, Turkle argues that this form of transactional communication means that users spend less time learning to see the world through the eyes of another person, which is an important skill that fosters empathy.

Drawing together these various threads of arguments, Turkle believes we are in a robotic moment in which we long for, and in some cases, we even prefer AI-based robotic companionship to that of other humans. For example, some people enjoy having conversations with the Siri virtual assistant on their iPhones because they don't fear being judged by this device, which is highlighted by a series of Siri ads that feature celebrities talking to their phones. This is problematic for Turkle because these devices can only respond as if they understand the conversation. However, AI-based devices are limited to understanding the literal meanings of the data stored on the device. They can, for example, understand the content of calendars and emails that reside on phones, but they cannot actually understand what any of this data means to the user. For an AI-based device, there is no significant difference between a calendar appointment for car maintenance and one for chemotherapy. Entangled in a variety of these robotic relationships with an increasing number of devices, a person can forget what it means to have an authentic conversation with another human.

While *Reclaiming Conversation* reports on eroding conversation skills and shrinking levels of empathy, it also strikes a hopeful note. Because people are experiencing growing dissatisfaction in their relationships, there may yet be the possibility of reclaiming the important role of face-to-face human communication. Turkle's solutions emphasize decreasing the amount of time that one uses a cell phone, but the role of AI in this relationship is also of importance. Users must acknowledge that the relationships they have with their virtual assistants cannot replace face-to-face relationships. This will require being more deliberate about the way one uses devices, intentionally prioritizing in-person interactions over the quicker and easier interactions provided by AI-enabled devices.

*J. J. Sylvia*

*See also:* Caregiver Robots; Cognitive Psychology, AI and.

**Further Reading**
Turkle, Sherry. 1995. *Life on the Screen: Identity in the Age of the Internet.* New York: Simon and Schuster.

Turkle, Sherry. 2005. *The Second Self: Computers and the Human Spirit*. Cambridge, MA: MIT Press.

Turkle, Sherry. 2015. *Reclaiming Conversation: The Power of Talk in a Digital Age*. New York: Penguin Press.

## *2001: A Space Odyssey* (1968)

*2001: A Space Odyssey* was simultaneously developed as a novel and a motion picture, with both released in 1968. Director Stanley Kubrick and author Arthur C. Clarke worked closely to expand Clarke's 1948 short story "The Sentinel" into a grander narrative of space travel, discovery, and humanity's relative primitiveness. While the novel can be considered a classic, it is overshadowed by Stanley Kubrick's visual masterpiece. Collectively, both texts reveal two of the most preeminent representations of artificial intelligence in popular culture: the HAL 9000 computer.

In the film and novel, HAL's primary function is to operate and regulate the controls of the *Discovery One*—a spacecraft on an interplanetary mission to Jupiter in the film (Saturn in the novel). Initially, HAL is depicted as an essential member of the crew. He empathetically communicates with David Bowman and Frank Poole (the crew's operating astronauts), plays chess with them, and gives them aesthetic judgments. The book explains how Bowman and Poole "could talk to HAL as if he were a human being, and he would reply in the perfect idiomatic English he had learned during the fleeting weeks of his electronic childhood." In the film, HAL is introduced as the subject of a BBC interview, during which he is established as an exceptionally advanced artificial intelligence system. He is able to freely and seamlessly answer the interview questions as a human would.

As the plot of the film and novel progress, however, HAL's benign functioning is thrown into deep suspicion. HAL, whose operating systems are portrayed as infallible, informs the crew of the imminent failure of a satellite control device. Frank Poole investigates but finds nothing wrong. Bowman and Poole show concern over the (mal)functioning of HAL. HAL insists his calculations cannot be wrong. HAL suggests that the inability to identify the failing satellite is the result of human error, not its miscalculations. It is possible to read this episode in at least two ways. HAL could be deliberately deceiving the astronauts as part of a master plan to cut off base communications and bait the astronauts outside the shuttle where it can more easily kill them. A second option is that HAL is legitimately confused about the state of the satellite. Either way, Poole and Bowman discuss shutting off HAL. HAL, however, reads the lips of both men and severs the oxygen to Poole's flight pod. HAL also turns off the life support functions to the crew in suspended animation. Bowman, now the lone remaining human crew member, eventually deactivates HAL's circuits.

The most common reading of this scene, that HAL deliberately acted maliciously toward Poole and Bowman, dovetails into the most prevalent theme in the representation of artificial intelligence in popular culture. That theme is that of complex, "intelligent" machines catastrophically malfunctioning or violently rebelling against humanity. As a result, HAL has commonly been read as a cautionary tale of our fear of the blurring boundaries between human and machine and the expanding autonomy and sophistication of technology.

HAL can also be read less pessimistically, as a representation of the goals, methods, and dreams of the artificial intelligence field. Indeed, one of the founders of artificial intelligence, Marvin Minsky, served as a consultant for the film. Accordingly, HAL depicted (and still depicts) a plausible representation of future artificial intelligence systems. HAL integrates many subfields of artificial intelligence work, including but not limited to visual processing, natural language acuity, and chess playing.

HAL's advanced visual processing is demonstrated in numerous ways. The film contains frequent cuts to HAL's camera eye and gives a subjective view through the camera. HAL is also able to recognize the face of crewmembers in one of Bowman's drawings. He even moves beyond simple face-object recognition and gives aesthetic judgment to Bowman's drawing style. In so doing, HAL demonstrates visual processing capabilities that still surpass what we are able to accomplish today.

One of the most significant signs of intelligence is language, which has interested the field of artificial intelligence since its inception. Throughout the film and the novel, HAL shows an impressive array of language competencies. He is able to understand and make sense of sentences and conversations, and he can generate appropriate responses to social interactions. HAL is further capable of taking part in conversations ranging from simple commands—such as displaying Poole's parents' birthday message—to complex interactions as in expressing inner conflicts and, perhaps, telling cunning lies. In other words, HAL is capable of deciphering the connotative meanings of human interactions, something that remains out of reach for current artificial intelligence systems. Moreover, HAL's language acuity would allow the computer to easily pass the Turing Test. The Turing Test is a trial wherein a confederate is unable to determine whether they are communicating with a human or a robot.

Game playing, especially chess, has been a perennial issue in artificial intelligence. Chess was singled out in early artificial intelligence research because of the game's difficulty, a game for intelligent people. Thus, it was assumed that if computers could play chess, they must also be intelligent. At the time of *2001*'s release, there was plenty of optimism in artificial intelligence research given the early successes in designing machines capable of playing chess. HAL's ability to play the game surpassed artificial intelligence systems from the time. Consequently, the chess playing scenes in the novel and the movie, where HAL is virtually unbeatable, coincided with optimism that in the foreseeable future, there would be chess-playing systems superior to any human player. Today, artificial intelligence has caught up to science fiction with chess-playing supercomputers such as Deep Blue and advanced AI programs such as AlphaZero.

Over fifty years after their release, the novel and film retain the power to inspire awe. In 1991, the film was deemed "culturally, historically, or aesthetically" significant enough to be preserved by the National Film Registry.

*Todd K. Platts*

*See also:* Deep Blue; Minsky, Marvin; Robot Ethics; Turing Test.

**Further Reading**

Brammer, Rebekah. 2018. "Welcome to the Machine: Artificial Intelligence on Screen." *Screen Education* 90 (September): 38–45.

Kolker, Robert, ed. 2006. *Stanley Kubrick's* 2001: A Space Odyssey: *New Essays*. New York: Oxford University Press.

Krämer, Peter. 2010. *2001: A Space Odyssey*. London: British Film Institute.

# U

## Unmanned Ground and Aerial Vehicles

Unmanned vehicles are machines that operate without a human operator physically present onboard. There are a diverse variety of such vehicles that can work in different environments including ground, underground, underwater, airspace, and outer space. Unmanned vehicles are either preprogrammed by algorithms or remotely controlled by a human operator, and they can have varying degrees of autonomy in their operations. Typically, the overall system of an unmanned vehicle consists of sensors that collect data about the surrounding environment and actuators that enable the vehicle to maneuver. After the human operator or the machine program receives the information gathered by sensors, the actuators guide the vehicle in accordance with the commands received from the operator. A major premise of unmanned vehicles is to reach places that are partially or fully unreachable or dangerous for humans and to perform potentially dangerous, dirty, and dull (three D) tasks for humans. Unmanned vehicles also bear the potential of lowering labor costs while increasing work and time efficiency in both military and civilian settings.

The history of unmanned vehicle technology goes back to the nineteenth century. In 1898, the Serbian-American engineer and inventor Nikola Tesla presented a remotely controlled boat at an exhibition at Madison Square Garden in New York City. The miniature boat he built had hull, keel, rudder, electric motor, battery, and an antenna receiver. He could remotely control the vehicle's speed, position, and direction through the radio signals of a radio-transmitting control box. He called his invention teleautomaton and received patent the same year (*Method of an Apparatus for Controlling Mechanism of Moving Vessels or Vehicles*).

Even though Tesla's invention remained largely unnoticed by the military at the time, unmanned aerial and ground vehicles have been widely used in the military for strategic and tactical purposes such as reconnaissance, surveillance, and target acquisition since the early twentieth century. With the beginning of World War I (1914–1918), the idea of remotely controlled unmanned vehicles gained popularity among armies as they were seeking ways both to increase the efficiency of military operations and also to reduce the human costs of combat. Today, diverse military and civilian applications of unmanned vehicles are being developed in a wide array of industries such as agriculture, manufacturing, mining, emergency response, transportation of goods and people, and security operations of police and the military.

Unmanned Ground Vehicles (UGVs) are land-based vehicles that operate on the ground without a human driver inside. They are commonly described as mobile robots. Typically, a UGV consists of sensors, utility and power platform, infrared equipment, cameras, and a communication interface that enables human-machine interaction. Generally, UGVs are either preprogrammed to act autonomously or remotely operated by a human controller or by a computer algorithm. In an ordinary operation, a UGV can gather data about the surrounding environment, sense and avoid obstacles, and follow a desired path to perform a desired mission.

The history of UGVs goes back to the early twentieth century. In 1904, Spanish civil engineer and mathematician Leonardo Torres y Quevedo presented a radio-controlled ground vehicle at the Center for Essays in Aeronautics in Madrid, Spain. The unmanned tricycle he built was a successor of an earlier invention of his from the year before, a small electric boat remotely controlled by a telekino (also known as telekine) system. The remotely controlled tricycle was equipped with an antenna and radiator that received guidance through electromagnetic waves generated by a remote wireless transmitter. Torres y Quevedo's telekino is considered as an early form of today's common control.

A major point of interest in the history of UGV research development after World War I has been its potential to mitigate life-threatening risks on the battlefield. UGVs held the potential to revolutionize military capabilities with their mobility, endurance, maneuverability, and long-distance navigation and communication. Development of UGVs equipped with weapons could carry explosives to the enemy lines. An important moment in the history of UGV development was the early 1920s when the U.S. Army developed a radio-controlled driverless automobile.

Later, in 1958, authorized by President Dwight D. Eisenhower, the United States' Department of Defense started a development program called the Advanced Research Projects Agency (ARPA), which later became the Defense Advanced Research Projects Agency (DARPA). Since then, DARPA has sponsored UGV development and research projects for the military. The agency funded the development of Shakey the Robot in the late 1960s. Developed as a part of Stanford Research Institute's research, Shakey was a robot with a TV camera, a triangulating range finder, and bump sensors. Connected to computers with radio and video links, Shakey had programs for perception, world modeling, and acting. Today, Shakey is considered by many as the baseline for function and performance of robotics with AI, which includes autonomous vehicle development.

In addition to DARPA, according to the National Research Council's Consensus Study Report, several other government agencies have also been using and developing UGVs in the United States. The report states:

> The National Aeronautics and Space Administration (NASA) uses UGVs for planetary exploration; the Department of Energy needs UGVs for nuclear site maintenance and coal mining; the Department of Homeland Security employs UGVs for search and rescue; and the Department of Transportation is developing UGVs as cars, trucks, and buses that can drive themselves or assist a human driver. (National Research Council 2005, 135)

Today, UGVs are used in a wide range of operations in both indoor and outdoor environments. While the military uses UGVs in modern warfare, commercial applications of UGV technology are also being rapidly developed. UGVs are widely used in factories and warehouses across the world. They also have logistical applications in areas such as agriculture and mining. Research and development of self-driving cars (also known as driverless cars) for everyday mobility has been on the rise in the past few decades.

Unmanned Aerial Vehicles (UAVs), popularly known as drones, are aircraft operating without a human pilot on board. UAVs function in an interface called Unmanned Aircraft System (UAS), which refers to the complex system that consists of UAV, remote operator, and the communication link between the two. UAVs are either remotely controlled by an operator on the ground or preprogrammed to fly autonomously. The typical components of a UAV are sensors, camera(s), propellers, flight controller, landing gear, receiver, and transmitter. While smaller UAVs are usually powered with lithium polymer batteries, larger ones have aircraft engines powered with gasoline.

The history of UAVs can be traced back to the late 1910s, only a few years after the Wright brothers' first documented successful piloted heavier-than-air human flight in 1903. The beginning of World War I sparked interest in unmanned aircraft technology across the world to be employed for airdropping, aerial surveillance, and aerial bombing purposes. In the United States, Elmer Sperry and Peter Hewitt developed the pilotless Hewitt-Sperry Automatic Airplane, also known as the Flying Bomb, in 1917. The remote-controlled airplane could carry explosives to targets along enemy lines. In 1918, Orville Wright and Charles F. Kettering from Dayton-Wright Airplane Company developed the pilotless Kettering Aerial Torpedo (also known as Kettering Bug) to carry explosives for aerial bombing. World War I ended shortly after the completion of more than twenty of these Bugs, and so they remained unused in combat. The project eventually was discontinued.

UAV research gained renewed impetus during World War II (1939–1945) as the military worked to reduce the economic and human costs of aerial warfare. In 1946, the U.S military founded the U.S. Air Force Pilotless Aircraft Branch, dedicated to developing UAVs. In the following decades, as UAV and remote sensors technologies became more sophisticated, UAVs found their way into U.S. Air Force and Navy operations. The U.S. military used UAVs extensively during the Vietnam War (1955–1975) to airdrop military supplies, conduct reconnaissance, and engage in target tracking. Nowadays, UAVs are a crucial part of the U.S. military, especially in overseas operations of the War on Terror campaign following the terrorist attacks of September 11, 2001. Predator and Reaper drones are widely used in such military operations.

Recreational and commercial uses of UAVs are a relatively recent development. UAVs available for commercial sale are also cheaper, smaller, and easier to operate than military UAVs. Consumer UAVs' small size, durability, and maneuverability make them ideal for a variety of commercial and social purposes such as aerial imaging, natural resource development, search and rescue operations, wildlife protection, freight delivery, infrastructure maintenance, and crop monitoring. UAV users and developers are still discovering potential commercial and social uses of UAVs. Currently, with its Part 107 Small Unmanned Aircraft Regulations, the Federal Aviation Administration is overseeing commercial, recreational, and government uses of UAVs in the United States.

*Fatma Derya Mentes*

*See also:* Battlefield AI and Robotics; Driverless Cars and Trucks.

## Further Reading

Cheney, Margaret. 1981. *Tesla, Man Out of Time.* New York: Prentice-Hall.

Everett, H. R. 2015. *Unmanned Systems of World Wars I and II.* Cambridge, MA: MIT Press.

Gettinger, Dan, Arthur Holland Michel, Alex Pasternack, Jason Koebler, Shawn Musgrave, and Jared Rankin. 2014. *The Drone Primer: A Compendium of the Key Issues.* Annandale-on-Hudson, NY: Bard College, Center for the Study of the Drone.

National Research Council. 2002. *Technology Development for Army Unmanned Ground Vehicles.* Washington, DC: National Academies Press.

National Research Council. 2005. *Autonomous Vehicles in Support of Naval Operations.* Washington, DC: National Academies Press.

Ni, Jun, Jibin Hu, and Changle Xiang. 2018. *Design and Advanced Robust Chassis Dynamics Control for x-by-wire Unmanned Ground Vehicle.* Williston, VT: Morgan & Claypool.

Rothstein, Adam. 2015. *Drone.* London: Bloomsbury Academic.

# W

## Warwick, Kevin (1954–)

Kevin Warwick, aka Captain Cyborg, is an engineer, researcher, and educator in the fields of robotics, control systems, artificial intelligence, and human-machine interfaces. Before undertaking university studies, Warwick worked for six years as a technician for British Telecom. In 1982, he received his PhD from Imperial College, London, where he was supervised by Panos Antsaklis and John Westcott. His doctoral thesis was on self-tuning controllers, and he has published extensively on control theory. Warwick has held positions at many universities, including Oxford and Reading, where he was Professor of Cybernetics. Warwick is Emeritus Professor and former Deputy Vice Chancellor of Research at Coventry University. Warwick is a member of the European Academy of Sciences and Arts. He has received many awards and honorary doctoral degrees, including the Institution of Engineering and Technology Mountbatten Medal for his work to bring science to the public.

Some of the formative influences on Warwick's research were his father's illness, a severe agoraphobia that was successfully treated by brain surgery, as well as the writings of Michael Crichton, especially *The Terminal Man* (1972), and the film *The Terminator* (1984). Warwick remains fascinated by direct brain interfacing, stimulation, and modification, and he has been a persistent voice of caution about the potential of robotic and AI threats to human life and liberty.

Warwick is a well-published author, having written, coauthored, and edited hundreds of texts. He has produced a steady stream of academic journal articles throughout his career, writing on topics that range from the highly technical to the speculative and philosophical. His publications include textbooks on engineering and control systems, such as the 1995 volume, *Neural Network Applications in Control*, which Warwick edited along with G. W. Irwin and K. J. Hunt. Many of Warwick's book length works are written for broader audiences, such as *QI: The Quest for Intelligence* (2000) and *Artificial Intelligence: The Basics* (2011), and with Huma Shah, *Turing's Imitation Game* (2016).

Warwick has participated in a wide range of media programs and films. Some of his appearances include the *BBC News*, *The Science of Doctor Who*, and the 2006 documentary by Ken Gumbs, *Building Gods*. In 2012, Kevin and Irena Warwick recorded an hour-long episode for the British Library Listening Project, in which they discussed their relationship and collaboration in human-machine interfacing experiments. Warwick also conducted a Reddit AMA (Ask Me Anything) in 2012. The AMA is a well-established mechanism through which prominent figures from any field can field questions directly from users of the popular internet message board. During his AMA, Warwick discussed his movement into and out of the status of a cybernetic entity, as his claim to be the first cyborg has often been contested.

Warwick's status as the first cyborg is a matter of definition. Implantable therapeutic devices such as the pacemaker were developed mid-twentieth century. Steve Mann, another contender for the title of first cyborg, has experimented with wearable sensory enhancement technology since the 1970s. Artist Eduardo Kac implanted himself with a pet registration microchip in 1997. Warwick's first implant in 1998 provided location data to sensors in his lab and office, while his more extensive 2002 implant transmitted signals across networked computers. Warwick acknowledges that his claim to be a cybernetic entity is much stronger in the latter case.

The 1998 Project Cyborg involved the implantation of a chip in his arm. The chip was a simple transmitter, of a type called a radio-frequency identification device (RFID). This chip was placed millimeters deep in his left arm, and the experiment lasted nine days, after which the chip was removed. While the chip was implanted, it communicated with sensors in Warwick's laboratory and office to control environmental conditions such as electric lights. The operation was conducted by Dr. George Boulos and his medical team in Reading. Warwick has written and spoken extensively about experiments he and his collaborators have performed on his body and the body of Irena Warwick, Kevin's wife and partner, in the BrainGate (also known as Utah Array) experiment. In *I, Cyborg* (2002), Warwick provides an autobiographical account of being implanted with a computer chip, and an overview of his research to that point attempts to integrate the human body with machines.

In 2002, a group of neurosurgeons at Radcliffe Infirmary, Oxford, led by Amjad Shad and Peter Teddy, implanted a device connected to the median nerve of Warwick's left arm. The device was the brainchild of Mark Gasson, who earned his PhD under Warwick, and his research team. The implant in Warwick was connected to a computer at Columbia University, in a New York lab, and sent signals to a robot hand in the United Kingdom, at what was then Warwick's home institution of the University of Reading. The hand could be made to open or close following Warwick's movements, and Warwick could receive limited sensory data from the hand.

Warwick's wife Irena has objected to characterizations of her participation as being done under compulsion, emphasizing her determination and enthusiasm in being part of cybernetic experiments, and she was aware of the risks to Kevin and herself. Warwick described the most significant part of the project being that they were able through network transmission to extend his body across the ocean with the New York-Reading experiment. Warwick has spoken with intensity of the extraordinary intimacy of the linkage of his and Irena's nervous systems and expressed a desire to experiment with direct brain connection. Warwick has likened brain linkage to sexual intimacy and suggested that brain-to-brain connections could be influenced by human sexual preferences and aversions. Warwick had the interface in his body for three months prior to the experiment. Some wires remained in his arm from the BrainGate experiment.

Warwick and his collaborators have designed several robots, including Morgui, a skull-like robot that was restricted to visitors over the age of eighteen on the grounds that it was frightening; Hissing Sid, a robotic cat that made headlines when it was barred from air travel by British Airways; and Roger, a robot designed for long-distance running.

Warwick began working on rat brain cells as robotic controllers in 2007. Rat embryo brain cells are harvested to serve as parts in order to study how brain cell networks can be used to control mechanics. The Animat project involved connecting cells to electrodes, with those electrodes transmitting signals from robot sensors.

In 2014, Warwick participated in organizing the Turing Test of a chatbot called Eugene Goostman. In the test, a computer producing a facsimile of human speech attempts to fool judges into evaluating it as a fellow human. The Royal Society-hosted event in London led to bombastic headlines in the press about robot overlords and the test being passed as a huge milestone. Warwick was criticized for participating in the hype when he claimed the test should be considered passed, though he also cautioned that predictions based on the results should be greeted with profound skepticism.

Warwick has been the subject of reporting in *The Telegraph, The Atlantic, The Register, Wired Magazine, Forbes, Scientific American, The Guardian*, and many other publications. Warwick appeared on the cover of the February 2000 issue of *Wired*. He has been covered most intensively by *The Register*, which gave him his nickname, Captain Cyborg. *The Register* has persistently reported on Warwick's experiments, public appearances, and projections about the future of technology, though it was always to tease, heckle, and lampoon. In the course of their reporting, *The Register* has run an assortment of quotes critical of Warwick from scientists in related disciplines. Their criticisms characterize Warwick as a showman and self-promoter, not a serious scientist. Warwick has dismissed such criticisms as unfounded or the result of jealousy, pointing out that people who attempt to make science accessible and exciting to the larger public often suffer similar critiques. Warwick has also been criticized for the uncritical use of the categories of black and white races in his discussion of intelligence in *QI*.

Warwick has raised concerns about the potential for humans to be overtaken by technology; for example, he has warned that designing military robots with enhanced intelligence might overwhelm humanity if they decide to do so. Warwick argues that those who do not make use of implantable tech in the future will be disadvantaged or left behind. This view is predicated on the idea that very powerful computers will be running human societies in the near future and that implantable technology will provide the most effective mechanism of communication with these superintelligent machines. Warwick has made expansive claims about the fate of unaugmented humans in the future, imagining they will become one of a kind like cows, a form of life regarded as fundamentally inferior.

Warwick is a fan of the transhumanist movement, especially the biohackers who work outside of traditional institutions. Beyond his warnings about how humanity might become undone by machines, Warwick anticipates remarkable expansions of the human sensorium through mechanical enhancement. His fondest dream is to transcend speech through technology, merge human minds together to create new intimacies, and create new refined forms of communication.

*Jacob Aaron Boss*

*See also:* Cybernetics and AI.

**Further Reading**

O'Shea, Ryan. 2017. "Kevin Warwick on Project Cyborg." *The Future Grind*, November 24, 2017. https://futuregrind.org/podcast-episodes/2018/5/17/ep-10-kevin-warwick-on-project-cyborg.

Stangroom, Jeremy. 2005. *What Scientists Think*. New York: Routledge.

Warwick, Kevin. 2000. *QI: The Quest for Intelligence*. London: Piatkus.

Warwick, Kevin. 2002. *I, Cyborg*. Champaign: University of Illinois Press.

Warwick, Kevin. 2004. *March of the Machines: The Breakthrough in Artificial Intelligence*. Champaign: University of Illinois Press.

Warwick, Kevin. 2012. *Artificial Intelligence: The Basics*. New York: Routledge.

# Workplace Automation

The term "automation" is derived from the Greek word *automatos,* meaning "acting of one's own will." In a modern economic context, the term is used to describe any digital or physical process that has been designed to perform tasks requiring minimal human input or intervention. While technological development has been continuous over human history, the pervasive level of automation in modern society is a relatively recent development. The replacement of humans by machines performing routine manual labor roles is now the de facto norm.

The modern debate on the pros and cons associated with automation has shifted away from industrial automation (i.e., robotic assembly lines) to the future impact of artificial intelligence—a general-purpose technology with the potential to redefine the nature of human labor. Technologies built upon AI and machine learning are increasingly impacting all sectors of the economy. Autonomous driving systems are expected to make travel in all domains (e.g., land, sea, and air) much safer and more efficient. In health care, artificial intelligence will be able to make disease diagnoses, perform surgeries, and enhance our knowledge of gene editing at a speed and efficiency greater than any human doctor. Financial services are also affected by the introduction of many AI-related innovations, such as robo-advisors—intelligent robots that provide financial advice and investment management services to clients online.

There are two opposing opinions on the long-term effect that the development of AI-based technologies will have on automation and human labor. They can be summarized as follows:

NEGATIVE: Automation due to AI leads to widespread unemployment as increasingly more complex human labor is replaced by machines.

POSITIVE: Automation due to AI replaces some human labor, but simultaneously creates higher-quality jobs in new sectors, which results in a net positive societal outcome.

The types of human labor can be qualitatively described as falling under one of the following categories: (1) routine manual, (2) nonroutine manual, (3) routine cognitive, and (4) nonroutine cognitive (Autor et al. 2003). The routine/nonroutine division refers to how much a job follows a set of predetermined steps, while the manual/cognitive division refers to whether a job requires physical or creative input. As expected, routine manual jobs (e.g., machining, welding, and painting) are the easiest to automate, while nonroutine manual jobs (e.g., operating a vehicle) and nonroutine cognitive jobs (e.g., white-collar work) have traditionally been considered more difficult to automate. But the advent of AI-based technologies, which combine the speed and efficiency of a machine with the agency of a human, is now putting pressure on employment in typically human applications—jobs that require some level of creative decision-making.

For example, a large percentage of the U.S. population is currently employed to drive a vehicle in some capacity. With the current advances in self-driving technology, these jobs are imminently automatable. More generally, Frey and Osbourne (2017) predict that 47 percent of all jobs in the U.S. economy fall in the category of easily automatable over the next two decades.

Meanwhile, 33 percent of jobs fall in the nonroutine cognitive category, which are considered relatively protected from automation (Frey and Osbourne 2017). But even such jobs are under pressure from recent advancements in AI-based technologies, such as deciphering handwriting, decision-making in fraud detection, and paralegal research. For example, in the legal profession, advanced text analysis algorithms are increasingly able to read and detect relevant information among thousands of pages of complex case documents such as license agreements, employment agreements, customer contracts, and common law precedents. This makes human labor, allocated to information processing and synthesizing, redundant. Similar trends can be seen in other white-collar professions where human agency has traditionally been required to perform tasks.

The fear is that the falling costs of robotic technology, coupled with developments in AI-based technology, will accelerate the loss of routine jobs and nonroutine jobs typically allocated to humans across all economic sectors. The process of automation is not inherently negative, but if this widespread replacement of labor is not met by an adequately robust growth in new jobs, the result will be mass unemployment.

On the other hand, a significant number of technology experts and other societal stakeholders argue that such doomsday scenarios in the labor markets will not materialize. In their opinion, the effects of the AI-driven technological revolution will not be dissimilar to those of the industrial revolutions that preceded it. For example, during the Second Industrial Revolution—which saw the introduction of the steam engine, electricity, and associated advanced machinery—many routine jobs, especially in the textile industry, were rendered redundant. This rapidly shifting employment landscape led to the organization of disenfranchised English textile workers, known as the Luddites, to protest the changes. Despite these short-term upheavals, the labor market ultimately adapted rather successfully, which led to the creation of new jobs as a result of these new technologies.

The proponents of AI-driven technological change consider the modern situation to be no different. Susskind and Susskind (2015), among other academics, suggest that the new technologies resulting from an AI-driven technological revolution will lead to a beneficial Schumpeterian innovation effect: The economy will be stimulated with the increased output coming from automation, which will unlock latent demand for professional services as they become cheaper. Acemoglu and Restrepo (2019) mention that while new technologies lead to displacement of labor (i.e., the displacement effect) they may also increase overall productivity (i.e., the productivity effect). This could increase demand for labor in tasks that are not automated or allow labor to move into a broader range of possibilities (i.e., the reinstatement effect). Hence a technological displacement of labor in the short term will be canceled out by the new tasks and possibilities that AI and machine learning will introduce into the labor markets. More fundamentally, Autor (2015) predicts that AI-based technologies are simply too far from being able to replicate human behavior in the foreseeable future. Thus, mass technological unemployment will not become an issue in society.

No matter on which side of the debate one sits, one thing is for certain: the way that universities, apprenticeship systems, social security schemes, and national governments adapt to the rising tide of automation from AI-based technologies will determine the participation of the human workforce in the labor markets of the future. Recent proposals, such as the universal basic income, are just responses to the potential effects of rising unemployment and stagnating wages resulting from automation. Failure to adapt will result in more political divisiveness and economic inequality that will test the cohesion of our societies.

*Edvard P. G. Bruun and Alban Duka*

*See also:* Autonomous and Semiautonomous Systems; Autonomy and Complacency; Brynjolfsson, Erik; Ford, Martin; "Software Eating the World."

**Further Reading**

Acemoglu, Daron, and Pascual Restrepo. 2019. "Automation and New Tasks: How Technology Displaces and Reinstates Labor." *Journal of Economic Perspectives* 33, no. 2 (Spring): 3–30.

Autor, David H. 2015. "Why Are There Still So Many Jobs? The History and Future of Workplace Automation." *Journal of Economic Perspectives* 29, no. 3 (Summer): 3–30.

Autor, David H., Frank Levy, and Richard J. Murnane. 2003. "The Skill Content of Recent Technological Change: An Empirical Exploration." *Quarterly Journal of Economics* 118, no. 4 (November): 1279–1333.

Bruun, Edvard P. G., and Alban Duka. 2018. "Artificial Intelligence, Jobs, and the Future of Work: Racing with the Machines." *Basic Income Studies* 13, no. 2: 1–15.

Brynjolfsson, Erik, and Andrew McAfee. 2011. *Race Against the Machine: How the Digital Revolution Is Accelerating Innovation, Driving Productivity, and Irreversibly Transforming Employment.* Lexington, MA: Digital Frontier Press.

Frey, Carl Benedikt, and Michael A. Osbourne. 2017. "The Future of Employment: How Susceptible Are Jobs to Computerization?" *Journal of Technological Forecasting and Social Change* 114: 254–80.

Manyika, James, Michael Chui, Jacques Bughin, Richard Dobbs, Peter Bisson, and Alex Marrs. 2013. *Disruptive Technologies: Advances that Will Transform Life, Business, and the Global Economy.* McKinsey Global Institute Technical Report.

Smith, Aaron, and Janna Anderson. 2014. "AI, Robotics, and the Future of Jobs." Pew Research Center Report. http://www.pewinternet.org/2014/08/06/future-of-jobs/.

Susskind, Richard E., and Daniel Susskind. 2015. *The Future of the Professions: How Technology Will Transform the Work of Human Experts.* New York: Oxford University Press.

# Y

## Yudkowsky, Eliezer (1979–)

Eliezer Yudkowsky is an artificial intelligence theorist, blogger, and autodidact best known for his commentaries on friendly artificial intelligence. He is cofounder and research fellow at the Machine Intelligence Research Institute (MIRI), founded in 2000 as the Singularity Institute for Artificial Intelligence. He is also a founding director of the World Transhumanist Association. Yudkowsky often writes on the topic of human rationality on the community blog Less Wrong.

Yudkowsky's inspiration for his life's work is the concept of an intelligence explosion first invoked by British statistician I. J. Good in 1965. In his essay "Speculations Concerning the First Ultraintelligent Machine," Good explained that a machine with intellectual capacities surpassing human intelligence should be capable of rewriting its own software to improve itself, presumably until reaching superintelligence. At this point, the machine would design artificial intelligence improvements far beyond the capabilities of any human being. Such a machine would be the last invention of humankind.

Yudkowsky has examined expectations for a coming intelligence explosion in depth. Greater-than-human intelligence, he asserts, depends on creating a machine capable of general intelligence beyond that of human beings. Whole brain emulation or mind uploading experiments such as the Blue Brain Project is considered as one possible outcome. Another is biological cognitive enhancement, which would improve the mental capacities of human beings by genetic or molecular modification. Augmented intelligence by direct brain-computer interfaces is a third possibility. A final approach to the intelligence explosion involves an artificial general intelligence built from neural networks or genetic algorithms, which could, for example, recursively self-improve on themselves. Yudkowsky has said that the conditions and groundwork for superintelligence could be in place by the year 2060.

Yudkowsky has argued that a superintelligent machine might disrupt computer networks, exploit infrastructure vulnerabilities, create copies of itself for purposes of global domination, or even eliminate humanity as a potential threat to its own survival. As a rival, an artificial superintelligence could overturn civilization in the blink of an eye. Thus, superintelligence represents an existential threat to human beings.

On the other hand, a superintelligent machine might resolve intractable human problems such as disease, famine, and war. It might discover ways to take humans to the stars or give them the ability to participate in their own evolution or discover the biological bases for immortality. It would be very difficult to assess the motivations of a superintelligent AI because the machine's ultimate motives would exist beyond human understanding. Yudkowsky has noted "the AI does not hate you, nor does it love you, but you are made out of atoms which it can use for something else" (Yudkowsky 2008, 27).

For this reason, Yudkowsky believes it is important to build safety mechanisms or basic machine morality into artificial intelligences. He calls this research perspective Friendly AI: How do you design machines that will remain friendly to humanity beyond the point of superintelligence? Yudkowsky admits the difficulties in approaching this problem. He notes in particular the superpowers such a machine might be expected to possess and the dangerous and unexpected literal ways it might go about fulfilling its programming.

To address these concerns about a superintelligent AI destroying humanity either inadvertently or by conscious choice, Yudkowsky devised the AI-box experiment. With no real AI oracle available for a test in a virtual prison, Yudkowsky cast himself as the imprisoned AI. He reports that during text-based terminal conversations with human gatekeepers, he has twice experimentally convinced them to let him out of the box. He has not yet published his winning tactics, but it stands to reason that a true superintelligence would be even more persuasive. Even if a first superintelligent AI could be checked, others would likely emerge from other labs in quick succession. It is unlikely that they could all be contained. It is possible that an advanced AI could even exploit currently unknown physics to route around automatic fail-safe control mechanisms. An AI-box experiment forms the basic premise of the 2014 science fiction film *Ex Machina*.

Yudkowsky and MIRI researchers do not hold out much hope that programming basic rules of benevolence into superintelligent machines will prevent catastrophe. A truly advanced machine will realize that programmed constraints (for example, Asimov's Three Laws) are obstacles to the achievement of its goals. It is unlikely that human designers would be able to outthink a rapidly improving machine. Or it might be that the machine achieves its goals in a way that minimizes harm to humans by subverting the basic human condition. If, for example, an AI is programmed to avoid inflicting pain, the caretaking AI might one day devise a way to remove all of the pain receptors from human beings. It is not likely that humans can anticipate every eventuality and write specific rules or goals that prevent infliction of every possible harm or that satisfy every human inclination.

Yudkowsky and his think tank group also believe it is unlikely that machine learning can be used to teach a superintelligence moral behavior. Humans themselves disagree on the morality of individual cases present in human society. A superintelligence may make incorrect decisions in a radically reshaped world, or it may not properly classify the unique sources of data in the ways originally intended from given human judgment datasets.

Instead, Yudkowsky suggests pursuing coherent extrapolated volition as a partial solution to these problems. He advances the proposition that a seed AI itself be given the task of exploring and generalizing from the vast storehouse of present human values in order to determine or make recommendations about where they converge and diverge. The machine would be tasked with identifying and drawing conclusions from our best natures and our best selves. Yudkowsky acknowledges that it may be that there are no places where human wishes for moral clarity and progress coincide.

Not surprisingly, Yudkowsky is interested in the consequences of superintelligence on society, framing the issue around the concept of a coming Singularity, the moment beyond which we live in a world of smarter-than-human intelligences. Yudkowsky frames claims about the Singularity in terms of three major schools of thought. The first is I. J. Good's concept of an Intelligence Explosion leading to a superintelligent AI. A second school, advocated by math professor and science fiction author Vernor Vinge, features an Event Horizon in technological progress where all bets are off and the future becomes truly unpredictable. A third school, Yudkowsky says, is Accelerating Change. In this school, he places Ray Kurzweil, John Smart, and (possibly) Alvin Toffler. The school of Accelerating Change claims that while we expect change to be linear, technological change can be exponential and is therefore predictable.

Yudkowsky has a number of academic publications. His "Levels of Organization in General Intelligence" (2007) uses neural complexity and evolutionary psychology to explore the foundations of what he calls Deliberative General Intelligence. In "Cognitive Biases Potentially Affecting Judgment of Global Risks" (2008), he uses the framework of cognitive psychology to systematically compile all human heuristics and biases that could be of value to appraisers of existential risk. "Artificial Intelligence as a Positive and Negative Factor in Global Risk" (2008) and "Complex Value Systems in Friendly AI" (2011) discuss the complex challenges involved in thinking about and building a Friendly AI. "The Ethics of Artificial Intelligence" (2014), cowritten with Nick Bostrom of the Future of Humanity Institute, assesses the state of the art in machine learning ethics, AI safety, and the moral status of AIs and speculates on the moral quandaries involved in advanced forms of superintelligence.

Yudkowsky was born in Chicago, Illinois, in 1979. In his autobiography, Yudkowsky explains how his interest in advanced technologies emerged from a steady diet of science fiction stories and a borrowed copy of *Great Mambo Chicken and the Transhuman Condition* (1990) by Ed Regis, which he read at age eleven. He remembers being struck by a passage in Vernor Vinge's cyberpunk novel *True Names* (1981) hinting at the Singularity, which led to his self-declaration as a Singularitarian. Yudkowsky dropped out of high school after completing eighth grade. For several years, he tried his hand at programming—including attempts to create a commodities trading program and an AI—before moving to Atlanta in June 2000 to cofound the Singularity Institute with Brian Atkins and Sabine Stoeckel. The Singularity Institute attracted the attention of entrepreneurs and investors Peter Thiel and Jaan Tallinn, and Yudkowsky moved with the institute to the San Francisco Bay Area in 2005. In fan fiction circles, Yudkowsky is well known as the author of the hard fantasy novel *Harry Potter and the Methods of Rationality* (2010), which recasts wizardry as a scientific method in order to advance the cause of rationality over magic.

*Philip L. Frana*

*See also:* *Ex Machina*; General and Narrow AI; Superintelligence; Technological Singularity.

## Further Reading

Horgan, John. 2016. "AI Visionary Eliezer Yudkowsky on the Singularity, Bayesian Brains, and Closet Goblins." *Scientific American*, March 1, 2016. https://blogs .scientificamerican.com/cross-check/ai-visionary-eliezer-yudkowsky-on-the -singularity-bayesian-brains-and-closet-goblins/.

Packer, George. 2011. "No Death, No Taxes." *New Yorker*, November 21, 2011. https:// www.newyorker.com/magazine/2011/11/28/no-death-no-taxes.

Yudkowsky, Eliezer. 2007. "Levels of Organization in General Intelligence." In *Artificial General Intelligence*, edited by Ben Goertzel and Cassio Pennachin, 389–501. New York: Springer.

Yudkowsky, Eliezer. 2008. "Artificial Intelligence as a Positive and Negative Factor in Global Risk." In *Global Catastrophic Risks*, edited by Nick Bostrom and Milan M. Ćirković, 308–45. Oxford, UK: Oxford University Press.

Yudkowsky, Eliezer. 2011a. "Cognitive Biases Potentially Affecting Judgment of Global Risks." In *Global Catastrophic Risks*, edited by Nick Bostrom and Milan M. Ćirković, 91–119. Oxford, UK: Oxford University Press.

Yudkowsky, Eliezer. 2011b. "Complex Value Systems in Friendly AI." In *Artificial General Intelligence: 4th International Conference, AGI 2011, Mountain View, CA, USA, August 3–6, 2011*. New York: Springer.

# Bibliography

Anderson, Susan L. 2008. "Asimov's 'Three Laws of Robotics' and Machine Metaethics." *AI & Society* 22, no. 4 (April): 477–93.

Ashby, W. Ross. 1952. *Design for a Brain: The Origin of Adaptive Behavior.* New York: Wiley.

Asimov, Isaac. 1982. *The Complete Robot.* New York: Doubleday.

Asimov, Isaac, Patricia S. Warrick, and Martin H. Greenberg. 1984. *Machines That Think: The Best Science Fiction Stories about Robots and Computers.* New York: Holt, Rinehart, and Winston.

Barrat, James. 2013. *Our Final Invention: Artificial Intelligence and the End of the Human Era.* New York: Thomas Dunne.

Bekey, George A. 2005. *Autonomous Robots: From Biological Inspiration to Implementation and Control.* Cambridge, MA: MIT Press.

Berkeley, Edmund. 1949. *Giant Brains, or Machines That Think.* New York: Wiley.

Bhuta, Nehal, Susanne Beck, Robin Geiss, Hin-Yan Liu, and Claus Kress, eds. 2016. *Autonomous Weapons: Law, Ethics, Policy.* Cambridge, UK: Cambridge University Press.

Boden, Margaret A., and Ernest A. Edmonds. 2009. "What Is Generative Art?" *Digital Creativity* 20, no. 1–2: 21–46.

Bonabeau, Eric, Marco Dorigo, and Guy Theraulaz. 1999. *Swarm Intelligence: From Natural to Artificial Systems.* New York: Oxford University Press.

Boring, Edwin G. 1946. "Mind and Mechanism." *American Journal of Psychology* 59 (April): 173–92.

Bostrom, Nick. 2005. "A History of Transhumanist Thought." *Journal of Evolution and Technology* 14, no. 1: 1–25.

Bostrom, Nick. 2014. *Superintelligence: Paths, Dangers, Strategies.* Oxford, UK: Oxford University Press.

Brammer, Rebekah. 2018. "Welcome to the Machine: Artificial Intelligence on Screen." *Screen Education* 90: 38–45.

Brockman, John, ed. 2019. *Possible Minds: 25 Ways of Looking at AI.* London: Penguin.

Brooks, Rodney. 2002. *Flesh and Machines: How Robots Will Change Us.* New York: Pantheon.

Brown, Richard, and Kevin S. Decker, eds. 2009. *Terminator and Philosophy: I'll Be Back Therefore I Am.* Hoboken, NJ: John Wiley.

Brynjolfsson, Erik, and Andrew McAfee. 2016. *The Second Machine Age: Work, Progress, and Prosperity in a Time of Brilliant Technologies*. New York: W. W. Norton.

Calo, Ryan. 2017. "Artificial Intelligence Policy: A Primer and Roadmap." *University of California, Davis Law Review* 51: 399–435.

Clarke, Arthur C. 1968. *2001: A Space Odyssey*. London: Hutchinson.

Clarke, Neil. 2017. *More Human Than Human: Stories of Androids, Robots, and Manufactured Humanity*. San Francisco: Night Shade Books.

Colby, Kenneth M. 1975. *Artificial Paranoia: A Computer Simulation of Paranoid Processes*. New York: Pergamon Press.

Conklin, Groff. 1954. *Science-Fiction Thinking Machines: Robots, Androids, Computers*. New York: Vanguard Press.

Cope, David. 2001. *Virtual Music: Computer Synthesis of Musical Style*. Cambridge, MA: MIT Press.

Crevier, Daniel. 1993. *AI: The Tumultuous Search for Artificial Intelligence*. New York: Basic Books.

Crowther-Heyck, Hunter. 2005. *Herbert A. Simon: The Bounds of Reason in Modern America*. Baltimore: Johns Hopkins University Press.

De Garis, Hugo. 2005. *The Artilect War: Cosmists vs. Terrans: A Bitter Controversy Concerning Whether Humanity Should Build Godlike Massively Intelligent Machines*. ETC Publications.

Dennett, Daniel. 1998. *Brainchildren: Essays on Designing Minds*. Cambridge, MA: MIT Press.

Dick, Philip K. 1968. *Do Androids Dream of Electric Sheep?* New York: Doubleday.

Diebold, John. 1995. *Transportation Infostructures: The Development of Intelligent Transportation Systems*. Westport, CT: Greenwood.

Dreyfus, Hubert. 1965. *Alchemy and Artificial Intelligence*. Santa Monica, CA: RAND Corporation.

Dreyfus, Hubert. 1972. *What Computers Can't Do*. New York: Harper & Row.

Dupuy, Jean-Pierre. 2000. *The Mechanization of the Mind: On the Origins of Cognitive Science*. Princeton, NJ: Princeton University Press.

Feigenbaum, Edward, and Julian Feldman. 1963. *Computers and Thought*. New York: McGraw-Hill.

Ferguson, Andrew G. 2017. *The Rise of Big Data Policing: Surveillance, Race, and the Future of Law Enforcement*. New York: New York University Press.

Foerst, Anne. 2005. *God in the Machine: What Robots Teach Us About Humanity and God*. New York: Plume.

Ford, Martin. 2016. *Rise of the Robots: Technology and the Threat of a Jobless Future*. New York: Basic Books.

Ford, Martin. 2018. *Architects of Intelligence: The Truth about AI from the People Building It*. Birmingham, UK: Packt Publishing.

Freedman, David H. 1994. *Brainmakers: How Scientists Are Moving Beyond Computers to Create a Rival to the Human Brain*. New York: Simon & Schuster.

Funge, John D. 2004. *Artificial Intelligence for Computer Games: An Introduction*. Boca Raton, FL: CRC Press.

Gallagher, Shaun. 2005. *How the Body Shapes the Mind*. Oxford, UK: Oxford University Press.

Garcia, Megan. 2016. "Racist in the Machine: The Disturbing Implications of Algorithmic Bias." *World Policy Journal* 33, no. 4 (Winter): 111–17.

Gardner, Howard. 1986. *The Mind's New Science: A History of the Cognitive Revolution*. New York: Basic Books.

Geraci, Robert M. 2008. "Apocalyptic AI: Religion and the Promise of Artificial Intelligence." *Journal of the American Academy of Religion* 76, no. 1: 138–66.

Goertzel, Ben. 2002. *Creating Internet Intelligence: Wild Computing, Distributed Digital Consciousness, and the Emerging Global Brain*. New York: Springer.

Goertzel, Ben, and Stephan V. Bugaj. 2006. *The Path to Posthumanity: 21st-Century Technology and Its Radical Implications for Mind, Society, and Reality*. Bethesda, MD: Academica.

Good, I. J. 1966. "Speculations Concerning the First Ultraintelligent Machine." In *Advances in Computers*, vol. 6, edited by Franz Alt and Morris Ruminoff, 31–88. Cambridge, MA: Academic Press.

Gunkel, David J. 2012. *The Machine Question: Critical Perspectives on AI, Robots, and Ethics*. Cambridge, MA: MIT Press.

Hanson, Robin. 2016. *The Age of Em: Work, Love, and Life When Robots Rule the Earth*. Oxford, UK: Oxford University Press.

Hebb, Donald O. 1949. *The Organization of Behavior*. New York: Wiley.

Hubbard, F. Patrick. 2010. "Do Androids Dream: Personhood and Intelligent Artifacts." *Temple Law Review* 83: 405–74.

Ishiguro, Hiroshi, and Fabio D. Libera, eds. 2018. *Geminoid Studies: Science and Technologies for Humanlike Teleoperated Androids*. New York: Springer.

Kasparov, Garry. 2018. *Deep Thinking: Where Machine Intelligence Ends and Human Creativity Begins*. London: John Murray.

Kelly, John, and Steve Hamm. 2013. *Smart Machines: IBM's Watson and the Era of Cognitive Computing*. New York: Columbia University Press.

Kline, Ronald R. 2017. *The Cybernetics Moment: Or Why We Call Our Age the Information Age*. Baltimore: Johns Hopkins University Press.

Knight, Heather. 2014. *How Humans Respond to Robots: Building Public Policy Through Good Design*. Washington, DC: Brookings Institution.

Kurzweil, Ray. 1999. *The Age of Spiritual Machines: When Computers Exceed Human Intelligence*. New York: Penguin.

Kurzweil, Ray. 2005. *The Singularity Is Near: When Humans Transcend Biology*. New York: Viking.

Lewis, Seth C., Andrea L. Guzman, and Thomas R. Schmidt. 2019. "Automation, Journalism, and Human-Machine Communication: Rethinking Roles and Relationships of Humans and Machines in News." *Digital Journalism* 7, no. 4: 409–27.

Lin, Patrick, Keith Abney, and George A. Bekey. 2012. *Robot Ethics: The Ethical and Social Implications of Robotics*. Cambridge, MA: MIT Press.

Lin, Patrick, Ryan Jenkins, and Keith Abney. 2017. *Robot Ethics 2.0: New Challenges in Philosophy, Law, and Society*. New York: Oxford University Press.

Lipson, Hod, and Melba Kurman. 2016. *Driverless: Intelligent Cars and the Road Ahead*. Cambridge, MA: MIT Press.

McCarthy, John. 1959. "Programs with Common Sense." In *Mechanisation of Thought Processes: Proceedings of the Symposium of the National Physics Laboratory*, 77–84. London: Her Majesty's Stationery Office.

McCarthy, John, and Patrick J. Hayes. 1969. "Some Philosophical Problems from the Standpoint of Artificial Intelligence." In *Machine Intelligence*, vol. 4, edited by Donald Michie and Bernard Meltzer, 463–502. Edinburgh, UK: Edinburgh University Press.

McCorduck, Pamela. 1979. *Machines Who Think: A Personal Inquiry into the History and Prospects of Artificial Intelligence*. San Francisco: W. H. Freeman.

McCorduck, Pamela. 1990. *Aaron's Code: Meta-Art, Artificial Intelligence, and the Work of Harold Cohen*. New York: W. H. Freeman.

McCulloch, Warren, and Walter Pitts. 1943. "A Logical Calculus of the Ideas Immanent in Nervous Activity." *Bulletin of Mathematical Biophysics* 5: 115–37.

Menges, Achim, and Sean Ahlquist. 2011. *Computational Design Thinking: Computation Design Thinking*. Chichester, UK: Wiley.

Mindell, David A. 2015. *Our Robots, Ourselves: Robotics and the Myths of Autonomy*. New York: Viking.

Minsky, Marvin. 1961. "Steps toward Artificial Intelligence." *Proceedings of the IRE* 49, no. 1 (January): 8–30.

Minsky, Marvin. 1982. "Why People Think Computers Can't." *AI Magazine* 3, no. 4 (Fall): 3–15.

Minsky, Marvin. 1986. *The Society of Mind*. New York: Simon & Schuster.

Moravec, Hans. 1988. *Mind Children: The Future of Robot and Human Intelligence*. Cambridge, MA: Harvard University Press.

Moravec, Hans. 1999. *Robot: Mere Machine to Transcendent Mind*. Oxford, UK: Oxford University Press.

Newell, Allen. 1990. *Unified Theories of Cognition*. Cambridge, MA: Harvard University Press.

Newell, Allen, and Herbert A. Simon. 1961. "Computer Simulation of Human Thinking." *Science* 134, no. 3495 (December 22): 2011–17.

Nilsson, Nils. 2009. *The Quest for Artificial Intelligence: A History of Ideas and Achievements*. Cambridge, UK: Cambridge University Press.

Nocks, Lisa. 2008. *The Robot: The Life Story of a Technology*. Baltimore: Johns Hopkins University Press.

Norvig, Peter, and Stuart J. Russell. 2020. *Artificial Intelligence: A Modern Approach*. Fourth Edition. Upper Saddle River, NJ: Prentice Hall.

Pasquale, Frank. 2016. *The Black Box Society: The Secret Algorithms that Control Money and Information*. Cambridge, MA: Harvard University Press.

Pfiefer, Rolf, and Josh Bongard. 2007. *How the Body Shapes the Way We Think*. Cambridge: MIT Press.

Pias, Claus. 2016. *The Macy Conferences, 1946–1953: The Complete Transactions*. Zürich, Switzerland: Diaphanes.

Pinker, Steven. 1997. *How the Mind Works*. New York: W. W. Norton.

Reddy, Raj. 1988. "Foundations and Grand Challenges of Artificial Intelligence." *AI Magazine* 9, no. 4: 9–21.

Reese, Byron. 2018. *The Fourth Age: Smart Robots, Conscious Computers, and the Future of Humanity*. New York: Atria Books.

Saberhagen, Fred. 1967. *Berserker*. New York: Ballantine.

Sammon, Paul S. 2017. *Future Noir: The Making of Blade Runner*. New York: Dey Street.

Searle, John. 1984. *Mind, Brains, and Science*. Cambridge, MA: Harvard University Press.

Searle, John. 1990. "Is the Brain a Digital Computer?" *Proceedings and Addresses of the American Philosophical Association* 64, no. 3 (November): 21–37.

Selbst, Andrew D., and Solon Barocas. 2018. "The Intuitive Appeal of Explainable Machines." *Fordham Law Review* 87, no. 3: 1085–1139.

Shanahan, Murray. 2015. *The Technological Singularity*. Cambridge, MA: MIT Press.

Simon, Herbert A. 1991. *Models of My Life*. New York: Basic Books.

Singer, Peter W. 2009. *Wired for War: The Robotics Revolution and Conflict in the 21st Century*. London: Penguin.

Sladek, John. 1980. *The Complete Roderick*. New York: Overlook Press.

Søraa, Roger A. 2017. "Mechanical Genders: How Do Humans Gender Robots?" *Gender, Technology, and Development* 21, no. 12: 99–115.

Sparrow, Robert. 2007. "Killer Robots." *Journal of Applied Philosophy* 24, no. 1: 62–77.

Tambe, Milind, and Eric Rice. 2018. *Artificial Intelligence and Social Work*. Cambridge, UK: Cambridge University Press.

Tegmark, Max. 2017. *Life 3.0: Being Human in the Age of Artificial Intelligence*. New York: Knopf.

Togelius, Julian. 2019. *Playing Smart: On Games, Intelligence, and Artificial Intelligence*. Cambridge, MA: MIT Press.

Townsend, Anthony. 2013. *Smart Cities: Big Data, Civic Hackers, and the Quest for a New Utopia*. New York: W. W. Norton.

Turing, Alan. 1950. "Computing Machinery and Intelligence." *Mind* 59, no. 236 (October): 433–60.

Turkle, Sherry. 2011. *Alone Together: Why We Expect More from Technology and Less from Each Other*. New York: Basic Books.

Turner, Jacob. 2018. *Robot Rules: Regulating Artificial Intelligence*. London: Palgrave Macmillan.

Ulam, Stanislaw. 1958. "Tribute to John von Neumann." *Bulletin of the American Mathematical Society* 64, no. 3, pt. 2 (May): 1–49.

Vinge, Vernor. 1993. "The Coming Technological Singularity: How to Survive in the Post-Human Era." In *Vision 21: Interdisciplinary Science and Engineering in the Era of Cyberspace*, 11–22. Cleveland, OH: NASA Lewis Research Center.

Weaver, John F. 2014. *Robots Are People Too: How Siri, Google Car, and Artificial Intelligence Will Force Us to Change Our Laws*. Santa Barbara, CA: Praeger.

Weizenbaum, Joseph. 1976. *Computer Power and Human Reason: From Judgment to Calculation*. San Francisco: W. H. Freeman.

Wiener, Norbert. 1948. *Cybernetics, or Control and Communication in the Animal and the Machine*. New York: Wiley.

Williamson, Jack. 1947. "With Folded Hands." *Astounding Science Fiction* 39, no. 5 (July): 6–44.

Wosk, Julie. 2015. *My Fair Ladies: Female Robots, Androids, and Other Artificial Eves*. New Brunswick, NJ: Rutgers University Press.

Yudkowsky, Eliezer. 2008. "Artificial Intelligence as a Positive and Negative Factor in Global Risk." In *Global Catastrophic Risks*, edited by Nick Bostrom and Milan M. Ćirković, 308–45. New York: Oxford University Press.

Zuboff, Shoshana. 2018. *The Age of Surveillance Capitalism: The Fight for a Human Future at the New Frontier of Power*. London: Profile Books.

# About the Editors

**Philip L. Frana,** PhD, is Associate Professor of Interdisciplinary Liberal Studies and Independent Scholars and Associate Dean and Academic Unit Head of the Honors College at James Madison University, Harrisonburg, VA. He is the current president of the Virginias Collegiate Honors Council and past executive secretary-treasurer of the Southern Regional Honors Council. He is also science programming lead for the Museum of Science Fiction's Escape Velocity Conference. He has published more than a dozen articles on the history of computing, medicine, and public health. He is currently finishing a project related to the politics of big data and beginning a history of comparisons between computing and devilry.

Dr. Frana teaches interdisciplinary honors seminars on the past, present, and future of artificial intelligence, robotics, and automation. He is also the faculty lead on James Madison University's honors area of Emphasis Course Sequence on Creativity and Innovation. He posts daily on subjects related to AI and computational creativity at twitter.com/ArtificialOther.

Dr. Frana received a doctorate in the history of technology and science from Iowa State University and completed postdoctoral training at the Charles Babbage Institute for Computing, Information & Culture at the University of Minnesota-Twin Cities. At CBI, he also managed the four-year National Science Foundation-funded project "Building a Future for Software History." He received his bachelor's degrees in history and economics from Wartburg College in Waverly, Iowa.

**Michael J. Klein,** PhD, is Associate Professor of Writing, Rhetoric, and Technical Communication at James Madison University in Harrisonburg, VA. Dr. Klein directs the Cohen Center for the Humanities, a university center focused on humanistic inquiry across disciplines and the support of graduate education through scholarships, travel grants, and learning opportunities. He is also the founder and coordinator of the interdisciplinary minor in medical humanities, which comprises twelve academic units across three colleges. He teaches courses in technical communication, scientific and medical communication, and writing in the health sciences.

Dr. Klein's recent scholarship has focused on medical narratives and intercultural communication and the creation of graphic embodiment memoirs in an interdisciplinary writing course. His forthcoming book, *Effective Teaching of Technical Communication: Theory, Practice, and Application* (2021), is a collection of essays on innovative pedagogical strategies and approaches by wide-ranging group of educators in the field of technical communication (WAC Clearinghouse).

Dr. Klein holds a doctorate in science and technology studies (Virginia Tech) and master's degrees in rhetoric and composition (University of Arizona) and technical communication (Rensselaer). His dissertation on the human cloning debate analyzed how the media's references to literature and film served as anti-cloning tropes.

# List of Contributors

Rachel Adams
*Human Sciences Research Council (South Africa)*

Sigrid Adriaenssens
*Princeton University*

Hamza Ahmad
*McGill University (Canada)*

Vincent Aleven
*Carnegie Mellon University*

Antonia Arnaert
*McGill University (Canada)*

Enrico Beltramini
*Notre Dame de Namur University*

Jacob Aaron Boss
*Indiana University*

Mat Brener
*Penn State University*

Edvard P. G. Bruun
*Princeton University*

Juliet Burba
*University of Minnesota*

Angelo Gamba Prata de Carvalho
*University of Brasília (Brazil)*

Shannon N. Conley
*James Madison University*

Zoumanan Debe
*McGill University (Canada)*

Kanta Dihal
*University of Cambridge*

Yeliz Doker
*Bournemouth University (United Kingdom)*

Evan Donahue
*Duke University*

Alban Duka
*University of New York Tirana (Albania)*

Jason R. Finley
*Fontbonne University*

Batya Friedman
*University of Washington*

Fatma Güneri
*Lille Catholic University (France)*

David J. Gunkel
*Northern Illinois University*

Andrea L. Guzman
*Northern Illinois University*

Heiko Hamann
*University of Lübeck (Germany)*

Mihály Héder
*Budapest University of Technology and Economics (Hungary)*

Kenneth Holstein
*Carnegie Mellon University*

Laci Hubbard-Mattix
*Washington State University*

Ming-Yu Bob Kao
*Queen Mary University of London (United Kingdom)*

Argyro Karanasiou
*University of Greenwich (United Kingdom)*

Oliver J. Kim
*University of Pittsburgh*

Roman Krzanowski
*The Pontifical University of John Paul II (Poland)*

Victoriya Larchenko
*National Technical University "Kharkiv Polytechnic Institute" (Ukraine)*

Brenda Leong
*Future of Privacy Forum*

John Liebert
*Private Practice of Psychiatry (USA)*

Jeffrey Andrew Thorne Lupker
*Western University (Canada)*

Wing Kwan Man
*Bournemouth University (United Kingdom)*

M. Alroy Mascrenghe
*Informatics Institute of Technology (Sri Lanka)*

Crystal Matey
*University of North Georgia*

Fatma Derya Mentes
*Duke University*

Raymond J. Miran
*Northern Illinois University*

Jason H. Moore
*University of Pennsylvania*

Samantha Noll
*Washington State University*

Amanda K. O'Keefe
*Citigroup*

Anna Stephanie Elizabeth Orchard
*Bournemouth University (United Kingdom)*

Jacob D. Oury
*Indiana University*

William R. Patterson
*U.S. Department of State*

Todd K. Platts
*Piedmont Virginia Community College*

Pauline Quenescourt
*Lille Catholic University (France)*

Gwenola Ricordeau
*Cal State University–Chico*

Frank E. Ritter
*Penn State University*

Diane M. Rodgers
*Northern Illinois University*

Raphael A. Rodriguez
*Penn State University*

Joan Spicci Saberhagen
*Independent Scholar*

Konstantinos Sakalis
*University of Athens (Greece)*

David Schafer
*Western Connecticut State University*

Craig I. Schlenoff
*National Institute of Standards and Technology*

David M. Schwartz
*Penn State University*

J. J. Sylvia
*Fitchburg State University*

Farnaz Tehranchi
*Penn State University*

Michael Thomas
*Cruise LLC*

Christopher Tozzi
*Rensselaer Polytechnic Institute*

Stefka Tzanova
*York College*

Ikechukwu Ugwu
*Bournemouth University (United Kingdom)*

Steven Umbrello
*University of Torino (Italy)*

Elisabeth Van Meer
*College of Charleston*

Brett F. Woods
*American Public University System*

Robin L. Zebrowski
*Beloit College*

# Index

Page numbers in **bold** indicate the location of main entries.

AARON, **1–2**
Aaronson, Scott, 279
Accidents and Risk Assessment, **2–4**, 32, 131–133, 239, 326
Action-based robotics, 55
ACT-R (Adaptive Control of Thought-Rational), 83–85
Actroid robot, 195–197
Adaptive AI, 140
Advanced Soldier Sensor Information Systems and Technology (ASSIST), **4–7**
Advice Taker, 220
Aegis Combat System (ACS), 34–35
Agents, 124–126, 156, 179, 191–192, 202, 205, 225–230, 256–258
Agriculture, 76, 183–185, 265
*A.I. Artificial Intelligence*, 78–79, 236
AI-box experiment, 349
AI for Social Good, 41, 317
AI in Education (AIED), 189
AI Safety, 258, 350
AI Winter, **7–8**
Air traffic control, **9–10**
Air Travel Information System (ATIS), 246
Akscyn, Robert, 193
Alchemy and Artificial Intelligence, **10–12**, 211–212
Aldebaran Nao robot (SoftBank Group), 198–200, 304
Algorithmic bias and error, **12–14**, 297. *See also* Bias
Algorithmic black box, 13, 148
Algorithmic composition and music, 91–93, 142, 168–170
All of Us Research Program, 269
Allen, Colin, 284
Alpha-beta pruning, 110, 211, 291

al-Rifaie, Mohammad Majid, 93
Alvarez-Rodriguez, Unai, 279
Amazon, 12, 59; Alexa, 112, 157–158, 227, 229, 262; Comprehend, 72; Dot, 299; Echo, 229, 296
Anantharaman, Thomas, 110
Anderson, John, 84–85
Anderson, Susan Leigh, 255, 283
Andreessen, Marc, 300–301
Androids, 48–50, 194–197, 253–255, 262, 306
Animal consciousness, **14–17**; personhood, 16–17, 253
Animat project, 343
Anime, 65, 70, 303
Ant colony optimization (ACO), 126
Anthropomorphism, 66, 198–200, 229, 261–262
Apple, 48, 71, 75, 159; Siri, 73, 86, 112, 227–229, 333
Architecture, 75, 164–167
ARMOR software, 317
Art, 1–2, 165–166
Artificial brains, 106–109, 234
Artificial evolution, 14, 53, 78, 150, 170, 223; Berserkers, 42; de Garis, Hugo, 106–109; self-evolution, 308, 320; Sloman, Aaron, 293–294; swarm robotics, 127. *See also* Evolution; Genetic algorithms
Artificial general intelligence (AGI), 11, 160–162, 179, 238, 318, 325; Goertzel, Ben, 172–174; Omohundro, Steve, 256–258; Yudkowsky, Eliezer, 348
Artificial intelligence in medicine (AIM), 175–176, 222–223, 231. *See also* Medicine

Artificial life, 57, 78, 109, 123, 150, 178, 217; artificial quantum life, 279

Artificial neural networks (ANNs), 106, 112, 120, 225, 236, 277; cybernetics, 102–103; machine translation, 217; medicine, 222. *See also* Neural networks

Ashby, W. Ross, 217–218

Asimov, Isaac, **17–20**, 39–40, 254, 262, 283, 305, 317, 324

Assistive technology, 63–70

Atkinson and Shiffrin model, 89

ATTENDING system, 100

Austin, George, 89

Automata, 102, 303

Automated machine learning (AutoML), **20–23**

Automated multiphasic health testing (AMHT), **23–24**

Automated narrative generation systems, 91, 244

Automated trading software, 3, 86, 114, 122, 150, 350

Automatic film editing, **24–26**

Automatic Language Processing Advisory Committee (ALPAC), 216

Autonomous and semiautonomous systems, **26–30**

Autonomous capitalism, 258

Autonomous gaming agents, 212

Autonomous robotics, 27–28

Autonomous vehicles. *See* Driverless cars and trucks

Autonomous weapons systems (AWS), ethics of, 3, 28, **30–32**, 35–36, 158, 207–210. *See also* Lethal autonomous weapons systems

Autonomy and complacency, **32–33**

Autoverse, 123

Avatars, 229

Bach, Johann Sebastian, 91, 142, 170

Backpropagation, 120, 180, 214, 222

Backward chaining, 145, 241

Bar-Hillel, Yehoshua, 215

Barnett, G. Octo, 193

Basic AI drives, 257

Bateson, Gregory, 217–218

Battlefield AI and robotics, **34–37**

*Battlestar Galactica*, 158, 306

Bayes, Thomas, 37

Bayesian inference, **37–39**; classifiers, 86; cognitive models, 97–98, 190;

conditional probability, 193–194; estimation, 100; optimization, 20

BEAM (biology, electronics, aesthetics, and mechanics) robotics, 323–324

Beam search, 246

Beauchamp, James, 142

BECCA (brain-emulating cognition and control architecture), 84

Behavioral economics, 61, 150, 256–257

Behavior-based robotics and AI, 55–57

Behaviorism, 88–89

Beneficial AI, **39–40**

Berger-Wolf, Tanya, **41–42**

Berkeley, Edmund, 88, 170–172

Bernstein, Ethan, 277

Berserkers, **42–44**

Bertillon, Alphonse, 44

Bias, 12–14, 131, 147, 152, 251; gender bias, 157–159; policing, 272–273, 297

Bible concordance and translation, 175, 216

Bina48 robot, 122, 305

Biometric privacy and security, **44–47**

Biometric technology, 12, **47–48**

Biomorphic robots, 323–324

BIONET, 231

BioRC (Biomimetic Real-Time Cortex) Project, 121

Bishop, Mark, 93

Blackboard architectural model and system architectures, 145, 282

*Blade Runner*, **48–50**, 158, 262

Blockchain AI, 172, 258

Blois, Marsden, 175

Blue Brain Project, **50–51**, 315, 348

BlueGene supercomputer, 51, 315

Bobrow, Daniel, 225–226

Boden, Margaret, 90, 294

Bongard, Josh, 137

Boole, George, 39, 313

Boolean algebra, 171, 218, 313–314

Booth, Andrew D., 215

Boring, Edwin, 312

Bostrom, Nick, **52–55**, 257, 308–310, 320, 350

Boucher, Anthony, 305

Boulez, Pierre, 169

Boulos, George, 342

Bounded rationality, 292

Brain-computer interfacing (BCI), 174, 238, 348

BrainGate (Utah Array) experiment, 342

BrainScaleS, 181
Breazeal, Cynthia, 56, 199, 229
Breslow, Lester, 23
British House of Lords Artificial
    Intelligence Committee, 19
Broadbent, Donald, 88
Brooks, Rodney, **55–58**, 135–137, 324
Browne, Mike, 110
Bruner, Jerome, 89
Brutlag, Douglas, 231
Brynjolfsson, Erik, **58–60**, 266
Buchanan, Bruce, 240–241, 291
Buddhism, 303
Bunraku, 302
Buolamwini, Joy, 159

Cage, John, 168–169
Calo, Ryan, **61–62**
Calverley, David, 254
Cambridge Declaration on Consciousness,
    14, 16
Campaign to Stop Killer Robots, 30,
    **62–63**, 209
Campbell, Murray, 110–111
CAPS (Collaborative, Activation-based,
    Production System), 84
Carboncopies, 122
Card, Stuart, 249
Caregiver robots, **63–71**, 194
Carley, Kathleen, 250
Carnevale, Ted, 51
Cartesian dualism, 116, 309, 329
Case-based reasoning, 25, 146
Caselles-Dupré, Hugo, 92
CASNET expert system, 145
CASPER affect management agent, 68
Cellular automata, 107, 123
Cellular Automata Machine, 107
Certainty factors, 241
Champernowne, D. G., 328
Chatbots, **71–74**, 147, 227–228, 245,
    260–261, 298, 343
Checkers, 232
Cheng, Lili, **74–75**
Chess, 11, 91, 105, 117, 218, 221, 291, 309;
    ChipTest, 110; Deep Blue, 91, 110–112,
    179, 211; HAL 9000, 334–335; Mac
    Hack, 11, 211–212; Stockfish, 180;
    Turochamp, 328
CheXpert, 101
Chicago Police Department Strategic
    Subject List (SSL), 271, 273

China-Brain Project, 109
Chinese Room argument, 92, 330
Chomsky, Noam, 89, 170
Christian Transhumanist Association, 304
Christianity, 303–306
Church, Alonzo, 225, 327
Church of Perpetual Life, 305
Church-Turing thesis, 277
Circumscription, 221
CLARION (Connectionist Learning with
    Adaptive Rule Induction On-line), 83
Clarke, Arthur C., 122, 334
Classification, 5–7, 21, 81–82, 97, 113–114,
    159, 201, 277
CLAVIER expert system, 291
Clerwall, Christer, 91
Climate crisis, **75–79**, 188, 317
Clinical decision support systems, **79–83**,
    193. *See also* Artificial intelligence in
    medicine; Computer-assisted diagnosis;
    Medicine
CLIPS expert system, 145
Clowes, Max, 294
Clynes, Manfred, 123
Coeckelbergh, Mark, 285
Cog robot, 56, 136–137, 151, 304
Cognitive science, 134, 163–164, 189, 218,
    223–226, 229, 248–250; android
    science, 195–197; animal cognition,
    14–17; architectures, 83–90; behavior-
    based, 55–57; biological information
    processing, 293–294; computational
    neuroscience, 95; computing, 85–89;
    cybernetics, 103, 105; embodiment, 136,
    138; interaction for agents, 191–192;
    neuroscience, 178–179; philosophy,
    117–118, 212; symbol grounding
    problem, the, 311; symbol manipulation,
    225, 260, 311–313, 314; SyNAPSE,
    314–316. *See also* Agents; Cognitive
    psychology
Cognitive agents, interaction for, 191–192
Cognitive architectures and models,
    **83–85**, 89, 96–97, 174, 191–192, 257,
    291, 311
Cognitive computing, 76, **85–88**, 315–316
Cognitive psychology, **88–90**, 95, 217, 224,
    292, 312, 350
Cognitive Revolution, 15, 88–89
Cohen, Harold, 1–2
Cohen, Michael F., 25
Colby, Kenneth, 73, 135, 260–261, 312

Collen, Morris F., 23
Colton, Simon, 90, 92–93
Combinatorial explosion, 21, 127
Comedy, 198–199
CommonKADS, 202
Companionship, 65, 67–69, 118, 153, 196–197, 200, 332–333
Complexity theory, 105, 107, 109, 248, 258, 277, 292
Composition, 90–93, 142, 168–170
Computational creativity, **90–95**, 165–167
Computational linguistics, 215, 225, 243, 247
Computational neuroscience (CNS), 15, 51–52, 90, **95–98**, 122, 159, 179, 315
Computational sustainability, 76–77
Computer-aided design (CAD), 166
Computer analysis of natural scenes, 282
Computer-assisted diagnosis, 38, **98–102**
Computer-generated art, 90
Computer Graphic exhibit, 165
Computer vision, 41, 112–113, 131, 225, 234–235, 239, 295
*Computers and Automation*, 171–172
Configurable role limiting methods (CRLM), 201
Connectionism, 90, 135, 330
Connectionist systems, 83–84, 102, 120, 217, 311
Connectomics, 124
Consciousness, 57, 137, 254; animal, 14–17; fringe, 11; machine, 205, 236, 252, 321–322; material origins, 120; mind uploading, 122–123, 234; philosophy, 116–118; sentience, 14, 17
Conway, John, 123
Cope, David, 91–94, 141–142, 170
Creativity, 90–95, 165–167, 257
Crime Anticipation System, 271
CrimeScan, 271
Cryptocurrency, 3, 173, 258
Cryptography, 258, 277, 327–328
Cuban, Mark, 264
Cupp, Richard, 17
CyBeRev (Cybernetic Beingness Revival) project, 121–122
Cybernetics, **102–104**, 105–106, 217–219, 221, 332, 341–344
Cyberpunk, 268, 287–289, 350
Cyborgs, 43, 103, 107–108, 123, 138, 174, 303, 341–344
Cyc project, 86

Dance, 199, 286
Darling, Kate, 285
Dartmouth AI conference (Summer Research Project), 103, **105–106**, 218, 220, 224, 312
Data analytics, 20, 24, 45, 74, 77, 295, 299; bias in big data, 250–252
Davis, Marc, 25
Dayton-Wright Airplane Company, 339
de Barros, Joao, 44
de Garis, Hugo, **106–110**
De Santo Caro, Hugo, 175, 303
Decision analysis, 99–100
Decision theory, 218
Decision tree algorithms, 21, 99, 145, 213–214, 228
Declarative Camera Control Language (DCCL), 25
Deep Blue, 91, **110–112**, 179, 211
Deep learning, 24–25, 41, 101, **112–114**, 178–180, 247, 317; deep neural networks (DNN), 101, 180; generative adversarial networks (GANs), 46, 77, 92, 170
Deep Thought, 110–111, 211
DeepDream, 165–166
Deepfakes, 46–47
DeepMind, 24; AlphaFold, 180; AlphaGo, 91, 93, 178–180; AlphaStar, 83, 180; AlphaZero, 180
Defense Advanced Research Projects Agency (DARPA), 56, 134, 145, 158, 221, 245–246, 315–316, 338
Deliberative General Intelligence, 256, 350
DENDRAL expert system, **114–115**, 146, 230, 241, 291
Dennett, Daniel, 17, **115–118**
DESIRE (design and specification of interacting reasoning components) language, 203
Deutsch, David, 276–277
Diamandis, Peter, **118–119**
Dice games, 142, 168
DiCenso, Alba, 80
Dick, Philip K., 48, 50
Digital immortality, 54, **119–124**, 205, 288, 304
Dijkstra's algorithm, 186
Distributed and swarm intelligence, **124–128**
Djerassi, Carl, 114–115
DOCTOR chatbot, 134–135
Drexler, K. Eric, 226, 265, 267

Dreyfus, Hubert L., 10–11, 91, 211–212
Driverless cars and trucks, 27, **128–131**, 132–133, 235–236, 239, 301, 338; Automated and Electric Vehicles Act, 133; automatic collision avoidance, 3–4, 9, 27, 57, 187, 295; connected vehicles, 128, 295; intelligent driver feedback, 76; obstacle detection, 29; platooning, 76, 187; Trolley Problem, 326–327
Driverless vehicles and liability, **131–133**
Drones, 30–31, 34, 62, 207–208, 257, 268, 317, 339
Dubinsky, Donna, 122
Dudley, Homer, 245
DXplain expert system, 193

Ecology, 41, 76–79, 150, 217, 265. *See also* Environment
Economics, 58–61, 153–155, 248, 256–258, 344–347; complexity economics, 150; post-scarcity, 264–269
Edelman, David, 16
Edmonds, Dean, 224
Education, 74, 87, 154–155, 189–191, 205, 227–228
Electronic medical records (EMR), 23–24, 296
Electronic persons, 255, 285
ELIZA chatbot, 73, **134–135**, 260, 329
Embodiment, 55, 127, **135–138**, 152, 159, 304, 321, 325
Emergence and emergent systems, 1–2, 84, 138, 224–226, 331
Emergent gameplay and non-player characters, **138–141**
Emily Howell program, **141–142**
Emmy, 91–92, 94, 141–142
Empathy, 36, 43, 49–50, 70, 77, 81, 158, 333–334
Employment and unemployment, 59, 154, 184, 236, 320, 344–346
EMYCIN expert system, 240–241
Energy, 258, 267, 299, 338; renewable, 76–78, 204, 237–239, 264
Engrams, 87
Eno, Brian, 142, 169
Entertainment, 65, 69, 198–199, 227, 244, 286, 299, 325
Environment, 41, 299, 310, 317; agriculture, 183–184; climate crisis, 75–79, 320; environmental robotics (ecobots), 76

EPIC (Executive-Process/Interactive Control), 84
Ernst, George W., 163–164
Ernst, Henry, 225
Eschatology, 304–305
Ethics, 108, 229, 244, 250–255, 262–263, 272, 350; algorithmic bias, 12–13; Asimov's laws, 18–20, 40; autonomous weapons, 28–31, 34–36, 62–63, 207–208; caregiver robots, 68–70; chatbots, 73; gender, 157, 159; human enhancement, 54; Moral Turing Test, 233; religion and spirituality, 302–307; robot ethics, 283–286; superintelligence, 320; Trolley Problem, 325–327
EU General Data Protection Regulation (GDPR), 13, 148
Eugene Goostman chatbot, 343
Evidence-based medicine, 79–82, 270
Evolution, 14, 42, 53, 127, 205–206, 234, 236, 293–294; evolutionary algorithms, 78, 106–109; evolutionary biology, 116–118; evolutionary computation, 127, 222–223; evolutionary robotics, 78, 170; evolvable hardware, 106–107. *See also* Artificial evolution; Genetic algorithms
E-waste and recycling, 76–77, 268, 296
*Ex Machina*, **143–144**, 158, 349
Existential risks, 15, 44, 52–53, 107–108, 239–240, 304, 309–310, 348, 350
Experimental Music Studio, 142
Experiments in Musical Intelligence, 91, 141, 170
Expert systems, 37, 100, **144–147**, 222–223, 291; DENDRAL, 114–115; INTERNIST-I and QMR, 192–194; knowledge engineering, 200–204; MOLGEN, 230–232; MYCIN, 240–242
Explainable AI (XAI), **147–149**
Extraterrestrials, 14–15, 108, 114, 122, 226

FaceApp, 47
Facebook, 12, 46, 157; Messenger chatbot, 228
Face-recognition systems, 45–48, 65, 86, 131, 174, 196, 286, 295; FaceFINDER, 45
Farmer, J. Doyne, **150–151**
Fautrel, Pierre, 92
Feenberg, Andrew, 92
Feigenbaum, Edward, 114–115, 230–231, 240–241, 282, 291

Feynman, Richard, 276
Film, 46–47, 70, 78–79, 123, 158–159, 224, 236, 262, 270, 325, 341; automatic film editing, 24–25; *Blade Runner*, 48–50; *Ex Machina*, 143, 349; Robot Film Festival, 198; *Terminator*, 321–323; *2001: A Space Odyssey*, 334–335
Finance, 46, 150–151, 258, 344
Fingerprinting systems, 44–48
Fletcher, Harvey, 245
Flight, 9, 102, 339
Floridi, Luciano, 285
Flynn, Anita, 56
Foerst, Anne, **151–153**, 303–304
Foot, Philippa, 325
Ford, Martin, **153–155**
Forsythe, George, 221
Forward chaining, 145
Fractal branching ultra-dexterous robots (Bush robots), 235
Frame based languages, 203
Frame problem, the, **155–156**
*Frankenstein*, 143, 304
Frey, Carl Benedikt, 154
Friedland, Peter, 230–231
Friedman, Batya, 61
Friendly AI, 54, 238, 348–350
Froomkin, Michael, 61
Fukushima Daiichi nuclear plant, 57, 76
Fux, Johann Joseph, 168
Fuzzy electric lighting system, 147
Fuzzy expert systems, 222–223
Fuzzy logic, 147, 188

Gait recognition, 12, 45
Gallagher, Shaun, 138
Gambling, 150, 173, 260
Game AI, 139
Game of Life, 123
Game theory, 102, 142
Gasson, Mark, 342
Gates, Bill, 238, 240
Geddes, Norman Bel, 130
Gelernter, Herbert, 225
Geminoid robot, 196
Gender and AI, 12, **157–160**
General and Narrow AI, **160–162**
General Problem Solver, 25, 84, **162–164**, 248, 292, 312
Generative adversarial networks (GANs), 46, 77, 92, 170
Generative art, 165–166

Generative design, **164–168**
Generative music and algorithmic composition, 142, **168–170**
Generic tasks (GT), 201–202
GENET network, 230–231
Genetic algorithms, 106, 109, 168, 170, 223, 348
George, Dileep, 122
Geraci, Robert, 304
Giant Brains, 88, **170–172**
Gidayū, Takemoto, 302
Global Air Navigation Plan (GANP), 9
Goal-directed backward chaining, 241
God and Computers Project, 151, 303–304
Gödel, Kurt, 313–314
Goertzel, Ben, **172–175**, 325
Good, Irving John, 318, 348, 350
Good Old Fashioned AI (GOFAI), 136, 225, 311, 329–330
Goodnow, Jacqueline, 89
Google, 81, 157, 204, 305; AI Quantum, 277–278; Assistant, 71, 159, 227–228, 247, 297; Behavioral Advertising, 251; Cloud Advanced Solutions Lab, 78; DeepDream, 165; DeepMind, 24, 83, 91, 180; Duplex, 228; Home, 296; Now, 86; Translate, 217
GPS (global positioning system), 4, 129, 158, 184, 187
Grasshopper, 166
Gray, John Chipman, 253
Great Ape Project, 16, 253
Greenblatt, Richard, 211
Grok application, 122
Group Symbol Associator (GSA), 98–99, **175–177**
Grover, Lov, 277
Grover's search algorithm, 279
Grymes, Rosalind, 231
Gunkel, David J., 285
Gunn, Alexander, 222

HAM (Human Associative Memory), 85
Hanson, Robin, 124
Hardy, James, 99, 176
Harmonic Tone Generator/Beauchamp Synthesizer, 142
Harnad, Stevan, 310
Harpy, 246, 282
Harpy drone, 208
Harrison, Deborah, 158
Harrow, Aram, 277

Hartmann, Michael, 277–278
Hassabis, Demis, **178–181**
Hassidum, Avinatan, 277
Haugeland, John, 309, 329
Hawking, Stephen, 36, 54, 119, 173, 240
Hayes, Patrick, 155
He, Li-wei, 25
HEARSAY expert system, 145
Hearsay II, 282
Hebb, Donald, and Hebbian theory, 87, 103, 224
Heidegger, Martin, 10, 92, 137
Hematology, 38, 99, 176
Henry, Richard Edward, 44
Heuristic Programming Project (HPP), 230, 241
Heuristics, 91, 106, 150, 179, 201, 228, 246, 290–291; automated machine learning, 22; automatic film editing, 25; expert systems, 86, 115, 144–145, 162–164, 192–194
Hewitt, Peter, 339
Hewitt-Sperry Automatic Airplane, 339
Hidden Markov models, 38, 190, 247
Hiller, Lajaren, 142
Hilton hotels, 298–299
Hines, Michael, 51
Hirata, Oriza, 197
Hixon Symposium on Cerebral Mechanisms in Behavior, 219
Hoffman, Guy, 198
Hofstadter, Douglas, 91, 93–94, 142, 302
Holland, John, 223, 302
Hospitality, 153, 298–300
Hsu, Feng-Hsiung, 110
Hubbard, F. Patrick, 254
Human Brain Project (HBP), 51, **181–182**
Human-computer interaction (HCI), 68, 85, 92, 192, 249
Human Rights Watch, 30, 62
Humanity+ (World Transhumanist Association), 52, 54, 172, 348

IBM (International Business Machines), 105, 219–221, 224–225, 245–246; BlueGene, 51; Deep Blue, 91, 110–112, 179, 211; Green Horizon, 76; Q Experience, 278–279; Shoebox, 228; SyNAPSE, 314–316; Watson, 24, 86–87
Identity and the self, 46, 120–121, 124, 254–255, 331; mirror test of self-recognition, 15

IDIC film editing system, 25
Iliad expert system, 100, 193–194
Image recognition and classification, 5, 86, 113–114, 121, 160, 197
Imitation game, 328–330
Immortality, 54, 119–124, 205, 254, 288, 304–305, 348
Inanimates, 15, 261–262, 302
Information and communication technologies (ICT), 67, 181, 295
Information processing, 10–11, 84, 88–89, 95–98, 171, 212, 311
Information theory, 88, 105, 219
Instrumental convergence thesis, 257
Instrumental theory of technology, 92
Integrated Automated Fingerprint Identification System (IAFIS), 45
Intel Loihi chip, 87
IntelliCorp, 230–231, 241
Intelligence explosion, 309, 318, 348, 350
IntelliGenetics, 231
Intelligent cinematography, 24–25
Intelligent sensing agriculture, 76, **183–185**
Intelligent transportation, 130, **185–189**, 295; electronic road pricing (ERP), 187; electronic toll and traffic management system (ETTM), 187; V2X (vehicle to everything), 129. *See also* Driverless cars and trucks
Intelligent tutoring systems, **189–191**
Interaction for Cognitive Agents, **191–192**
Internet of Things (IoT), 295, 298–299
INTERNIST-I and QMR expert systems, **192–194**
IPL language, 164, 292
Iris recognition systems, 45–46, 48
iRobot Corporation, 34, 55–57, 67, 76
Ishiguro, Hiroshi, **194–198**
Issacson, Leonard, 142
Istvan, Zoltan, 305
Itskov, Dmitry, 122

Jackson, Peter, 145
Johnson, Mark, 138
Just War Theory, 30–31

Kac, Eduardo, 342
Karpathy, Andrej, 239
Kasparov, Garry, 91, 110–111, 179
Kassirer, Jerome, 99
Kedes, Larry, 231

Kelso, Louis, 264
Kerner, Winifred, 142
Kerr, Ian, 61
Kettering, Charles F., 339
Kettering Aerial Torpedo ("Kettering Bug"), 339
Keystroke level model, 249
Kismet robot, 56, 151, 304
Klein, Jonathan, 68
Kline, Nathan, 123
K-nearest neighbor algorithms, 21
Knight, Heather, **198–200**
Knowledge acquisition and documentation structuring (KADS), 200–203
Knowledge based systems (KBS), 170, 200–203
Knowledge engineering (KE), **200–204**
Knowledge engineering environment (KEE), 231
Koch, Christof, 16
Koene, Randal, 122
Kohno, Tadayoshi, 61
Kronrod, Alexander, 211
Kubrick, Stanley, 224, 236, 334
Kurzweil, Ray, 1, 119, 173, **204–206**, 302, 319, 350

Lakoff, George, 137–138
Landscape of human competence, 235
Larsen, Bent, 110
Larson, Steve, 142
Law enforcement, 44–46, 48–49, 270–273, 295
Lawrence, David, 254
Laws of war, 36, 208
Leabra, 84
Lederberg, Joshua, 114–115, 230, 291
Ledley, Robert, 37, 39
Legg, Shane, 179
Lehman, Joel, 78
Lethal autonomous weapons systems (LAWS), 3, 35, 62–63, 158, **207–210**. *See also* Autonomous weapons systems, ethics of
Levandowski, Anthony, 205
Li, Fei-Fei, 159
Licklider, J.C.R., 221, 224
LIDAR sensors, 129, 184, 239
Life extension, 54, 122, 173
Lifenaut study, 122
Lilly, John, 15
Lin, Patrick, 284

Lipkin, Martin, 38, 99, 176
LISP (List Processing or List Programming), 141, 146, 225, 292
Loebner, Hugh, 226
Loebner Prize, 71–74, 226
Logic Theorist (Logic Theory Machine), 105, 163, 248, 290, 292, 311–312, 314
Logistic regression models, 190
Loneliness, 65, 69, 73, 153
Loving AI research project, 174
Low, Philip, 16
Lusted, Lee, 37, 39

M40 fingerprint matching algorithm, 45
Ma, Jack, 77
Mac Hack, 11, **211–212**
MacDonald, Bruce, 66
Machine learning regressions, **212–214**
Machine translation (MT), **215–217**
MACSYMA program, 225
Macy Conferences on Cybernetics, 103, **217–219**
Manga, 65, 70, 197, 303
Manin, Yuri, 276
Mann, Steve, 342
Marino, Lori, 15
Markram, Henry, 51
Marr, David, 95
Marriott hotels, 298–299
Masarie, Fred E., Jr., 192
Mascelli, Joseph, 25
Mass spectrometry, 114–115, 230, 291
Matarić, Maja, 68
MATHia® (Cognitive Tutor), 189
Mauldin, Michael, 73
Mayes, Jason, 161
McAfee, Andrew, 59–60
McCarthy, John, 57, 102–103, 218, **219–222**, 225, 237, 294; Dartmouth AI conference, 105–106, 224, 312; frame problem, the, 155–156
McCracken, Donald, 193
McCulloch, Warren, 103, 217–218, 222, 224
Mead, Carver, 121
Means-ends analysis, 162–163, 292
Mechanisation of Thought Processes conference, 106, 176, 220
Medical HouseCall, 100, 194
Medicine, **222–223**; cybernetics, 103; decision-making, 37, 176, 291; medical informatics, 23, 98–100, 181. *See also*

Artificial intelligence in medicine;
Clinical decision support systems;
Computer-assisted diagnosis
Meditel expert system, 193–194
Meditel Pediatric System, 193
Mental health and depression, 65, 81, 101
Merleau-Ponty, Maurice, 10, 137
Messiaen, Olivier, 142
META-DENDRAL, 146
Metadynamics AI algorithms, 150
Meta-Morphogenesis Project, 293
Meteorology, 9, 77, 232, 295
Micro-macro problem, 127
Microsoft, 74–75, 77, 78, 157, 228, 238,
304; Aether, 159; AI for Earth program,
41; Cortana, 157–158, 227; QuArC, 276;
Social Computing group, 75; Speech
Application Programming Interface
(SAPI), 246; Tay, 3, 72–73; Translator,
216–217
MIDAS assembly language, 211
Miikkulainen, Risto, 78
Miller, George, 89, 224
Miller, Kenneth D., 123–124
Miller, Randolph A., 192
Mind uploading, 54, 120–121, 123, 288,
304–305, 307, 348
Mineta, Norman, 186
Minsky, Marvin, 7, 102–103, 105–106,
134, 219–220, **223–227**, 335
Mobile recommendation assistants, **227–230**
Model based and incremental knowledge
engineering (MIKE), 202
Model human processor, 249
Modha, Dharmendra, 316
Molecular biology, 230–231
MOLGEN expert system, 115, **230–232**, 241
Molyneux, Peter, 178
Monte Carlo simulation methods, **232**
Monty Hall problem, 37, 39
Monzaemon, Chikamatsu, 302
Moon, Youngme, 68
Moore, John W., 51
Moore's Law, 107, 302, 319
Moral Machines, 284, 326
Moral Turing Test (MTT), **233–234**
Moran, Thomas, 249
Moravec, Hans, 120, 123, **234–237**, 302
Moravec's paradox, 234
Mordvintsev, Alexander, 165–166
Morgenstern, Oskar, 217, 257
Mori, Masahiro, 303

Mormon Transhumanist Association, 304
Mozart, Wolfgang Amadeus, 91, 142, 168, 170
Mukai, Toshiharu, 66
Music, 91–93, 105, 141–142, 168, 204
Music videos, 168–170, 200
*Musikalisches Würfelspiel (Musical Dice
Game)*, 142, 168
Musk, Elon, 54, 174, **237–240**
My Artificial Muse, 166
My Real Baby, 57, 332
MYCIN expert system, 144–145, 230,
**240–242**, 291
Myers, Jack D., 192

Nagel, Ernst, 314
Nagel, Thomas, 14
Nake, Frieder, 165
Nanosocialism, 264
Nanotechnology, 53, 121–122, 265, 267,
288, 320
Narrow AI, 160–162, 173
Nash, Firmin (F. A.), 98–99, 175–177
Natural language generation (NLG), 91,
**243–244**
Natural language processing (NLP) and
speech understanding, 81, 204, 228,
**244–248**; chatbots, 72–73; cognitive
architectures and computing, 84–86.
*See also* Machine translation; Natural
language generation; Speech
recognition
Nees, Georg, 165
Nejat, Goldie, 67
Neural networks, 86–87, 97–98, 105–107,
112, 146, 213–214, 224–225;
convolutional neural networks, 41, 81,
87, 97, 101, 113, 166; generative
adversarial networks (GANs), 77;
quantum neural networks, 277;
recurrent neural networks, 114;
stochastic neural networks, 224
Neuralink, 174, 238–239
Neurobiology, 90, 95, 97, 120
Neuroengineering, 107
Neuromorphic electronics, 78, 87, 121–122,
181, 314–315
NEURON, 51
Neven, Hartmut, 278
Newell, Allen, 91, 95, 105, 245, **248–250**,
314; General Problem Solver, 163–164;
Herbert Simon, collaboration with,
290–292; symbol manipulation, 311–312

Newman, James, 314
Next Generation Identification system, 45
Neyman, Jerzy, 23
Nissenbaum, Helen, **250–252**
Noll, A. Michael, 165
Nonhuman Rights and Personhood, 16, **252–255**
Nonhuman Rights Project, 14, 16
Non-player characters (NPCs), 138–141
Norvig, Peter, 83
Nowatzyk, Andreas, 110

Obfuscation tools, 251–252
Object oriented languages, 258
Object-oriented systems, 146
Obvious collective, 92
OI-KSL (Object Inference Knowledge Specification Language), 146
Omohundro, Steve, **256–259**, 309
OPAL system, 146
Open learner models (OLMs), 190
OpenAI, 238
OpenCog, 172–174, 325
OpenNMT, 217
Operations research, 102, 105, 246
Osborne, Michael, 154
Otterbach, Johannes, 276

Packard, Norman, 150, 258
PackBot, 34, 56–57, 76
Painting, 1–2, 92–93, 166, 197
Painting Fool, The, 92–93
Paperclip Maximizer, 320
Papert, Seymour, 11, 211, 225
Parker, Alice, 121
PARRY chatbot, 73, 135, **260–261**, 312
Particle swarm optimization (PSO), 126
Pathetic fallacy, **261–262**
Pattern recognition, 37, 48, 85, 89, 101, 225, 247
Pattern Recognition Theory of Mind, 204
Patterson, Dan, 144
Pauker, Stephen, 99, 193
Pearce, David, 54
Perceptrons, 103, 225
*Person of Interest*, **262–263**
Personhood, 14, 16–17, 124, 151–152, 252–255
Petrov, Stanislav, 3
Pfeifer, Rolf, 137
Phenomenology, 10, 137–138
Philosophy of AI, 116

Philosophy of mind, 90, 116, 122
Phylogenetics, 37
Physical symbol system hypothesis, 135–136
Picard, Rosalind, 68
Pinker, Steven, 234
Pitts, Walter, 103, 222
Platt, Charles, 123
Policing, 44, 46, 270–273, 295, 297, 317, 337
Pollution, 76–77, 310
Pople, Harry, 21, 192
Poplog system, 294
Post, Emil, 290
Posthumanism, 54, 205
Post-scarcity, **264–269**
Power laws, 249, 292
Precision medicine, **269–270**
Predator, 339
Predictive policing, **270–273**, 297
*Principia Mathematica*, 105, 290, 311, 313–314
Privacy, 40, 61–62, 69–70, 229, 250–252, 297; biometric, 44–48; vehicular, 131, 185–186
Problem-solving methods (PSM), 201–203
Product liability, 100, 132–133, **273–275**
Production rules, 84, 201, 241, 291
Production systems, 84, 97, 145, 249, 290
Productivity, 58–60, 154, 183–184, 186, 239, 264, 296, 346
Prolog (Programming in Logic), 146–147
PROSPECTOR expert system, 291
Psycholinguistics, 89
Psychology, 103, 262, 312, 331, 350; associationism, 10; cognitive psychology, 88–90, 292–293; concept attainment, 89; ELIZA, 134–135; Minsky, Marvin, 224–225; Newell, Allen, 249–250
Public health, 23, 135, 317
PUFF expert system, 291

Qualia, 116–117
Quantum AI, 107, 109, **276–280**
Quantum ontology, 123
Quick Medical Reference (QMR) expert system, 192
QuoVADis Project, 67

R1 expert system, 291
R1-Soar, 85

Race, 73, 131, 159, 251, 272–273
Radiology, 24, 37, 101, 222
Ramakrisha, Varun, 199
RAND Corporation, 10–11, 211, 248
Random forest algorithms, 21, 213–214
Real-time processing and analytics, 9, 46, 86, 190, 196, 199, 246; driving, 158, 186, 188, 239
Reaper, 339
Recombinacy, 91
RECONSIDER, 175
Reddy, Raj, 246, 249, **281–282**
*Regina* v. *Adams* (1996), 39
Region of interest (ROI), 234
Regis, Ed, 350
Regulation, 13, 31, 63, 129, 148, 209, 239–240, 251
Reinforcement learning, 97, 162, 179–180, 190, 247
Reiss, Diana, 15
ReliefF, 20
Religion and spirituality, 121, 205, 302–307
Renal disease, 145
Replicants, 12, 49–50, 262
Replika, 73
Responsible Robotics, 62, 284
Retailing, 59, 153, 266
Reynolds, Craig W., 126
Riemann Hypothesis Catastrophe, 226
Rin, Kagamine, 158
RoboCup soccer, 125
Roboethics, 283–285
Robot ethics (machine ethics), 19, 70, **283–286**
Robot Ethics Charter, 19
Robot Film Festival, 198
Robot nannies, 64, 70
Robot rights, 16, 283, 285
RoboThespian, **286–287**
Rochester, Nathaniel, 105, 219, 224–225
Role limiting methods (RLM), 201
Roomba robot, 34, 55, 57, 76, 262, 296
Rosenblatt, Frank, 103, 225
Rosenbloom, Paul, 249
Rosenblueth, Arturo, 218
Rothblatt, Bina Aspen, 121
Rothblatt, Gabriel, 305
Rothblatt, Martine, 121, 305
Rucker, Rudy, **287–289**
Ruskin, John, 261
Russell, Bertrand, 105, 290, 313–314

Russell, Stuart, 83
Ryder, Richard, 14

Saberhagen, Fred, 42–44
Sack, Warren, 25
Safe AI, 258
SAFE program, 231
Safety, 3, 9, 35, 57, 131, 256, 273, 295
SAINT (Symbolic Automatic INTegrator), 225
Salesin, David H., 25
Samuel, Arthur, 212
Sanders, J. W., 285
Sarao, Navinder Singh, 3
SATBOT autonomous spacecraft, 324
Sather language, 258
Satie, Erik, 169
Satisficing, 248, 292
Satkin, Scott, 199
Savage, Leonard, 217–218
Schank, Roger, 7, 87
Schoenberg, Arnold, 169
Schumpeter, Joseph, 59, 300, 346
Schwartz, Eric, 95
Science fiction, 70, 78–79, 226, 236, 262, 270; Asimov, Isaac, 17–20, 283; Berserkers, 42–44; digital immortality, 122–123; post-scarcity, 265–268; Rucker, Rudy, 287–289; spiritual robots, 305–306; Vinge, Vernor, 205, 350
Searle, John, 15, 92–93, 309, 330
Security, 185, 229–230, 317; biometric security, 44–48; cybersecurity, 3, 73, 131; robots, 2, 68
Sedol, Lee, 91, 93, 179–180
Seed AI, 308–309, 349
Seinendan theatre company, 197
Self-Informing Universe, 294
Self-organizing systems, 67, 84, 126–127, 224
Self-replicating machines, 42–43, 236, 264–265, 320
Selfridge, Oliver, 225, 248
Semiautonomous systems, 26–30, 35, 128
Semiautonomous weapons, 28
Sensors, 65–68, 76, 129, 136–137, 195–196, 199, 229; agriculture, 183–185; automated systems, 26; biometrics, 46, 48; obstacle detection, 29, 56–57; smart cities and buildings, 295, 299; transportation, 187–188, 239; wearable, 4–7, 342–343

SEQ program, 231
Shad, Amjad, 342
Shakey the Robot, 137, 338
Shanahan, Murray, 321
Shannon, Claude, 88, 102–103, 105, 171, 219–220, 224, 313
Sharkey, Amanda, 68–70
Sharkey, Noel, 68–70
Shaw, Cliff, 290, 311
Shelley, Mary, 143
Shibata, Takanori, 65
Shor, Peter, 277
Shor's algorithm, 277–279
Shortliffe, Edward, 291
Simon, Herbert, 84, 91, 163, 248, 250, **290–293**, 311–312, 314
Simulation argument, 52–53
Simultaneous localization and mapping (SLAM), 197
Singer, Peter, 14, 254
Singularity, 78, 161, 172–174, 205–206, 294, 302, 318–321, 350
SingularityNET, 172–174
6S Hierarchy of Evidence-Based Resources, 80
Skinner, B. F., 89
Skynet, 262, 322
Slagle, James Robert, 225
Sloman, Aaron, **293–294**
Smart cities and homes, 46, 67, 128, 130, 188, **295–298**
Smart governance, 296
Smart hotel rooms, **298–300**
Sninsky, John, 231
Soar, 83–86, 249–250
Socher, Richard, 247
Social media, 3, 25, 46, 63, 74, 228, 238, 332
Social robots, 64–65, 68, 122, 198
Society of Mind, 86, 225–226
Softbots, 68
Software Eating the World, **300–302**
Solum, Lawrence, 254
Sophia robot, 174, 255, 262
Southern Baptist Convention Ethics and Religious Liberty Commission, 304
Space, 122–123, 193, 204, 334–335, 338; Apollo-Soyuz Test Project, 216; Diamandis, Peter, 118–119; rovers, 28, 56, 235; SpaceX, 237–238
Speech recognition, 5, 9, 195, 204, 228, 249, 282. *See also* Natural language processing and speech understanding

Sperry, Elmer, 339
Sphinx I/II, 246, 282
SpiNNaker (Spiking Neural Network Architecture), 181
Spiritual robots, **302–307**
Srinivasa, Narayan, 316
*Star Trek*, 25, 158, 317
StarLisp, 258
Stefik, Mark, 230–231
Stiehl, Walter Dan, 68, 199
Stigmergy, 136
Stochastic Neural-Analog Reinforcement Calculator (SNARC), 224
Stochastic search algorithms, 20–22
Stockhausen, Karlheinz, 169
Stoeckel, Sabine, 350
Stop Killer Robots, 30, 62–63, 209
Strong AI, 116, 118, 314
Strout, Joe, 120
StyleGAN, 46
Subsumption architecture, 55–56, 136–137
Suleyman, Mustafa, 179
Superintelligence, 40, 78, 161, 236, 268, **307–310**, 343; Artilect War, 107–108; Bostrom, Nick, 53–54; Goertzel, Ben, 173; Kurzweil, Ray, 204–206; Riemann Hypothesis Catastrophe, 226; spiritual robots, 305; Technological Singularity, 319–321; Yudkowsky, Eliezer, 348–350
Super-Turing machine, 294
Supervised learning algorithms, 38, 212
Surveillance, 46, 185, 187, 271, 297, 339; caregiver robots, 68–69; Nissenbaum, Helen, 251–252; *Person of Interest*, 262–263
Sustainability, 76–77, 184, 238, 264, 297
Sutherland, Ivan, 226
Swarm intelligence, 124–128
Swarm robotics, 127
Swiss Robots, 137
Symbol grounding problem, the, **310–311**
Symbol manipulation, 225, 260, **311–313**, 314
Symbolic AI, 102, 106, 221, 225
Symbolic logic, 171–172, 312, **313–314**
Symbolic reasoning and representation, 55
SyNAPSE (Systems of Neuromorphic Adaptive Plastic Scalable Electronics), **314–316**
SYSTRAN (System Translation), 216

Tambe, Milind, **317–318**
Task-oriented computer architectures, 282

Task structures (TS), 202

Taylor, Charles, 253

Teaching, 75, 189–192, 294

Technological Singularity, 173–174, 205–206, 302, **318–321**

Teddy, Peter, 342

Tegmark, Max, 54, 123, 235, 240

TeKnowledge company, 240–241

Teleautomaton, 337

Telekino, 338

Telenoid R1 robot, 196–197

Television shows, 70, 158, 262–263, 306, 325

TensorFlow, 213

Terasem Movement, 121–122, 305

*Terminator, The*, 44, 108, 207, 262, **321–323**, 341

Termites, 116, 125

Tesla, 130, 237–239; Autopilot, 3, 131–132, 239; Tesla Grohmann Automation, 239

Theater, 197–198, 286–287

Theology, 151–153, 303–305, 307

Three Laws of Robotics, 18, 39–40, 208, 254, 283, 324

Tilden, Mark, **323–325**

Tillich, Paul, 151–152

Toma, Peter, 216

Topological quantum computing, 106–107, 109

Torres y Quevedo, Leonardo, 338

Toyota Partner Robots, 65–66

Toys, 57, 65, 234, 294, 323–325, 331–332

Traffic, 125, 132, 276, 295; intelligent transportation, 185–188; traffic optimization algorithms, 130; Traffic Simulator, 185. *See also* Air traffic control

Transhumanism, 54, 236, 304–305, 320, 343–344

Transhumanist Party, 305

Transparency, 147–148, 229, 272–273, 296

Trolley Problem, **325–327**

Tronto, Joan, 69

Trustworthy AI, 13

Tsugawa, Sadayuki, 130

Tsuji, Saburo, 197

Turing, Alan, 37, 55, 89, 293, **327–328**, 329–331. *See also* Turing Test

Turing Test, 43, 84, 91, 142–143, 261, 322, **329–331**, 335; Coffee Test, 161; Comparative Moral Turing Test (cMTT), 233; Ethical Turing Test, 233;

Flat Pack Furniture Test, 161; Loebner Prize, 226, 343; Moral Turing Test (MTT), 233–234; Total Turing Test, 195, 233

Turkle, Sherry, 69–70, **331–334**

Tweney, Dylan, 301

*2001: A Space Odyssey*, 159, 224, **334–336**

Unified theories of cognition (UTC), 191, 249

United Nations Convention on Certain Conventional Weapons (CCW), 63, 209

Universal basic income (UBI), 154, 236, 264, 346

Universal quantum Turing machine, 277–278

Unmanned Aircraft System (UAS), 339

Unmanned combat aerial vehicles (UCAV), 207

Unmanned ground and aerial vehicles, 34, 184, **337–340**

Unmanned underwater vehicles (UUVs), 34

Unsupervised learning, 38, 103, 179, 212, 278–279

U.S. Air Force Pilotless Aircraft Branch, 339

User-interface challenges, 29

van Melle, William, 241

Van Wynsberghe, Aimee, 69

Vazirani, Umesh, 277

Vernie, Gauthier, 92

Veruggio, Gianmarco, 283–284

Video games, 44, 138, 180, 238

Vienna Convention on Road Traffic, 132

Vinge, Vernor, 173, 205, 318, 350

Virtual Cinematographer, 25

Virtual personal assistants, 227

Virtual reality (VR), 77, 122–124, 140, 189, 236, 305

Vision Zero project, 295

Visual Prolog, 146

Voice recognition, 48, 228, 245

Voigt-Kampff test, 49–50

von Foerster, Heinz, 217–218

von Neumann, John, 43, 55, 205, 217, 219, 257, 318

Wallach, Wendell, 284

Ware, Andrew, 265

Warner, Homer, Jr., 38–39, 194

Warwick, Irena, 341–342
Warwick, Kevin, **341–344**
Watson, 8, 24, 86–87, 262
Waxman, Herbert, 193
Way of the Future church, 305
Weaver, Warren, 215
Weizenbaum, Joseph, 73, 134–135, 260, 329
Werbos, Paul, 222
Whitehead, Alfred North, 105, 290, 311, 313–314
Whole brain emulation, 120, 124, 304, 307–308, 348
Wiener, Norbert, 102–103, 217
Wiggins, Geraint A., 90, 92–93
Wildlife conservation, 41, 317, 339
Winograd, Terry, 329
Wise, Steven, 16

Wittek, Peter, 279
Wolfram, Stephen, 258
Wong, S. Y., 185
Workplace automation, 153, **344–347**
World War I, 337–339
World War II, 18, 102, 171, 207, 327, 339
Worley, William, 193
WowWee Toys, 325

Xenakis, Iannis, 142, 169

Yu, Victor, 192
Yudkowsky, Eliezer, 54, **348–351**

Zhou, Chongwu, 121
Zog knowledge management system, 193, 249
Zworykin, Vladimir, 99